# Lecture Notes in Computer Science

*Commenced Publication in 1973*
Founding and Former Series Editors:
Gerhard Goos, Juris Hartmanis, and Jan van Leeuwen

Martin Grohe   Rolf Niedermeier (Eds.)

# Parameterized and Exact Computation

Third International Workshop, IWPEC 2008
Victoria, Canada, May 14-16, 2008
Proceedings

 Springer

Volume Editors

Martin Grohe
Humbold Universität zu Berlin
Institut für Information
Unter den Linden 6, 10099 Berlin, Germany
E-mail: grohe@informatik.hu-berlin.de

Rolf Niedermeier
Friedrich-Schiller-Universität Jena
Institut für Informatik
Ernst-Abbe-Platz 2, 07743 Jena, Germany
E-mail: niedermr@minet.uni-jena.de

Library of Congress Control Number: 2008926222


CR Subject Classification (1998): F.2, F.1.3, F.1, E.1, I.3.5, G.2

LNCS Sublibrary: SL 1 – Theoretical Computer Science and General Issues

ISSN       0302-9743
ISBN-10    3-540-79722-X Springer Berlin Heidelberg New York
ISBN-13    978-3-540-79722-7 Springer Berlin Heidelberg New York

Springer is a part of Springer Science+Business Media

springer.com

© Springer-Verlag Berlin Heidelberg 2008

Typesetting: Camera-ready by author, data conversion by Scientific Publishing Services, Chennai, India
Printed on acid-free paper      SPIN: 12265862      06/3180      5 4 3 2 1 0

# Preface

The Third International Workshop on Parameterized and Exact Computation was held in Victoria, B.C. during May 14–16, 2008. The workshop was co-located with the 40th ACM Symposium on Theory of Computing, which took place in Victoria during May 17–20. Previous meetings of the IWPEC series were held in Bergen, Norway 2004 and Zürich, Switzerland 2006, both as part of the ALGO joint conference.

The International Workshop on Parameterized and Exact Computation covers research in all aspects of parameterized and exact computation and complexity, including but not limited to: new techniques for the design and analysis of parameterized and exact algorithms, parameterized complexity theory, relationship between parameterized complexity and traditional complexity classifications, applications of parameterized computation, implementation and experiments, high-performance computing and fixed-parameter tractability.

We received 32 submissions. Each submission was reviewed by at least 3, and on the average 3.9, Program Committee (PC) members. We held an electronic PC meeting using the EasyChair system. The committee decided to accept 17 papers. We would thoroughly like to thank the members of the PC:

Yijia Chen, Shanghai, China
Benny Chor, Tel Aviv, Israel
Fedor V. Fomin, Bergen, Norway
Jiong Guo, Jena, Germany
Gregory Gutin, London, UK
MohammadTaghi Hajiaghayi, AT&T, USA
Peter Jonsson, Linköping, Sweden
Iyad Kanj, Chicago, USA
Dieter Kratsch, Metz, France
Dániel Marx, Budapest, Hungary
Prabhakar Ragde, Waterloo, Canada
Kenneth W. Regan, Buffalo, USA
Ulrike Stege, Victoria, Canada
Stephan Szeider, Durham, UK
Todd Wareham, Newfoundland, Canada
Osamu Watanabe, Tokyo, Japan

and all external referees for the valuable work they put in the reviewing process.

We would like to thank the three invited speakers Jianer Chen (Texas A&M University), Erik Demaine (MIT), and Stephan Kreutzer (Oxford University) for their contribution to the program of the workshop and their contributions for this proceedings volume.

Special thanks go to Jiong Guo and Johannes Uhlmann for preparing the camera-ready version of the volume. We thank the members of the IWPEC

Steering Committee (Jianer Chen, Frank Dehne, Rodney G. Downey, Michael R. Fellows, Michael A. Langston, Venkatesh Raman) for their continuous support. Last but not least we would like to thank Ulrike Stege for the local organization of the IWPEC workshop in Victoria.

March 2008                                                    Martin Grohe
                                                         Rolf Niedermeier

# Organization

## External Reviewers

Isolde Adler
Mohammad H. Bateni
Hans L. Bodlaender
Ran Canetti
Peter Damaschke
Stefan Dantchev
Henning Fernau
Arik Friedman
Serge Gaspers
Jens Gramm
Fabrizio Grandoni
Sushmita Gupta
Ken-ichi Kawarabayashi
Christian Komusiewicz
Jan Kratochvil
Mathieu Liedloff
Barnaby Martin
Luke Mathieson
Egbert Mujuni
Michael Pelsmajer
Ljubomir Perkovic
Igor Razgon
Marko Samer
Saket Saurbah
Amin Sayedi
Marcus Schaefer
Allan Scott
Sagi Snir
Srinath Sridhar
Iain Stewart
Hisao Tamaki
Dimitrios Thilikos
Ioan Todinca
Yngve Villanger
Magnus Wahlström
Ge Xia
Takayuki Yato
Uri Zwick

# Table of Contents

# Randomized Disposal of Unknowns and Implicitly Enforced Bounds on Parameters*

Jianer Chen

Department of Computer Science
Texas A&M University
College Station, TX 77843-3112, USA
chen@cs.tamu.edu

**Abstract.** We study two algorithmic techniques that have turned out to be useful in the recent development of parameterized algorithms: randomized disposal of a small unknown subset of a given universal set, and implicitly enforced bounds on parameters in a branch-and-search process. These techniques are simple, effective, and have led to improved algorithms for a number of well-known parameterized problems.

## 1 Introduction

Parameterized algorithms have witnessed a tremendous growth in the last decade and have become increasingly important in dealing with NP-hard problems that arise from the world of practical computation.

Different from general (i.e., un-parameterized) algorithms, whose complexity is measured by a single parameter, the input length $n$, the complexity of parameterized algorithms is measured by multi-dimensional parameters, in particular assuming a parameter $k$ with small values, besides the input length $n$. This assumption characterizes an important feature of many NP-hard problems of practical significance. By taking the advantage of this parameter $k$ of small values, many parameterized algorithms become practically efficient when applied to NP-hard problems arisen from the real world of computing. On the other hand, the multi-dimensional parametrization has proposed new challenges to algorithmic research. Traditional algorithmic techniques seem not sufficiently effective and precise, and new design and analysis techniques have been in high demand in the development of parameterized algorithms. In the past years, many new algorithmic techniques for parameterized algorithms have been proposed and studied, including kernelization, branch-and-search, graph branch/tree decompositions, graph minor theory, graph crown rules, greedy localization, iterative compression, color coding, and many others [1,3,8,9,10,13,14].

In the current paper, we will investigate two more algorithmic techniques that have turned out to be useful in the recent development of parameterized algorithms: the randomized disposal of a small unknown subset of a given universal

---

* This work was supported in part by the National Science Foundation under the Grant CCF-0430683.

M. Grohe and R. Niedermeier (Eds.): IWPEC 2008, LNCS 5018, pp. 1–8, 2008.

set and the implicitly enforced bounds on parameters in a branch-and-search process.

## 2   Disposal of a Small Unknown Subset

The following have been well-known and extensively studied techniques in general algorithmic research:

- *Sorting* will significantly speedup later searching and ordering processes;
- *Divide-and-Conquer* is very useful in developing efficient recursive algorithms;
- *Dynamic Programming* helps effectively avoiding unnecessary recomputation.

These techniques, which dispose a given set, have proved to be very useful in developing efficient algorithms. In the research of parameterized algorithms, we are often seeking a small subset $S$ of size $k$ in a universal set $U$ of size $n$, where $k \ll n$. It will be nice that the above algorithmic techniques can be applied to the *unknown* subset $S$ and speedup the running time of parameterized algorithms. However, the subset $S$ is unknown and, how do we dispose an unknown subset?

By simple probability analysis, we note the following facts:

**D1.** A random permutation of the universal set $U$ will give the subset $S$ any pre-specified order with probability $1/k!$;

**D2.** A random partition of the university set $U$ will split the subset $S$ into any pre-specified partition with probability $1/2^k$; and

**D3.** A random coloring of the university set $U$ by $k$ colors will *not* assign the same color to any two elements in the subset $S$, with probability larger than $1/e^k$.

Therefore, although we do not know where is the subset $S$, by randomly disposing the universal set $U$, which is known and under our control, we can achieve a desired disposal of the unknown subset $S$, with a reasonable probability.

Observations D1 and D3 have allowed Alon, Yuster, and Zwick [1] to develop the first group of randomized parameterized algorithms for the $k$-PATH problem, with running time bounded by $O^*(k!)$ and by $O^*((2e)^k) = O^*(5.5^k)$, respectively[1]. In the following, we show how Observation D2 can be used in development of efficient parameterized algorithms.

Consider the $k$-PATH problem: given an undirected graph $G$ and a parameter $k$, decide whether the graph $G$ contains a $k$-path (i.e., a simple path of $k$ vertices). Suppose that the graph $G$ contains a $k$-path $P$. By randomly splitting the vertices of $G$ into two parts $V_1$ and $V_2$, with probability $1/2^k$, the $k$ vertices on the path $P$ are partitioned in such a way that the first $k/2$ vertices on $P$ are contained in $V_1$ and the last $k/2$ vertices on $P$ are contained in $V_2$. Therefore, if we recursively look for $(k/2)$-paths in the induced subgraph $G[V_1]$ and for $(k/2)$-paths in the

---

[1] We have followed the convention of using $O^*(f(k))$ to denote the bound $f(k)n^{O(1)}$.

induced subgraph $G[V_2]$, and consider all possible concatenations of $(k/2)$-paths in $G[V_1]$ with $(k/2)$-paths in $G[V_2]$, we will have a good chance to construct a $k$-path in the original graph $G$. This has led to $O^*(4^k)$ time algorithms for the $k$-PATH problem [6,11].

Next we consider the SET SPLITTING problem: given a collection $C$ of subsets of a universal set $U$, decide whether there is a partition $(U_1, U_2)$ of the universal set $U$, where $U_1 \cap U_2 = \emptyset$ and $U_1 \cup U_2 = U$, that splits at least $k$ subsets in the collection $C$ (the partition $(U_1, U_2)$ "splits" a subset $S$ of $U$ if $S \cap U_1 \neq \emptyset$ and $S \cap U_2 \neq \emptyset$). Suppose that such a partition $(U_1, U_2)$ of $U$ exists, then there are $k$ subsets $S_1, \ldots, S_k$ in the collection $C$ and elements $a_1, \ldots, a_k, b_1, \ldots, b_k$ in $U$ such that for each $i$, $a_i, b_i \in S_i$, $a_i \in U_1$ and $b_i \in U_2$ (note that it is possible that $a_i = a_j$ or $b_i = b_j$ for $i \neq j$, but no $a_i$ can be $b_j$ for any $i$ and $j$). Therefore, what we are looking for is the subset $P = \{a_1, \ldots, a_k, b_1, \ldots, b_k\}$ of no more than $2k$ elements in the universal set $U$, and a partition $(\{a_1, \ldots, a_k\}, \{b_1, \ldots, b_k\})$ of the subset $P$. By Observation D2, by a random partition of the universal set $U$, we will achieve the desired partition of the unknown subset $P$ with a probability at least $1/2^{2k}$. Note that with a partition of the universal set $U$, it is trivial to verify if the partition splits at least $k$ subsets in $C$. Therefore, by a straightforward probability analysis, with $O^*(4^k)$ times of iterations of this random partition, we will, with a high probability, "implement" the desired partition of the unknown subset $P$ that splits the $k$ subsets in the collection $C$. This leads to a randomized algorithm of running time $O^*(4^k)$ for the SET SPLITTING problem. A little bit more careful analysis [5] shows that this randomized algorithm in fact solves the SET SPLITTING problem in time $O^*(2^k)$.

We make two remarks on the above results. First, the above techniques can be used to deal with weighted problem instances without any essential changes. In consequence, we can construct in time $O^*(4^k)$ a $k$-path of the maximum weight in a given graph whose vertices and/or edges are assigned weights, and construct in time $O^*(2^k)$ a partition of a universal set $U$ that splits $k$ subsets of a given collection $C$ such that the sum of the weights of the $k$ split subsets is maximized.

The second remark is that the above randomized techniques can be de-randomized [1,5,6,12]. For example, the concept of *universal* $(n, k)$-*set* [12] has been introduced that de-randomizes the random partition process. A *universal* $(n, k)$-*set* $F$ is a collection of partitions of the universal set $U = \{1, 2, \ldots, n\}$ such that for every subset $S$ of $k$ elements in $U$ and for any partition $(S_1, S_2)$ of $S$, there is a partition $(U_1, U_2)$ of $U$ in $F$ that implements $(S_1, S_2)$, i.e., $S_1 \subseteq U_1$ and $S_2 \subseteq U_2$. In other words, any partition of any subset of $k$ elements in $U$ is implemented by at least one partition in $F$.

**Theorem 1.** ([12]) *There is a universal* $(n, k)$-*set* $F$ *that consists of* $O^*(2^{k+o(k)})$ *partitions of the universal set* $U = \{1, 2, \ldots, n\}$ *and can be constructed in time* $O^*(2^{k+o(k)})$.

Therefore, in order to achieve a specific partition of an unknown subset $S$ of $k$ elements in the universal set $U = \{1, 2, \ldots, n\}$, we can try each of the partitions in the universal $(n, k)$-set $F$ in Theorem 1 (there are only $O^*(2^{k+o(k)})$ such partitions), and are guaranteed that at least one of the partitions in $\mathcal{F}$ implements the desired partition of the unknown subset $S$.

# 3   Implicitly Enforced Bounds on Parameters

Branch-and-search has been a general approach in the development of parameterized algorithms. For example, to find a vertex cover of $k$ vertices in a graph $G$, a typical step is to pick a vertex $v$ of degree $d > 2$, then branch on $v$, by either including or excluding $v$ in the objective vertex cover, then recursively search in the resulting graph. In the branch of including $v$, we are looking for a vertex cover of $k - 1$ vertices in the resulting graph; while in the branch of excluding $v$, we must include all neighbors of $v$ in the objective vertex cover, thus are looking for a vertex cover of $k - d$ vertices in the resulting graph. This recursive process gives a recurrence relation $T(k) = T(k - 1) + T(k - d)$ for the size $T(k)$ of the search tree for a vertex cover of $k$ vertices, and leads to a parameterized algorithm of running time $O^*(c^k)$ for a small constant $c$ for the problem.

The above branch-and-search process seems to fail in dealing with some other parameterized problems. For example, suppose we are looking for an independent set of $k$ vertices in a given graph $G$. If we branch at a vertex $v$ by either including or excluding $v$ in the objective independent set, then in the branch of excluding $v$, we will not be able to directly include any vertex in the objective independent set and decrease the parameter value. Thus, the branch-and-search process does not lead to efficient parameterized algorithms for the INDEPENDENT SET problem. This difficulty seems essential, as the INDEPENDENT SET problem is $W[1]$-complete, and by the working hypothesis in parameterized complexity theory, the INDEPENDENT SET problem is not fixed-parameter tractable.

Similar difficulties have also arisen for other parameterized problems. Recall that a *feedback vertex set* (FVS) $F$ in a graph $G$ is a vertex subset such that $G - F$ is an acyclic graph, and that the FEEDBACK VERTEX SET problem is for a given undirected graph $G$ and a parameter $k$, to decide whether $G$ has an FVS of $k$ vertices.

By the technique of *iterative compression* [14], it is know that the FEEDBACK VERTEX SET problem can be reduced to the following more restricted version of the problem:

> FOREST BIPARTITION FVS: given an undirected graph $G = (V, E)$, a partition $(V_1, V_2)$ of the vertices of $G$ such that both induced subgraphs $G[V_1]$ and $G[V_2]$ are acyclic, and a parameter $h$, decide whether $G$ has an FVS $F$ of $h$ vertices such that $F \subseteq V_1$.

Consider the following branch-and-search process for the FOREST BIPARTITION FVS problem:

1. pick a vertex $v$ in $V_1$ that has at least two neighbors in $V_2$;[2]

---

[2] It can be shown [2] that such a vertex always exists unless the problem is trivially solvable.

2. **if** $v$ has two neighbors in the same connected component in $G[V_2]$
   **then** include $v$ in the objective FVS and recursively work on the instance $(G - v, V_1 - v, V_2, h - 1)$;
3. **else** branch at the vertex $v$:
   (B1) include $v$ in the objective FVS and recursively work on the instance $(G - v, V_1 - v, V_2, h - 1)$,
   (B2) exclude $v$ from the objective FVS and recursively work on the instance $(G, V_1 - v, V_2 + v, h)$.

The above branch-and-search process seems to encounter the same difficulty as we had for the INDEPENDENT SET problem: in the branch of excluding the vertex $v$, we cannot directly decrease the parameter value $h$. However, in this case, we have an implicitly enforced bound on the depth of the search tree. Observe the following simple facts:

**F1.** If a vertex $v$ in $V_1$ has two neighbors in the same connected component in $G[V_2]$, then $v$ *must* be included in the objective FVS;

**F2.** If a vertex $v$ in $V_1$ has two neighbors that are in different connected components in $G[V_2]$, then moving $v$ from $V_1$ to $V_2$ makes the number of connected components of $G[V_2 + v]$ strictly less than that of $G[V_2]$.

Therefore, if we start with a vertex subset $V_2$ of $l$ vertices, then the number of connected components in the induced subgraph $G[V_2]$ is bounded by $l$. Thus, each computational path in the search tree for the process passes through at most $l$ branches that exclude a vertex $v$ from the objective FVS by moving $v$ from $V_1$ to $V_2$. In consequence, each computational path in the search tree passes through at most $h + l$ branches, of which $h$ reduce the parameter value and $l$ reduce the number of connected components in $G[V_2]$. This concludes that the total running time of the above branch-and-search process is bounded by $O^*(2^{h+l})$.

It has been shown [2] that to determine whether a given graph $G$ has a feedback vertex set of $k$ vertices, we need to solve $\binom{k+1}{j}$ instances $(G', V_1', V_2', k - j)$ of the FOREST BIPARTITION FVS problem, for all $j$, $0 \le j \le k + 1$, where $|V_2'| = k - j + 1$. By the above analysis, each of such instances can be solved in time $O^*(4^{k-j})$. Therefore, the FEEDBACK VERTEX SET problem can be solved in time $O^*(5^k)$ (see [2] for more details).

We give another more sophisticated example to further illustrate the power of this technique. Let $G$ be a graph and let $T_1, \ldots, T_l$ be disjoint vertex subsets in $G$ (the vertex subsets $T_1, \ldots, T_l$ will be called *terminal sets* in the following discussion). A vertex subset $S$ is a *multi-way cut* for the terminal sets $T_1, \ldots, T_l$ if there is no path in the graph $G - S$ from any vertex in $T_i$ to any vertex in $T_j$ for any $i \ne j$. Consider the following problem:

MULTI-WAY CUT problem: given a graph $G$, and terminal sets $T_1, \ldots, T_l$, and a parameter $k$, decide whether there is a multi-way cut $S$ of $k$ vertices for the terminal sets $T_1, \ldots, T_l$.

In the following discussion, we fix an instance $(G; T_1, \ldots, T_l; k)$ of the MULTI-WAY CUT problem. A vertex $u$ is a *critical vertex* if $u \notin \bigcup_{i=1}^{l} T_i$ and $u$ is adjacent to a vertex in $T_1$ but not adjacent to any vertex in any $T_i$ for $i \neq 1$. It can be proved [4] that the instance $(G; T_1, \ldots, T_l; k)$ can be trivially reduced without branching if it contains no critical vertices.

**Lemma 1.** ([4]) *Let $u$ be a critical vertex, and let $S$ be a vertex subset in the graph $G$ such that $u \notin S$. Then $S$ is a multi-way cut for $T_1$, $T_2$, $\ldots$, $T_l$ if and only if $S$ is a multi-way cut for $T_1 + u$, $T_2$, $\ldots$, $T_l$.*

**Lemma 2.** ([4]) *Let $u$ be a critical vertex. Let $m_1$ be the size of a minimum cut between the two sets $T_1$ and $\bigcup_{i=2}^{l} T_i$, and let $m_2$ be the size of a minimum cut between the two sets $T_1 + u$ and $\bigcup_{i=2}^{l} T_i$. Then $m_1 \leq m_2$. Moreover, if $m_1 = m_2$, then there is a multi-way cut of $k$ vertices for the terminal sets $T_1$, $\ldots$, $T_l$ if and only if there is a multi-way cut of $k$ vertices for the terminal sets $T_1 + u$, $T_2$ $\ldots$, $T_l$.*

We first explain how Lemmas 1 and 2 are used in our algorithm. If we decide to exclude a critical vertex $u$ in the objective multi-way cut for the terminal sets $T_1$, $\ldots$, $T_l$, then we are looking for a multi-way cut of $k$ vertices that does not include $u$. By Lemma 1, this can be reduced to looking for a multi-way cut of $k$ vertices for the terminal sets $T_1 + u$, $T_2$, $\ldots$, $T_l$. Similarly, if the conditions in Lemma 2 hold true, then we can work, without branching, on a multi-way cut of $k$ vertices for the terminal sets $T_1 + u$, $T_2$, $\ldots$, $T_l$.

Consider the the following branch-and-search process:

1. pick a critical vertex $u$;
2. compute the size $m_1$ of a minimum cut between $T_1$ and $\bigcup_{i=2}^{l} T_i$ and the size $m_2$ of a minimum cut between $T_1 + u$ and $\bigcup_{i=2}^{l} T_i$;
3. **if** $m_1 = m_2$
   **then** recursively work on the instance $(G, \{T_1 + u, T_2, \ldots, T_l\}, k)$;
4. **else** branch at $u$:
   (B1) include $u$ in the objective multi-way cut and recursively work on the instance $(G - u, T_1, T_2, \ldots, T_l; k - 1)$;
   (B2) exclude $u$ from the objective multi-way cut and recursively work on the instance $(G, T_1 + u, T_2, \ldots, T_l; k)$.

The correctness of step 3 follows from Lemma 2, while the correctness of the branching (B2) follows from Lemma 1. Again, the branch at step 4 seems to have the difficulty that in the branching (B2) of excluding the critical vertex $u$, the parameter value $k$ is not decreased. In this case, the implicitly enforced bound on the parameter is the size of the minimum cut between the two sets $T_1$ and $\bigcup_{i=2}^{l} T_i$. By Lemma 2, in step 4 we must have $m_1 < m_2$. Therefore, at each branch, we will either decrease the parameter value $k$ (branching (B1)), or increase the size of a minimum cut between the two sets $T_1$ and $\bigcup_{i=1}^{l} T_i$ (branching (B2)). Note that the size of a minimum cut between $T_1$ and $\bigcup_{i=1}^{l} T_i$ must be bounded from above by $k$: by definition the size of a minimum cut between $T_1$ and $\bigcup_{i=1}^{l} T_i$ cannot be larger than that of a multi-way cut for the terminal sets $T_1$, $\ldots$, $T_l$. Therefore, again, each computational path in the search tree of this process can pass through at most $2k$ branches, of which $k$ decrease the parameter value and $k$ increase the size of a minimum cut between $T_1$ and $\bigcup_{i=1}^{l} T_i$. In consequence, the search-tree for the above branch-and-search process contains at most $2^{2k}$ leaves, and the process solves the MULTI-WAY CUT problem in time $O^*(4^k)$.

## 4   Conclusion

Parameterized algorithmic research is an exciting but still young research area in which new techniques and methodology are being explored. In this paper, we have studied two algorithmic techniques of general interest that have been useful in the recent development of parameterized algorithms. The techniques are simple and effective, and have led to improved algorithms for a number of parameterized problems, including some well-known ones that have been extensively studied in the past and are with an impressive list of gradually improved algorithms.

## References

1. Alon, N., Yuster, R., Zwick, U.: Color-coding. Journal of the ACM 42, 844–856 (1995)
2. Chen, J., Fomin, F., Liu, Y., Lu, S., Villanger, Y.: Improved algorithms for the feedback vertex set problems. In: Dehne, F., Sack, J.-R., Zeh, N. (eds.) WADS 2007. LNCS, vol. 4619, pp. 422–433. Springer, Heidelberg (2007); Journal version is to appear in Journal of Computer and System Sciences
3. Chen, J., Friesen, D., Kanj, I., Jia, W.: Using nondeterminism to design efficient deterministic algorithms. Algorithmica 40, 83–97 (2004)
4. Chen, J., Liu, Y., Lu, S.: An improved paraeterized algorithm for the minimum node multiway cut problem. In: Dehne, F., Sack, J.-R., Zeh, N. (eds.) WADS 2007. LNCS, vol. 4619, pp. 495–506. Springer, Heidelberg (2007); Journal version is to appear in Algorithmica
5. Chen, J., Lu, S.: Improved algorithms for weighted and unweighted set splitting problems. In: Lin, G. (ed.) COCOON 2007. LNCS, vol. 4598, pp. 537–547. Springer, Heidelberg (2007); Journal version is to appear in Algorithmica
6. Chen, J., Lu, S., Sze, S.-H., Zhang, F.: Improved algorithms for path, matching, and packing problems. In: Proc. 18th Annual ACM-SIAM Symposium on Discrete Algorithms (SODA 2007), pp. 298–307 (2007)
7. Dehne, F., Fellows, M., Rosamond, F., Shaw, P.: Greedy localization, iterative compression, modeled crown reductions: New FPT techniques, and improved algorithm for set splitting, and a novel $2k$ kernelization of vertex cover. In: Downey, R.G., Fellows, M.R., Dehne, F. (eds.) IWPEC 2004. LNCS, vol. 3162, pp. 127–137. Springer, Heidelberg (2004)
8. Dorn, F., Fomin, F., Thilikos, D.: Subexponential parameterized algorithms. In: Arge, L., Cachin, C., Jurdziński, T., Tarlecki, A. (eds.) ICALP 2007. LNCS, vol. 4596, pp. 15–27. Springer, Heidelberg (2007)
9. Downey, R., Fellows, M.: Parameterized Complexity. Springer, New York (1999)

10. Fellows, M.: Blow-ups, win/win's, and crown rules: Some new directions in FPT. In: Bodlaender, H.L. (ed.) WG 2003. LNCS, vol. 2880, pp. 1–12. Springer, Heidelberg (2003)
11. Kneis, J., Molle, D., Richter, S., Rossmanith, P.: Divide-and-color. In: Fomin, F.V. (ed.) WG 2006. LNCS, vol. 4271, pp. 58–67. Springer, Heidelberg (2006)
12. Naor, M., Schulman, L., Srinivasan, A.: Splitters and near-optimal derandomization. In: Proc. 36th IEEE Symp. on Foundations of Computer Science (FOCS 1995), pp. 182–190 (1995)
13. Niedermeier, R.: Invitation to Fixed-Parameter Algorithms. Oxford University Press, Oxford (2006)
14. Reed, B., Smith, K., Vetta, A.: Finding odd cycle transversals. Oper. Res. Lett. 32, 299–301 (2004)

# Algorithmic Graph Minors and Bidimensionality

Erik D. Demaine

MIT Computer Science and Artificial Intelligence Laboratory,
32 Vassar Street, Cambridge, MA 02139, USA

**Abstract.** Robertson and Seymour developed the seminal Graph Minor Theory over the past two decades. This breakthrough in graph structure theory tells us that a very wide family of graph classes (anything closed under deletion and contraction) have a rich structure similar to planar graphs. This structure has many algorithmic applications that have become increasingly prominent over the past decade. For example, Fellows and Langston showed in 1988 that it immediately leads to a wealth of (nonconstructive) fixed-parameter algorithms.

One recent approach to algorithmic graph minor theory is "bidimensionality theory". This theory provides general tools for designing fast (constructive, often subexponential) fixed-parameter algorithms, and approximation algorithms (often PTASs), for a wide variety of NP-hard graph problems in graphs excluding a fixed minor. For example, some of the most general algorithms for feedback vertex set and connected dominating set are based on bidimensionality. Another approach is "deletion and contraction decompositions", which split any graph excluding a fixed minor into a bounded number of small-treewidth graphs. For example, this approach has led to some of the most general algorithms for graph coloring and the Traveling Salesman Problem on graphs. I will describe these and other approaches to efficient algorithms through graph minors.

M. Grohe and R. Niedermeier (Eds.): IWPEC 2008, LNCS 5018, p. 9, 2008.

# Algorithmic Meta-theorems

Stephan Kreutzer

Oxford University Computing Laboratory
stephan.kreutzer@comlab.ox.ac.uk

Algorithmic meta-theorems are algorithmic results that apply to a whole range of problems, instead of addressing just one specific problem. This kind of theorems are often stated relative to a certain class of graphs, so the general form of a meta theorem reads "every problem in a certain class $\mathcal{C}$ of problems can be solved efficiently on every graph satisfying a certain property $\mathcal{P}$". A particularly well known example of a meta-theorem is Courcelle's theorem that every decision problem definable in monadic second-order logic (MSO) can be decided in linear time on any class of graphs of bounded tree-width [1].

The class $\mathcal{C}$ of problems can be defined in a number of different ways. One option is to state combinatorial or algorithmic criteria of problems in $\mathcal{C}$. For instance, Demaine, Hajiaghayi and Kawarabayashi [5] showed that every minimisation problem that can be solved efficiently on graph classes of bounded tree-width and for which approximate solutions can be computed efficiently from solutions of certain sub-instances, have a PTAS on any class of graphs excluding a fixed minor. While this gives a strong unifying explanation for PTAS of many problems on H-minor free graphs, the class of problems it defines is not very natural. In particular, it may require some work to decide if a given problem belongs to this class or not.

Another approach to define meta-theorems is therefore based on definability in logical systems, e.g. to consider the class of problems that can be defined in first-order logic. For instance, related to the above example, a result by Dawar, Grohe, Kreutzer and Schweikardt [4] states that every minimisation problem definable in first-order logic has an EPTAS on every class of graphs excluding a minor. While the actual complexity bounds obtained in this way may not live up to bounds derivable for each individual problem, the class of problems described in this way is extremely natural and for many problems their mathematical formulation already shows that they are first-order definable. Consider, e.g., the definition of a dominating set.

Such meta-theorems based on definability in a given logic have received much attention in the literature. For instance, for the case of decision problems, it is has been shown that every problem definable in monadic second-order logic can be decided in polynomial time on graph classes of bounded clique width (see e.g. [2]). For first-order logic (FO), Seese [14] showed that every FO-definable decision problem is solvable in linear time on graph classes of bounded degree. This has then been extended to planar graphs and, more generally, graph classes

M. Grohe and R. Niedermeier (Eds.): IWPEC 2008, LNCS 5018, pp. 10–12, 2008.

of bounded local tree-width by Frick and Grohe [7] and to H-minor free graphs by Flum and Grohe [6]. Finally, Dawar, Grohe, Kreutzer generalised these results even further to classes of graphs locally excluding a minor [3].

For optimisation problems, well-known meta-theorems employing logic can been found in the work on MAXSNP by Papadimitriou, Yannakakis and also Kolaitis, Thakus (see e.g. [13,10,11]) or in other syntactical approaches to meta-theorems as in, e.g., [9].

While meta-theorems based on logical definability give strong algorithmic results for natural classes of problems, this often comes at the price of sacrificing efficiency in the algorithms derived from them. For instance, while the linear time bound $\mathcal{O}(n)$ of Courcelle's theorem on MSO-properties of graph classes of bounded tree-width is optimal, the constants hidden in the $\mathcal{O}$-notation are horrific and prevent any practical applications of the theorem. It is clear, thus, that the benefit from algorithmic meta-theorems does not lie within immediate applications or (practically) efficient algorithms. Instead they help understand the algorithmic theory of certain graph classes and form an efficient tool for establishing that a problem is tractable on a certain class of graphs or structures.

In this talk I will survey algorithmic meta-theorems with emphasis on motivation and future directions. While the focus will be on decision problems, I will also discuss logical approaches to optimisation problems. Recent surveys on meta-theorems can be found in [8] and also [12].

# References

1. Courcelle, B.: Graph rewriting: An algebraic and logic approach. In: van Leeuwen, J. (ed.) Handbook of Theoretical Computer Science, pp. 194–242. Elsevier, Amsterdam (1990)
2. Courcelle, B., Makowsky, J., Rotics, U.: Linear time solvable optimization problems on graphs of bounded clique-width. Theory Comput. Systems 33, 125–150 (2000)
3. Dawar, A., Grohe, M., Kreutzer, S.: Locally excluding a minor. In: Proc. of the 22nd IEEE Symp. on Logic in Computer Science, pp. 270–279 (2007)
4. Dawar, A., Grohe, M., Kreutzer, S., Schweikardt, S.: Approximation schemes for first-order definable optimisation problems. In: Proc. of the 21st IEEE Symp. on Logic in Computer Science, pp. 411–420 (2006)
5. Demaine, E., Hajiaghayi, M., Kawarabayashi, K.: Algorithmic graph minor theory: Decomposition, approximation, and coloring. In: Symposium on Foundations of Computer Science (FOCS), pp. 637–646 (2005)
6. Flum, J., Grohe, M.: Fixed-parameter tractability, definability, and model checking. SIAM Journal on Computing 31, 113–145 (2001)
7. Frick, M., Grohe, M.: Deciding first-order properties of locally tree-decomposable structures. Journal of the ACM 48, 1184–1206 (2001)
8. Grohe, M.: Logic, graphs, and algorithms. In: Wilke, T., Flum, J., Grädel, E. (eds.) Logic and Automata History and Perspectives, Amsterdam University Press (2007)
9. Khanna, S., Motwani, R.: Towards a syntactic characterization of PTAS. In: Proc. of STOC 1996, pp. 329–337 (1996)
10. Kolaitis, P.G., Thakur, M.N.: Logical definability of NP optimization problems. Information and Computation 115(2), 321–353 (1994)

11. Kolaitis, P.G., Thakur, M.N.: Approximation properties of NP minimization classes. Journal of Computer and System Sciences 50, 391–411 (1995)
12. Kreutzer, S.: Finite model-theory of tree-like structures, `http://web.comlab.ox.ac.uk/oucl/work/stephan.kreutzer/publications.html`
13. Papadimitriou, C.H., Yannakakis, M.: Optimization, approximation, and complexity classes. Journal of Computer and System Sciences 43, 425–440 (1991)
14. Seese, D.: Linear time computable problems and first-order descriptions. Mathematical Structures in Computer Science 2, 505–526 (1996)

# Parameterized Complexity of the Smallest Degree-Constrained Subgraph Problem[*]

Omid Amini[1], Ignasi Sau[2,3], and Saket Saurabh[4]

[1] Max-Planck-Institut für Informatik
Omid.Amini@mpi-inf.mpg.de
[2] Mascotte join Project- INRIA/CNRS-I3S/UNSA- 2004
route des Lucioles - Sophia-Antipolis, France
Ignasi.Sau@sophia.inria.fr
[3] Graph Theory and Combinatorics group at Applied Mathematics IV
Department of UPC - Barcelona, Spain
[4] Department of Informatics, University of Bergen,
N-5020 Bergen, Norway
saket@ii.uib.no

**Abstract.** In this paper we study the problem of finding an induced subgraph of size at most $k$ with minimum degree at least $d$ for a given graph $G$, from the parameterized complexity perspective. We call this problem MINIMUM SUBGRAPH OF MINIMUM DEGREE $_{\geq d}$ (MSMD$_d$). For $d = 2$ it corresponds to finding a shortest cycle of the graph. Our main motivation to study this problem is its strong relation to DENSE $k$-SUBGRAPH and TRAFFIC GROOMING problems.

First, we show that MSMS$_d$ is fixed-parameter intractable (provided $FPT \neq W[1]$) for $d \geq 3$ in general graphs, by showing it to be $W[1]$-hard using a reduction from MULTI-COLOR CLIQUE. In the second part of the paper we provide *explicit* fixed-parameter tractable (FPT) algorithms for the problem in graphs with bounded local tree-width and graphs with excluded minors, *faster* than those coming from the meta-theorem of Frick and Grohe [13] about problems definable in first order logic over "locally tree-decomposable structures". In particular, this implies faster fixed-parameter tractable algorithms in planar graphs, graphs of bounded genus, and graphs with bounded maximum degree.

## 1 Introduction

Problems of finding subgraphs with certain degree constraints are well studied both algorithmically and combinatorially, and have a number of applications in network design [10,12,15,16,17]. In this paper we initiate the study of one such

---

[*] This work has been partially supported by European project IST FET AEOLUS, PACA region of France, Ministerio de Educación y Ciencia of Spain, European Regional Development Fund under project TEC2005-03575, Catalan Research Council under project 2005SGR00256, and COST action 293 GRAAL, and has been done in the context of the CRC CORSO with France Telecom.

problem, namely MINIMUM SUBGRAPH OF MINIMUM DEGREE $_{\geq d}$ (MSMD$_d$), in the realm of parameterized complexity. More precisely, the problem we study is defined as follows:

MINIMUM SUBGRAPH OF MINIMUM DEGREE $_{\geq d}$ (MSMD$_d$)
**Input:** A graph $G = (V, E)$ and a positive integer $k$.
**Parameter:** $k$.
**Question:** Does there exist a subset $S \subseteq V$, with $|S| \leq k$, such that $G[S]$ has minimum degree at least $d$?

For $d = 2$, MSMD$_2$ corresponds to finding a shortest cycle in the graph, which can be done in polynomial time. Hence for $d \geq 3$, the MSMD$_d$ problem can also be thought of as a generalization of the girth problem. Besides this, our motivations for studying this problem are the following: (a) the problem is closely related to the well studied DENSE $k$-SUBGRAPH problem and (b) it is motivated from practical applications because of its close connection to the TRAFFIC GROOMING problem. In the next two paragraphs we briefly explain these two connections.

**Connection to the DENSE $k$-SUBGRAPH problem:** The *density* $\rho(G)$ of a graph $G = (V, E)$ is defined as its edges-to-vertices ratio, that is $\rho(G) := \frac{|E|}{|V|}$. More generally, for any subset $S \subseteq V$, we denote its *density* by $\rho_G(S)$ or simply $\rho(S)$, and define it to be the density of the induced graph on $S$, i.e. $\rho(S) := \rho(G[S])$. The DENSE $k$-SUBGRAPH problem is formulated as follows:

DENSE $k$-SUBGRAPH (D$k$S)
**Input:** A graph $G = (V, E)$.
**Output:** A subset $S \subseteq V$, with $|S| = k$, such that $\rho(S)$ is maximized.

First of all, we remark that the NP-hardness of D$k$S easily follows from the NP-hardness of CLIQUE. On the other hand, if we do not fix the size of $S$, then finding a densest subgraph of $G$ reduces to an instance of the MAX-FLOW MIN-CUT problem, and hence it can be solved in polynomial time (see Chapter 4 of [18] for more details). The D$k$S problem has attracted a lot of attention, primarily in approximation algorithms [3,7,10,15].

Now we show how MSMD$_d$ is related to D$k$S. Suppose we are looking for an induced subgraph $G[S]$ of size at most $k$ and with density at least $\rho$. In addition, assume that $S$ is minimal, i.e. no subset of $S$ has density greater than $\rho(S)$. This implies that every vertex of $S$ has degree at least $\rho/2$ in $G[S]$. To see this, observe that if there is a vertex $v$ with degree strictly smaller than $\rho/2$, then removing $v$ from $S$ results in a subgraph of density greater than $\rho(S)$ and of smaller size, contradicting the minimality of $S$. Secondly, if we have an induced subgraph $G[S]$ of minimum degree at least $\rho$, then $S$ is a subset of density at least $\rho/2$. These two observations together show that, modulo a constant factor, looking for a densest subgraph of $G$ of size at most $k$ is equivalent to looking for the largest possible value of $\rho$ for which MSMD$_\rho$ returns YES for the parameter

$k$. As the degree conditions are more rigid than the global density of a subgraph, it is easier to work directly with $\text{MSMD}_d$. This is why we hope that a better understanding of the $\text{MSMD}_d$ problem will provide an alternative way to attack the outstanding open problem of the complexity of $D k S$.

**Connection to the** TRAFFIC GROOMING **problem:** Traffic grooming in optical networks refers to packing small traffic flows into larger units, which can then be processed as single entities. For example, in a network using both time-division and wavelength-division multiplexing, flows destined to a common node can be aggregated into the same wavelength, allowing them to be dropped by a single optical Add-Drop Multiplexer (ADM for short). The objectives of grooming are to improve bandwidth utilization and to minimize the equipment cost of the network. In WDM optical networks, the most accepted criterion is to minimize the number of electronic terminations, namely the number of SONET ADMs. See [8] for a general survey on grooming. It has been recently proved [2] that the TRAFFIC GROOMING problem in optical networks can be reduced (modulo polylogarithmic factors) to $D k S$, or equivalently to $\text{MSMD}_d$. Indeed, in graph theoretic terms, the problem can be translated to partitioning the edges of a given request graph into subgraphs with a constraint on their number of edges. The objective is to minimize the total number of vertices of the subgraphs of the partition. Hence, in this context of partitioning a given set of edges while minimizing the total number of vertices, is where $D k S$ and $\text{MSMD}_d$ come into play.

**Our Results:** We do a thorough study of the $\text{MSMD}_d$ problem in the realm of parameterized complexity. Our results can be classified into two categories:

*General Graphs:* In the first part of this paper, we show in Section 2 that $\text{MSMD}_d$ is not fixed-parameter tractable, provided $FPT \neq W[1]$, by proving it to be $W[1]$-hard for $d \geq 3$. In general, parameterized reductions are very stringent because of parameter-preserving requirements of the reduction, and require a lot of technical care. Our reduction is based on a new methodology emerging in parameterized complexity, called *multi-color clique edge representation*. This has proved to be useful in showing various problems to be $W[1]$-hard recently [4]. We first spell out step by step the procedure to use this methodology, which can be used as a template for future purposes. Then we adapt this methodology to the reduction for the $\text{MSMD}_d$ problem. Our reduction is robust, in the sense that many similar problems can be shown to be $W[1]$-hard with just minor modifications.

*Graphs with Bounded Local Tree-width and Graphs with Excluded Minors:* $\text{MSMD}_d$ problem can be easily defined in first-order logic, where the formula only depends on $k$ and $d$, both being bounded by the parameter. Frick and Grohe [13] have shown that first-order definable properties of graph classes of bounded local tree-width can be decided in time $n^{1+1/k}$ for all $k$, in particular in time $n^2$, and first-order model checking is FPT on $M$-minor free graphs. This immediately gives us the *classification result* that $\text{MSMD}_d$ for $d \geq 3$ is FPT in graphs with bounded local tree-width and graphs excluding a fixed graph $M$ as a minor. These classification results can be generalized to a larger class of graphs, namely graphs "locally excluding" a fixed graph $M$ as a minor, due to a recent result of Dawar et al. [5]. These results are very general and involve huge coefficients (in other words, huge dependence on $k$). Because of this, a natural problem arising in this context is the following: can we obtain an algorithm for $MSMD_d$ for $d \geq 3$ in these graph classes faster than the one coming from the meta-theorem of Frick and Grohe, using some specific properties of the problem?

In Section 3, we answer this question in the affirmative by giving a *faster* and *explicit* algorithm for $MSMD_d$, $d \geq 3$, in graphs with bounded local tree-width and graphs excluding a fixed graph $M$ as a minor. In particular, this gives us faster FPT algorithms on planar graphs and graphs of bounded genus. Though our algorithms use standard dynamic programming over graphs with bounded tree-width, and a few results concerning the clique decomposition of $M$-minor free graphs developed by Robertson and Seymour in their Graph Minor Theory [19], we needed to make a few non-trivial observations to get significant improvements in the time complexity of the algorithms. Finally, we would like to stress that our dynamic programming over graphs with bounded tree-width is very generic and can handle variations on degree-constrained subgraph problems with simple changes.

**Notations:** We use standard graph terminology. Let $G$ be a graph. We use $V(G)$ and $E(G)$ to denote vertex and the edge set of $G$, respectively. We simply write $V$ and $E$ if the graph is clear from the context. For $V' \subseteq V$, we denote the *induced subgraph* on $V'$ by $G[V'] = (V', E')$, where $E' = \{\{u, v\} \in E : u, v \in V'\}$. For $v \in V$, we denote by $N(v)$ (or $N_G(v)$) the *neighborhood* of $v$, namely $N(v) = \{u \in V : \{u, v\} \in E\}$. The *closed neighborhood* $N[v]$ of $v$ is $N(v) \cup \{v\}$. In the same way we define $N[S]$ (or $N_G[S]$) for $S \subseteq V$ as $N[S] = \cup_{v \in S} N[v]$, and $N(S) = N[S] \setminus S$. We define the *degree* of vertex $v$ in $G$ as the number of vertices incident to $v$ in $G$. Namely, $d(v) = |\{u \in V(G) : \{u, v\} \in E(G)\}|$.

## 2    Fixed-Parameter In-Tractability of $MSMD_d$ for $d \geq 3$

In this section we give a $W[1]$-hardness reduction for $MSMD_d$. We first define parameterized reductions.

**Definition 1.** *Let $\Pi, \Pi'$ be two parameterized problems, with instances $(x, k)$ and $(x', k')$, respectively. We say that $\Pi$ is (uniformly many:1) reducible to $\Pi'$ if there is a function $\Phi$, called a* parameterized reduction, *which transforms $(x, k)$ into $(x', g(k))$ in time $f(k)|x|^\alpha$, where $f, g : \mathbb{N} \to \mathbb{N}$ are arbitrary functions and $\alpha$ is a constant independent of $k$, so that $(x, k) \in \Pi$ if and only if $(x', g(k)) \in \Pi'$.*

Our reduction is from MULTI-COLOR CLIQUE, which is known to be $W[1]$-complete by a simple reduction from the ordinary CLIQUE [11], and is based on the methodology known as *multi-color edge representation*. The MULTI-COLOR CLIQUE problem is defined as follows:

MULTI-COLOR CLIQUE

**Input**: An graph $G = (V, E)$, a positive integer $k$, and a proper $k$-coloring of $V(G)$.

**Parameter**: $k$.

**Question**: Does there exist a clique of size $k$ in $G$ consisting of exactly one vertex of each color?

Consider an instance $G = (V, E)$ of MULTI-COLOR CLIQUE with its vertices colored with the set of colors $\{c_1, \cdots, c_k\}$. Let $V[c_i]$ denote the set of vertices of color $c_i$. We first replace each edge $e = \{u, v\}$ of $G$, with $u \in V[c_i]$, $v \in V[c_j]$ and $i < j$, with two arcs $e^f = (u, v)$ and $e^b = (v, u)$. By abuse of notation, we also call this digraph $G$. Let $E[c_i, c_j]$ be the set of arcs $e = (u, v)$, with $u \in V[c_i]$ and $v \in V[c_j]$, for $1 \le i \ne j \le k$. An arc $(u, v) \in E[c_i, c_j]$ is called *forward* (resp. *backward*) if $i < j$ (resp. $i > j$). We also assume that $|V[c_i]| = N$ for all $i$, and that $|E[c_i, c_j]| = M$ for all $i \ne j$, i.e. we assume that the color classes of $G$, and also the arc sets between them, have uniform sizes. For a simple justification of this assumption, we can reduce MULTI-COLOR CLIQUE to itself, taking the union of $k!$ disjoint copies of $G$, one for each permutation of the color sets.

In this methodology, the basic encoding bricks correspond to arcs of $G$, which we call **arc gadgets**. We generally have three kinds of gadgets, which we call **selection, coherence** and **match gadgets**. These are engineered together to get an overall reduction gadget for the problem. In an optimal solution to the problem (that is, a solution providing a YES answer), the selection gadget ensures that *exactly one* arc gadget is selected among arc gadgets corresponding to arcs going from a color class $V[c_i]$ to another color class $V[c_j]$. For any color class $V[c_i]$, the coherence gadget ensures that the out-going arcs from $V[c_i]$, corresponding to the selected arc gadgets, have a common vertex in $V[c_i]$. That is, all the arcs corresponding to these selected arc gadgets *emanate from the same vertex in $V[c_i]$*. Finally, the match gadget ensures that if we have selected an arc gadget corresponding to an arc $(u, v)$ from $V[c_i]$ to $V[c_j]$, then the arc gadget selected from $V[c_j]$ to $V[c_i]$ corresponds to $(v, u)$. That is, *both of $e^f$ and $e^b$ are selected together*. In what follows, we show how to particularize this general strategy to obtain a reduction from MULTI-COLOR CLIQUE to MSMD$_d$ for $d \ge 3$. To simplify the presentation, we first describe our reduction for the case $d = 3$ in Section 2.1 and then we describe the required modifications for the case $d \ge 4$ in Section 2.2.

## 2.1  $W[1]$-Hardness of MSMD$_d$ for $d = 3$

We now detail the construction of all the gadgets. Recall that an arc $(u, v) \in E[c_i, c_j]$ is forward if $i < j$, and it is backward if $i > j$. We refer the reader to Fig. 1 to get an idea of the construction.

***Arc Gadgets:*** For each arc $(u, v) \in E[c_i, c_j]$ with $i < j$ (resp. $i > j$) we have a cycle $\mathcal{C}_{e^f}$ (resp. $\mathcal{C}_{e^b}$) of length $3 + 2(k-2) + 2$, with vertex set:

- *selection vertices*: $e^f_{s1}$, $e^f_{s2}$, and $e^f_{s3}$ (resp. $e^b_{s1}$, $e^b_{s2}$, and $e^b_{s3}$);

- *coherence vertices:* $e^f_{ch1r}, e^f_{ch2r}$ (resp. $e^b_{ch1r}, e^b_{ch2r}$), for all $r \in \{1, \ldots, k\}$ and $r \neq i, j$; and
- *match vertices:* $e^f_{m1}$ and $e^f_{m2}$ (resp. $e^b_{m1}$ and $e^b_{m2}$).

**Selection Gadgets:** For each pair of indices $i, j$ with $1 \leq i \neq j \leq k$, we add a new vertex $A_{c_i, c_j}$, and connect it to all the selection vertices of the cycles $C_{e^f}$ if $i < j$ (resp. $C_{e^b}$ if $i > j$) for all $e \in E[c_i, c_j]$. This gadget is called *forward selection gadget* (resp. *backward selection gadget*) if $i < j$ (resp. $i > j$), and it is denoted by $S_{i,j}$.

That is, we have $k(k-1)$ clusters of gadgets: one gadget $S_{i,j}$ for each set $E[c_i, c_j]$, for $1 \leq i \neq j \leq k$.

**Coherence Gadgets:** For each $i$, $1 \leq i \leq k$, let us consider all the selection gadgets of the form $S_{i,p}$, $p \in \{1, \cdots, k\}$ and $p \neq i$. For any $u \in V[c_i]$, and any two indices $1 \leq p \neq q \leq k$, $p \neq i$, $q \neq i$ we add two new vertices $u_{pq}$ and $u_{qp}$, and a new edge $\{u_{pq}, u_{qp}\}$. For every arc $e = (u, v) \in E[c_i, c_p]$, with $u \in V[c_i]$, we pick the cycle $C_{e^x}$, $x \in \{f, b\}$ depending on whether $e$ is forward or backward, and add two edges of the form $\{e_{ch1q}, u_{pq}\}$ and $\{e_{ch2q}, u_{pq}\}$. Similarly, for an arc $e = (u, w) \in E[c_i, c_q]$, with $u \in V[c_i]$, we pick the cycle $C_{e^x}$, $x \in \{f, b\}$, and add two edges $\{e_{ch1p}, u_{qp}\}$ and $\{e_{ch2p}, u_{qp}\}$.

**Match Gadgets:** For any pair of arcs $e^f = (u, v)$ and $e^b = (v, u)$, we consider the two cycles $C_{e^f}$ and $C_{e^b}$ corresponding to $e^f$ and $e^b$. Now, we add two new vertices $e^*$ and $e_*$, a *matching edge* $\{e^*, e_*\}$, and all the edges of the form $\{e^f_{m1}, e^*\}$, $\{e^f_{m2}, e^*\}$, $\{e^b_{m1}, e_*\}$ and $\{e^b_{m2}, e_*\}$ where $e^f_{m1}, e^f_{m2}$ are match vertices on $C_{e^f}$, and $e^b_{m1}$, $e^b_{m2}$ are match vertices on $C_{e^b}$.

This completes the construction of the gadgets, and the union of all of them defines the graph $\mathscr{G}_G$ depicted in Fig. 1. Now, we prove that this construction yields the reduction through a sequence of simple claims.

**Claim 1.** *Let $G$ be an instance of* MULTI-COLOR CLIQUE, *and $\mathscr{G}_G$ be the graph we constructed above. If $G$ has a multi-colored $k$-clique, then $\mathscr{G}_G$ has a 3-regular subgraph of size $k' = (3k+1)k(k-1)$.*

*Proof.* Let $\omega$ be a multi-color clique of size $k$ in $G$. For every edge $e \in E(\omega)$, select the corresponding cycles $C_{e^f}, C_{e^b}$ in $\mathscr{G}_G$. Let us define $S$ as follows:

$$S = \bigcup_{e \in \omega, x \in \{f, b\}} N_{\mathscr{G}_G}[V(C_{e^x})] \, .$$

It is straightforward to check that $\mathscr{G}_G[S]$ is a 3-regular subgraph of $\mathscr{G}_G$. To verify the size of $\mathscr{G}_G[S]$, note that we have $2 \cdot \binom{k}{2}$ cycles in $\mathscr{G}_G[S]$ and each of them contribute $(3k-1)$ vertices (this includes vertices on the cycle themselves).    $\square$

**Claim 2.** *Any subgraph of $\mathscr{G}_G$ of minimum degree at least three should contain one of the cycles $C_{e^x}$, $x \in \{b, f\}$, corresponding to arc gadgets.*

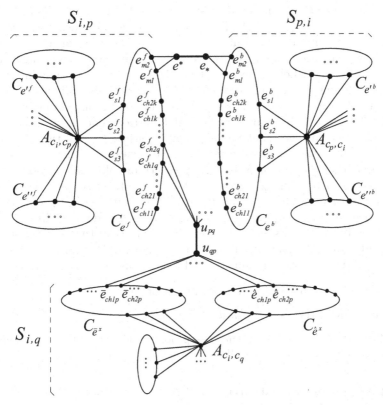

**Fig. 1.** Gadgets used in the reduction of the proof of Theorem 1 (we suppose $i < p$)

*Proof.* Note that if such a subgraph of $\mathscr{G}_G$ intersects a cycle $\mathcal{C}_{e^x}$, then it must contain all of its vertices. Further, if we remove all the vertices corresponding to arc gadgets in $\mathscr{G}_G$, then the remaining graph is a forest. These two facts together imply that any subgraph of $\mathscr{G}(G)$ of minimum degree at least three should intersect at least one cycle $\mathcal{C}_{e^x}$ corresponding to an arc gadget, and hence it must contain $\mathcal{C}_{e^x}$. □

**Claim 3.** *If $\mathscr{G}_G$ contains a subgraph of size $k' = (3k+1)k(k-1)$ and of minimum degree at least three, then $G$ has a multi-colored $k$-clique.*

*Proof.* Let $H = G[S]$ be a subgraph of size $k'$ of minimum degree at least three. Now, by Claim 2, $S$ must contain all the vertices of a cycle corresponding to an arc gadget. Furthermore, notice that to ensure the degree condition in $H$, once we have a vertex of a cycle in $S$ all the vertices of this cycle and their neighbors also are in $S$. Without loss of generality, let $\mathcal{C}_{e^f}$ be this cycle, and suppose that it belongs to the gadget $\mathcal{S}_{i,j}$, i.e., $e \in E[c_i, c_j]$ and $i < j$. Notice that, by construction, it forces some of the other vertices to belong to $S$. First, its match vertices force the cycle $\mathcal{C}_{e^b}$ of $\mathcal{S}_{j,i}$ to be in $S$. The coherence vertices of $\mathcal{C}_{e^f}$ force $S$ to contain at least one cycle in $\mathcal{S}_{i,l}$, for all $l \in \{1, \cdots, k\}$, $l \neq i$.

They in turn force $S$ to contain at least one cycle from the remaining gadgets $\mathcal{S}_{p,q}$ for all $p \neq q \in \{1, \ldots, k\}$. The selection vertices of each such cycle in $\mathcal{S}_{p,q}$ force $S$ to contain $A_{p,q}$. But because of our condition on the size of $S$ ($|S| = k'$), we can select *exactly one* cycle gadget from each of the gadgets $\mathcal{S}_{p,q}$, $p \neq q \in \{1, 2, \cdots, k\}$. Let $E'$ be the set of edges in $E(G)$ corresponding to arc gadgets selected in $S$. We claim that $G[V[E']]$ is a multi-color clique of size $k$ in $G$. Here $V[E']$ is a subset of vertices of $V(G)$ containing the end points of the edges in $E'$. First of all, because of the match vertices, once $e^f \in E'$, it forces $e^b \in E'$. To conclude the proof we only need to ensure that all the edges from a particular color class emanate from the same vertex. But this is ensured by the restriction on the size of $S$ and the presence of coherence vertices on the cycles selected in $S$ from $\mathcal{S}_{p,q}$, $p \neq q \in \{1, 2, \cdots, k\}$. To see this, let us take two arcs $e = (u, v) \in (E[c_i, c_p] \cap E')$ and $e' = (u', w) \in (E[c_i, c_q] \cap E')$. Now the 4 vertices $u_{pq}$, $u_{qp}$, $u'_{pq}$ and $u'_{qp}$ belong to $S$. If $u$ is different from $u'$ then it would imply that $S$ has at least 2 elements more than the expected size $k'$, which would contradict the condition on the size of $S$. All these facts together imply that $G[V(E')]$ forms a multi-colored $k$-clique in the original graph $G$.   $\square$

Claims 1 and 3 together yield the following theorem:

**Theorem 1.** $\mathrm{MSMD}_3$ *is* $W[1]$-*hard.*

We shall see in the next section that the proof of the Theorem 1 can be generalized for larger values of $d$, as well as for a few variants.

## 2.2   $W[1]$-Hardness of $\mathrm{MSMD}_d$ for $d \geq 4$

In this section we generalize the reduction given in Section 2 for $d \geq 4$. The main idea is to change the role of the cycles $\mathcal{C}_{e^x}$, $x \in \{b, f\}$, by $(d - 1)$-regular graphs of appropriate size. We show below all the necessary changes in the construction of the gadgets to ensure that the proof for $d = 3$ works for $d \geq 4$.

***Arc gadgets for*** $d \geq 4$: Let us take $\mathcal{C}$ to be a $(d - 1)$-regular graph of size $d + (d - 1)(k - 2) + (d - 1)$, if it exists (that is, if $(d - 1)$ is even or $k$ is odd). If it does not exist, we take a graph of size $(d + 1) + (d - 1)(k + 2) + (d - 1)$ and with regular degree $d - 1$ on the set $\mathcal{C}$ of $d + (d - 1)(k + 2) + (d - 1)$ vertices and degree $d$ on the last vertex $v$. As before, we replace each edge $e$ by two arcs $e^f$ and $e^b$. For each arc $e^x \in E[c_i, c_j]$, we add a copy of $\mathcal{C}$, that we call $\mathcal{C}_{e^x}$, with the following vertex set:

- *selection vertices:* $e^x_{s1}, e^x_{s2}, \cdots, e^x_{sd}$;
- *coherence vertices:* $e^x_{ch1r}, \cdots, e^x_{ch(d-1)r}$, for all $r \in \{1, \ldots, k\}$, $r \neq i, j$; and
- *match vertices:* $e^x_{m1}, \cdots, e^x_{m(d-1)}$.

***Selection gadgets for*** $d \geq 4$: Without loss of generality suppose that $x = f$. As before, we add a vertex $A_{c_i, c_j}$, and for every arc $e^f \in E[c_i, c_j]$ we add all the edges from $A_{c_i, c_j}$ to all the selection vertices of the graph $\mathcal{C}_{e^f}$. We call this gadget $\mathcal{S}_{i,j}$.

***Coherence gadgets for*** $d \geq 4$***:*** Fix an $i$, $1 \leq i \leq k$. Let us consider all the selection gadgets of the form $\mathcal{S}_{i,p}$, $p \in \{1, \cdots, k\}$ and $p \neq i$. For any $u \in V[c_i]$, and any two indices $p \neq q \leq k$, $p, q \neq i$, we add a new edge $\{u_{pq}, u_{qp}\}$. For every arc $e = (u, v) \in E[c_i, c_p]$, with $u \in V[c_i]$, we pick the graph $\mathcal{C}_{e^x}$, $x \in \{f, b\}$, depending on whether $e$ is forward or backward, and add $d - 1$ edges of the form $\{e_{ch1q}, u_{pq}\}, \{e_{ch2q}, u_{pq}\}, \ldots, \{e_{ch(d-1)q}, u_{pq}\}$. Similarly, for an edge $e = (u, w) \in E[c_i, c_q]$, with $u \in V[c_i]$, we pick the graph $\mathcal{C}_{e^x}$, $x \in \{f, b\}$, and add $d - 1$ edges of the form $\{e_{ch1p}, u_{qp}\}, \ldots, \{e_{ch(d-1)p}, u_{qp}\}$.

***Match gadgets for*** $d \geq 4$***:*** For the two arcs $e^f = (u, v)$ and $e^b = (v, u)$, we consider the two graphs $\mathcal{C}_{e^f}$ and $\mathcal{C}_{e^b}$ corresponding to $e^f$ and $e^b$. Now we add a matching edge $\{e^*, e_*\}$ and add all the edges of the form $\{e^f_{m1}, e^*\}, \ldots, \{e^f_{m(d-1)}, e^*\}$ and $\{e^b_{m1}, e_*\}, \ldots, \{e^b_{m1}, e_*\}$, where $e^f_{mi}$, $e^b_{mi}$ are match vertices of $\mathcal{C}_{e^f}$ and of $\mathcal{C}_{e^b}$, respectively.

This completes the construction of the gadgets, and the union of all of them defines the graph $\mathscr{G}_G$. It is not hard to see that a proof similar to that of Theorem 1 shows that $G$, an instance of multi-color clique, has a multi-colored clique of size $k$ if and only if $\mathscr{G}_G$ has a subgraph of size $k'$ with minimum degree $d$. Here $k'$ depends on the size of $(d-1)$-regular graph chosen in the construction of the arc gadget. For an example if we take a $(d-1)$-regular graph of size $d + (d-1)(k-2) + (d-1)$, then $k' = k(k-1)[(2d-1) + (k-2)(d-1) + (k-1)]$. The proof of the next theorem is along the lines of the proof of Theorem 1.

**Theorem 2.** $\mathrm{MSMD}_d$ *is* W[1]*-hard for all* $d \geq 3$.

If, instead of finding an induced subgraph of size at most $k$ of degree at least $d$ in $\mathrm{MSMD}_d$, we would like to find a $d$-regular induced subgraph (or subgraph) $H$ of $G$ of size at most $k$, we get MI-$d$-RSP and M-$d$-RSP problems respectively. Notice that the minimum subgraph of degree at least $d$ in the proofs of Theorems 1 and 2 turns out to be an induced subgraph of regular degree $d$ in $\mathcal{G}_G$. As a consequence we obtain the following corollary:

**Corollary 1.** MI-$d$-RSP *and* M-$d$-RSP *are* W[1]*-hard for all* $d \geq 3$.

# 3    Faster FPT Algorithms for Graphs with Bounded Local Tree-Width and Graphs with Excluded Minors

In this section we give a *fast* and *explicit* algorithm for $\mathrm{MSMD}_d$, $d \geq 3$, in graphs with bounded local tree-width and graphs excluding a fixed graph $M$ as a minor. In Section 3.1 we describe our algorithm for graphs with bounded local tree-width, and finally in Section 3.2 we give our parameterized algorithms for $\mathrm{MSMD}_d$ (for any $d \geq 3$) for classes of graphs excluding a fixed minor $M$. We need the definitions of local tree-width, clique-sum and $h$-nearly embeddable graphs to handle these graph classes.

The definition of tree-width, which has become quite standard now, can been generalized to take into account the local properties of $G$, and this is called *local tree-width*. To define it formally we first need to define the $r$-neighborhood of vertices of $G$. The *distance* $d_G(u, v)$ between two vertices $u$ and $v$ of $G$ is the length of a shortest path in $G$ from $u$ to $v$. For $r \geq 1$, a *r-neighborhood* of a vertex $v \in V$ is defined as $N_G^r(v) = \{u \mid d_G(v, u) \leq r\}$.

**Definition 2 (Local tree-width [14]).** *The* local tree-width *of a graph $G$ is a function $ltw^G : \mathbb{N} \to \mathbb{N}$ which associates to every integer $r \in \mathbb{N}$ the maximum tree-width of an $r$-neighborhood of vertices of $G$, i.e.*

$$ltw^G(r) = \max_{v \in V(G)} \{tw(G[N_G^r(v)])\}.$$

A graph class $\mathcal{G}$ has *bounded local tree-width* if there exists a function $f : \mathbb{N} \to \mathbb{N}$ such that for each graph $G \in \mathcal{G}$, and for each integer $r \in \mathbb{N}$, we have $ltw^G(r) \leq f(r)$. For a given function $f : \mathbb{N} \to \mathbb{N}$, $\mathcal{G}_f$ is the class of all graphs $G$ of local tree-width at most $f$, i.e. such that $ltw^G(r) \leq f(r)$ for every $r \in \mathbb{N}$. See [9] and [14] for more details.

Let us now provide the basics to understand the structure of the classes of graphs excluding a fixed graphs as a minor.

**Definition 3 (Clique-sum).** *Let $G_1 = (V_1, E_1)$ and $G_2 = (V_2, E_2)$ be two disjoint graphs, and $k \geq 0$ an integer. For $i = 1, 2$, let $W_i \subset V_i$ form a clique of size $h$ and let $G_i'$ be the graph obtained from $G_i$ by removing a set of edges (possibly empty) from the clique $G_i[W_i]$. Let $F : W_1 \to W_2$ be a bijection between $W_1$ and $W_2$. We define the $h$-clique-sum or the $h$-sum of $G_1$ and $G_2$, denoted by $G_1 \oplus_{h, F} G_2$, or simply $G_1 \oplus G_2$ if there is no confusion, as the graph obtained by taking the union of $G_1'$ and $G_2'$ by identifying $w \in W_1$ with $F(w) \in W_2$, and by removing all the multiple edges. The image of the vertices of $W_1$ and $W_2$ in $G_i \oplus G_2$ is called the* join *of the sum.*

Note that $\oplus$ is not well defined; different choices of $G_i'$ and the bijection $F$ can give different clique-sums. A sequence of $h$-sums, not necessarily unique, which result in a graph $G$, is called a *clique-sum decomposition* or, simply, a *clique-decomposition* of $G$.

**Definition 4 ($h$-nearly embeddable graph).** *Let $\Sigma$ be a surface with boundary cycles $C_1, \ldots, C_h$. A graph $G$ is $h$-nearly embeddable in $\Sigma$, if $G$ has a subset $X$ of size at most $h$, called* apices, *such that there are (possibly empty) subgraphs $G_0, \ldots, G_h$ of $G \setminus X$ such that*

1. $G \setminus X = G_0 \cup \cdots \cup G_h$;
2. $G_0$ *is embeddable in $\Sigma$ (we fix an embedding of $G_0$);*
3. $G_1, \ldots, G_h$ *are pairwise disjoint;*
4. *for $1 \leq \cdots \leq h$, let $U_i := \{u_{i_1}, \ldots, u_{i_{m_i}}\} = V(G_0) \cap V(G_i)$, $G_i$ has a path decomposition $(B_{ij})$, $1 \leq j \leq m_i$) of width at most $h$ such that*
   (a) *for $1 \leq i \leq h$ and for $1 \leq j \leq m_i$ we have $u_j \in B_{ij}$*
   (b) *for $1 \leq i \leq h$, we have $V(G_0) \cap C_i = \{u_{i_1}, \ldots, u_{i_{m_i}}\}$ and the points $u_{i_1}, \ldots, u_{i_{m_i}}$ appear on $C_i$ in this order (either walking through the cycles walk clockwise or counterclockwise).*

## 3.1   Graphs with Bounded Local Tree-Width

In order to prove our results we need the following lemma, which gives the time complexity of finding a smallest induced subgraph of degree at least $d$ in graphs with bounded tree-width.

**Lemma 1.** *Let $G$ be a graph on $n$ vertices with a tree-decomposition of width at most $t$, and let $d$ be a positive integer. Then in time $\mathcal{O}((d+1)^t(t+1)^{d^2}n)$ we can decide whether there exists an induced subgraph of degree at least $d$ in $G$ and if yes find one of the smallest size.*

As is usual in algorithms based on tree-decompositions, the proof uses the dynamic programming approach based on a given *nice* tree-decomposition, which at the end either produces a connected subgraph of $G$ of minimum degree at least $d$ and of size at most $k$, or decides that $G$ does not have any such subgraph. Given a tree-decomposition $(T, \mathcal{X})$, first we suppose that the tree $T$ is rooted at a fixed vertex $r$. A $\{0, 1, 2, 3, \ldots, d\}$-*coloring* of vertices in $X_i$ is a function $c_i : X_i \to \{0, 1, \ldots, d-1, d\}$. Let $supp(c) = \{v \in X_i \mid c(v) \neq 0\}$ be the *support* of $c$. For any such $\{0, 1, \ldots, d\}$-coloring $c$ of vertices in $X_i$, we denote by $a(i, c)$ the minimum size of an induced subgraph $H(i, c)$ of $G[X_i \cup \bigcup_{j \text{ a child of } i} X_j]$, which has degree $c(v)$ for every $v \in X_i$ with $c(v) \neq d$, and degree at least $d$ on its other vertices. Note that $H(i, c) \cap X_i = supp(c)$. If such a subgraph does not exist, we define $a(i, c) = +\infty$. We can then develop recursive formulas for $a(i, c)$, starting from the leaves of $T$. Looking at the values of $a(r, c)$ we can decide if such a subgraph exists in $G$. The complete proof of this lemma can be found in [1].

**Theorem 3.** *For any $d \geq 3$ and any function $f : \mathbb{N} \to \mathbb{N}$, MSMD$_d$ is fixed-parameter tractable on $\mathcal{G}_f$. Furthermore, the algorithm runs in time $\mathcal{O}((d+1)^{f(2k)}(f(2k)+1)^{d^2}n^2)$.*

*Proof.* Given the input graph $G = (V, E) \in \mathcal{G}_f$, that is, $G$ has a bounded local tree-width and the bound is given by the function $f$. We first notice that if there exists an induced subgraph $H \subseteq G$ of size at most $k$ and degree at least $d$, then $H$ can be supposed to be connected. Secondly, if we know a vertex $v$ of $H$, then $H$ is contained in $N_G^r[v]$, which has diameter at most $2k$. Hence there exists the desired $H$ if and only if there exists $v \in V$ such that $H$ is contained in $N_G^r[v]$. So to solve the problem we find in polynomial time a tree-decomposition of $N_G^r[v]$ for all $v \in V$ of width $f(2k)$, and then run the algorithm of Lemma 1 to obtain the desired result.                                                                    □

The function $f(k)$ is known to be $3k$, $C_g gk$, and $b(b-1)^{k-1}$ for planar graphs, graphs of genus $g$, and graphs of degree at most $b$, respectively [9,14]. Here $C_g$ is a constant depending only on the genus $g$ of the graph. As an easy corollary of Theorem 3 we have the following:

**Corollary 2.** *MSMD$_d$ can be solved in $\mathcal{O}((d+1)^{6k}(6k+1)^{d^2}n^2)$, $\mathcal{O}((d+1)^{2C_g gk}(2C_g gk+1)^{d^2}n^2)$ and $\mathcal{O}((d+1)^{2b(b-1)^{k-1}}(2b(b-1)^{k-1}+1)^{d^2}n^2)$ time in planar graphs, graphs of genus $g$, and graphs of degree at most $b$, respectively.*

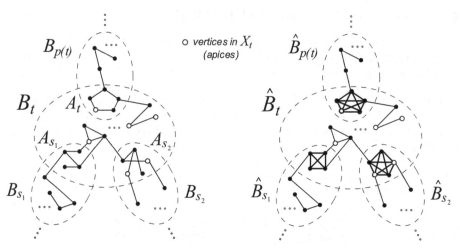

**Fig. 2.** Tree-decomposition of a minor free graph. The vertices in $X_t$ (i.e. the *apices*) are depicted by $\bigcirc$. Note that $B_{s_1}$ and $B_{s_2}$ could have non-empty intersection (in $B_t$).

### 3.2   *M*-Minor Free Graphs

In this section we give the results for the class of $M$-minor free graphs. To do so, we need the following theorem which gives the structural result about the class of graphs that are $h$-nearly embeddable in a fixed surface $\Sigma$. It says that they have linear local tree-width after removing the set of apices. More specifically, the result of Robertson and Seymour which was made algorithmic by Demaine et al. in [7], states the following:

**Theorem 4 ([19,7]).** *For every graph $M$ there exists an integer $h$, depending only on the size of $M$, such that every graph excluding $M$ as a minor can be obtained by clique sums of order at most $h$ from graphs that can be $h$-nearly embedded in a surface $\Sigma$ in which $M$ cannot be embedded. Furthermore, such a clique-decomposition can be found in polynomial time.*

Let $G$ be a $M$-minor free graph, and $(T, \mathcal{B} = \{B_t\})$ be a clique-decomposition of $G$ given by Theorem 4. Given this rooted tree $T$, we define $A_t := B_t \cap B_{p(t)}$ where $p(t)$ is the unique parent of the vertex $t$ in $T$, and $A_r = \emptyset$. Let $\hat{B}_t$ be the graph obtained from $B_t$ by adding all possible edges between the vertices of $A_t$ and also between the vertices of $A_s$, for each child $s$ of $t$, making $A_t$ and the $A_s$'s to induce cliques (see Fig. 2). In this way, $G$ becomes an $h$-clique sum of the graphs $\hat{B}_t$ according to the above tree $T$, and can also be viewed as a tree-decomposition given by $(T, \mathcal{B} = \{B_t\})$, where each $\hat{B}_t$ is $h$-nearly embeddable in a surface $\Sigma$ in which $M$ cannot be embedded. Let $X_t$ be the set of apices of $\hat{B}_t$. Then $|X_t| \leq h$, and $\hat{B}_t \setminus X_t$ has linear local tree-width by Theorem 4. We denote by $G_t$ the subgraph induced by all the vertices of $B_t \cup_s B_s$, $s$ being a descendant of $t$ in $T$.

Again to simplify the presentation, we give the proof for the case $d = 3$. Recall that we are looking for a subset of vertices $S$, of size at most $k$, which induces a graph $H = G[S]$ of minimum degree at least three.

Our algorithm consists of two levels of dynamic programming. The top level of dynamic programming runs over the clique-decomposition, and within each subproblem of this dynamic programming we focus on the induced subgraph of the vertices in $B_t$. Our first level of dynamic programming computes the size of a smallest subgraph of $G_t$, complying with degree constraints on the vertices of $A_t$. These constraints, as before, represent the degree of each vertex of $A_t$ in the subgraph $H_t := G_t[S_t]$, i.e. the *trace* of $H$ in $G_t$, where $S_t = S \cap V(G_t)$. This two level dynamic programming requires a combinatorial bound on the tree-width as a function of the parameter $k$ for each of the $B_t$'s, after removing the apices $X_t$ from $B_t$. This is done by making all possible choices in which they can interact with the desired solution. The next two lemmas are used later to obtain this combinatorial bound.

**Lemma 2.** *Let $H = G[S]$ be a connected induced subgraph of $G$. Then the subgraph $\hat{B}_t[S \cap B_t]$ is connected.*

The proof of Lemma 2 easily follows from the properties of a tree-decomposition and the fact that $A_t$ and $A_s$'s are cliques in $\hat{B}_t$, $s$ being a child of $t$ in $T$.

**Lemma 3.** *Let $H = G[S]$ be a smallest connected subgraph of $G$ of minimum degree at least three. Then the subgraph $\hat{B}_t[S_t \cap B_t \setminus X_t]$ has at most $3h + 1$ connected components, where $h$ is the integer given by Theorem 5.*

*Proof.* Let $C_1, \ldots, C_r$ be the connected components of $\hat{B}_t[S_t \cap B_t \setminus X_t] =: L$. We want to prove that $r \leq 3h + 1$. Assume to the contrary that $r > 3h + 1$. Now we form another solution $H'$ with size strictly smaller than $H$, which contradicts the fact that $H$ is of minimum size.

Let us now build the graph $H'$. For each vertex $v \in X_t \cap S_t$, let

$$b_v := \min\{d_{H_t}(v), 3\}$$

Then for each vertex $v \in X_t \cap S_t$, we choose at most $b_v$ connected components of $L$, covering at least $b_v$ neighbors of $v$ in $H_t$. We also add the connected component containing all the vertices of $A_t \setminus X_t$ (recall that $A_t$ induces a clique in $\hat{B}_t$). Let $A$ be the union of all the vertices of these connected components. Since $|X_t| \leq h$, $A$ has at most $3h + 1$ connected components. Also, since $A_s$ induces a clique in $\hat{B}_t$, for each child $s$ of $t$ such that $A_s \cap A \neq \emptyset$, we have that $A_s \setminus X_t \subset A$. We define $H'$ as follows

$$H' := G\left[\left(\bigcup_{\{s \,:\, A_s \cap A \neq \emptyset\}} S_s\right) \cup ((X_t \cup A) \cap S_t) \cup (S \setminus S_t)\right].$$

Clearly, $H' \subseteq H$. We have that $|H'| < |H|$ because, assuming that $r > 3h + 1$, there are some vertices of $H_t \subset H$ which are in some connected component $C_i$ which does not intersect $H'$.

Thus, it just remains to prove that $H'$ is indeed a solution of $\mathrm{MSMD}_3$, i. e., $H'$ has minimum degree at least 3. We prove it using a sequence of four simple claims:

**Claim 4.** The degree of each vertex $v \in (V(H') \cap X_t)$ is at least 3 in $H'$.

*Proof.* This is because each such vertex $v$ has degree at least $b_v$ in $H'_t$. If $d_v < 3$, then $v$ should be in $A_t$ (if not, $v$ has degree $d_v < 3$ in $H$, which is impossible), and hence $v$ is connected to at least $3 - d_v$ vertices in $S \setminus S_t$. But $S \setminus S_t$ is included in $H'$, and so every vertex of $X_t \cap V(H')$ has degree at least 3 in $H'$.                                                                  $\square$

**Claim 5.** The degree of each vertex in $(H \setminus H_t)$ is at least 3 in $H'$.

*Proof.* This follows because $A_t \cap H \subset H'$.                                  $\square$

**Claim 6.** The degree of each vertex in $A$ is at least 3 in $H'$.

*Proof.* Every vertex in $A$ has the same degree in both $H'$ and $H$. This is because $A$ is the union of some connected components, and no vertex of $A$ is connected to any other vertex in any other component.                           $\square$

**Claim 7.** Every other vertex of $H'$ also has degree at least 3.

*Proof.* To prove the claim we prove that the vertices of $H' \setminus (G[X_t] \cup (H \setminus H_t) \cup A)$ have degree at least 3 in $H'$. Remember that all these vertices are in some $S_s$, for some $s$ such that $A_s$ has a non-empty intersection with $A$. We claim that all these vertices have the same degree in both $H$ and $H'$. To prove this, note that $H' \cap A_s = H \cap A_s$ for all such $s$. Indeed, $(A_s \setminus X_t) \subset A$, and so $A_s \subset (A \cup X_t)$. Let $u$ be such a vertex. We can assume that $u \notin X_t$. If $u \in A_s$, then clearly $u \in A$, and so we are done. If $u \in (S_s \setminus A)$, then every neighbor of $u$ is in $H_s$. But $H_s \subset H'$, hence we are also done in this case.  $\square$

This concludes the proof of the lemma.                                            $\square$

We define a *coloring* of $A_t$ to be a function $c : A_t \cap S \to \{0, 1, 2, 3\}$. For $i < 3$, $c(v) = i$ means that the vertex $v$ has degree $i$ in the subgraph $H_t$ of $G_t$ that we are looking for, and $c(v) = 3$ means that $v$ has degree at least three in $H_t$. By $a(t, c)$ we denote the minimum size of a subgraph of $G_t$ with the prescribed degrees in $A_t$ according to $c$. We describe in what follows the different steps of our algorithm.

Recursively, starting from the leaves of $T$ to the root, for each node $t \in V(T)$ and for every coloring $c$ of $A_t$, we compute $a(t, c)$ from the values of $a(s, c)$, where $s$ is a child of $t$, or we store $a(t, c) = +\infty$ if no such subgraph exists. The steps involved in computing $a(t, c)$ for a fixed coloring $c$ are the following:

(i) We guess a subset $R_t \subseteq X_t \setminus A_t$ such that $R_t \subseteq S_t$. We have at most $2^h$ choices for $R_t$.

(ii) For each vertex $v$ in $R_t$, we guess whether $v$ is adjacent to a vertex of $B_t \setminus (R_t \cup A_t)$, i.e. we test all the 2-colorings $\gamma : R_t \to \{0, 1\}$ such that $\gamma(v) = 1$ means that $v$ is adjacent to a vertex of $B_t \setminus (R_t \cup A_t)$. The number of such colorings is at most $2^h$. For a fixed coloring $\gamma$, we guess one vertex in $B_t \setminus (R_t \cup A_t)$, which we suppose to be in $S_t$, for each of the vertices $v$ in $R_t$ such that $\gamma(v) = 1$. For each coloring $\gamma$, we have at most $n^h$ choices for the new vertices which could be included in $S_t$. If a vertex has $\gamma(v) = 0$, it is not allowed to be adjacent to any vertex of $B_t$ besides the vertices in $A_t \cup R_t$. Let $D_t^\gamma$ be the chosen vertices at this level.

(iii) Now we remove all the vertices of $X_t$ from $B_t$. Lemma 3 ensures that the induced graph $\hat{B}_t[S_t \cap B_t \setminus X_t]$ has at most $3h + 1$ connected components. Now we guess these connected components of $\hat{B}_t[S_t \cap B_t \setminus X_t]$ by guessing a vertex from these connected components in $B_t \setminus X_t$. Since we need to choose at most $3h + 1$ vertices this way, we have at most $(3h + 1)n^{3h+1}$ new choices. Let these newly chosen vertices be $F_t^\gamma$ and

$$R_t^\gamma = R_t \cup D_t^\gamma \cup F_t^\gamma \cup \{v \in A_t \setminus X_t \mid c(v) \neq 0\}.$$

Let $G_t^*$ be the graph induced by the $k$-neighborhood (vertices at distance at most $k$) of all vertices of $R_t^\gamma$ in $\hat{B}_t \setminus X_t$, i.e. $G_t^* = (\hat{B}_t \setminus X_t)[N^k(R_t^\gamma)]$.

(iv) Each connected component of $G_t^*$ has diameter at most $2k$ in $\hat{B}_t \setminus X_t$. As $\hat{B}_t \setminus X_t$ has bounded local tree-width, this implies that $G_t^*$ has tree-width bounded by a function of $k$. By the result of Demaine and Hajiaghayi [6], this function can be chosen to be linear.

(v) In this step we first find a tree-decomposition $(\mathcal{T}_\gamma, \{U_p\})$ of $G_t^*$. Since $A_s \cap G_t^*$ is a clique, it appears in a bag of this tree-decomposition. Let the node representing this bag in this tree be $p$. Now we make a new bag containing the vertices of $A_s \cap G_t^*$ and make it a leaf of the tree $\mathcal{T}_\gamma$ by adding a node and connecting this node to $p$. By abuse of notation, we denote by $s$ this distinguished leaf containing the bag $A_s \cap G_t^*$. We also add all the vertices of $A_t$ to all the bags of this tree-decomposition, increasing the bag size by at most $h$. Now we apply a dynamic programming algorithm similar to the one we used for the bounded local tree-width case. Remember that for each child $s$ of $t$, we have a leaf in this decomposition with the bag $A_s \cap G_s^*$. The aim is to find an induced subgraph which respects all the choices we made at earlier steps above, and with the minimum size.

   We start from the leaves of this decomposition $\mathcal{T}_\gamma$ and move towards its root. At this point we have all the values of $a(s, c')$ for all possible colorings $c'$ of $A_s$, where $s$ is a child of $t$ (because of the first level of dynamic programming). To compute $a(t, c)$ we apply the dynamic programming algorithm of Lemma 1 with the restriction that for each *distinguished* leaf $s$ of this decomposition, we already have all the values $a(s, c)$ for all colorings of $A_s \cap G_s^*$ (we extend this coloring to all $A_s$ by giving the zero values to the vertices of $A_s \setminus G_s^*$). Note that the only difference between this dynamic programming and the one of Lemma 1 is the way we initialize the leaves of the tree.

(vi) Return the minimum size of a subgraph with the degree constraint $c$ on $A_t$, among all the subgraphs we found in this way. Let $a(t, c)$ be this minimum.

(vii) If for some vertex $t$ and a colouring $c : A_t \to \{0, 3\}$, we have $1 \leq a(t, c) \leq k$, return YES. If not, we conclude that such a subgraph does not exist.

This completes the description of the algorithm. Now we discuss its time complexity. Let $C_M$ be the constant depending only on the linear local tree-width of the surfaces in which $M$ cannot be embedded. For each fixed coloring $c$, we need time $4^{\mathcal{C}_M k}(C_M k + 1)^9 n^{4h+1}$ to obtain $a(t, c)$, where $t \in T$. Since the number of colorings of each $A_t$ is at most $4^h$, and the size of the clique-decomposition is $O(n)$, we get the following theorem:

**Theorem 5.** *Let $\mathcal{C}$ be the class of graphs with excluded minor $M$. Then, for any graph in this class, one can find an induced subgraph of size at most $k$ with degree at least 3 in $\mathcal{C}$ in time $\mathcal{O}(4^{\mathcal{O}(k+h)}(\mathcal{O}(k))^9 n^{\mathcal{O}(1)})$, where the constants in the exponents depend only on $M$.*

Theorem 5 can be generalized to larger values of $d$ with slight modifications. We state the following theorem without a proof:

**Theorem 6.** *Let $\mathcal{C}$ be a class of graphs with an excluded minor $M$. Then, for any graph in this class, one can find an induced subgraph of size at most $k$ with degree at least $d$ in $\mathcal{C}$ in time $\mathcal{O}((d+1)^{\mathcal{O}(k+h)}(\mathcal{O}(k))^{d^2} n^{\mathcal{O}(1)})$, where the constants in the exponents depend only on $M$.*

## 4   Conclusions

In this paper we have introduced the $\text{MSMD}_d$ problem, a generalization of the problem of finding a shortest cycle in a graph, and we have studied its parameterized complexity. We have shown that $\text{MSMD}_d$ for $d \geq 3$ is $W[1]$-hard in general undirected graphs, and we have given explicit fixed-parameter tractable algorithms when the input graph is of bounded local tree-width or excludes a fixed minor $M$. These algorithms are faster than those coming from the meta-theorem of Frick and Grohe [13] about problems definable in first-order logic over "locally tree-decomposable structures". We believe that our algorithmic initiations will trigger further research on the problem. This will help us in understanding not only this problem, but also the closely related problems of DENSE $k$-SUBGRAPH and TRAFFIC GROOMING. The parameterized tractability of the TRAFFIC GROOMING problem still remains open.

**Acknowledgement.** The authors would like to thank M. Fellows , D. Lokshtanov and S. Pérennes for insightful discussions. We would especially like to thank D. Lokshtanov for his help in proving the hardness result and N. Misra for carefully reading the manuscript.

## References

1. Amini, O., Sau, I., Saurabh, S.: Parameterized Complexity of the Smallest Degree-Constrained Subgraph Problem, INRIA Technical Report 6237 (2007) (accessible in first author's homepage)
2. Amini, O., Pérennes, S., Sau, I.: Hardness and Approximation of Traffic Grooming. In: Tokuyama, T. (ed.) ISAAC 2007. LNCS, vol. 4835, pp. 561–573. Springer, Heidelberg (2007)

3. Andersen, R.: Finding large and small dense subgraphs (submitted, 2007), http://arXiv:cs/0702032v1
4. Chor, B., Fellows, M., Ragan, M.A., Razgon, I., Rosamond, F., Snir, S.: Connected coloring completion for general graphs: Algorithms and complexity. In: Lin, G. (ed.) COCOON 2007. LNCS, vol. 4598, pp. 75–85. Springer, Heidelberg (2007)
5. Dawar, A., Grohe, M., Kreutzer, S.: Locally Excluding a Minor. In: LICS, pp. 270–279 (2007)
6. Demaine, E., Haijaghayi, M.T.: Equivalence of Local Treewidth and Linear Local Treewidth and its Algorithmic Applications. In: SODA, pp. 840–849 (2004)
7. Demaine, E., Hajiaghayi, M.T., Kawarabayashi, K.C.: Algorithmic Graph Minor Theory: Decomposition, Approximation and Coloring. In: FOCS, pp. 637–646 (2005)
8. Dutta, R., Rouskas, N.: Traffic grooming in WDM networks: Past and future. IEEE Network 16(6), 46–56 (2002)
9. Eppstein, D.: Diameter and Tree-width in Minor-closed Graph Families. Algorithmica 27(3–4), 275–291 (2000)
10. Feige, U., Kortsarz, G., Peleg, D.: The Dense $k$-Subgraph Problem. Algorithmica 29(3), 410–421 (2001)
11. Fellows, M., Hermelin, D., Rosamond, F.: On the fixed-parameter intractability and tractability of multiple-interval graph properties. Manuscript (2007)
12. Goemans, M.X.: Minimum Bounded-Degree Spanning Trees. In: FOCS, pp. 273–282 (2006)
13. Frick, M., Grohe, M.: Deciding first-order properties of locally tree-decomposable structures. J. ACM 48(6), 1184–1206 (2001)
14. Grohe, M.: Local Tree-width, Excluded Minors and Approximation Algorithms. Combinatorica 23(4), 613–632 (2003)
15. Khot, S.: Ruling out PTAS for graph min-bisection, densest subgraph and bipartite clique. In: FOCS, pp. 136–145 (2004)
16. Klein, P.N., Krishnan, R., Raghavachari, B., Ravi, R.: Approximation algorithms for finding low-degree subgraphs. Networks 44(3), 203–215 (2004)
17. Könemann, J.: Approximation Algorithms for Minimum-Cost Low-Degree Subgraphs, PhD Thesis (2003)
18. Lawler, E.L.: Combinatorial Optimization: Networks and Matroids, Holt, Rinehart and Winston (1976)
19. Robertson, N., Seymour, P.: Graph minors XVI, Excluding a non-planar graph. J. Comb. Theory, Series B 77, 1–27 (1999)

# Fixed Structure Complexity

Yonatan Aumann and Yair Dombb

Department of Computer Science
Bar-Ilan University
Ramat Gan 52900, Israel
`aumann@cs.bu.ac.il, yairbiu@gmail.com`

**Abstract.** We consider a non-standard parametrization, where, for problems consisting of a combinatorial structure and a number, we parameterize by the combinatorial structure, rather than by the number. For example, in the *Short-Nondeterministic-Halt* problem, which is to determine if a nondeterministic machine $M$ accepts the empty string in $t$ steps, we parameterize by $|M|$, rather than $t$. We call such parametrization *fixed structure parametrization*. Fixed structure parametrization not only provides a new set of parameterized problems, but also results in problems that do not seem to fall within the classical parameterized complexity classes. In this paper we take the first steps in understanding these problems. We define fixed structure analogues of various classical problems, including graph problems, and provide complexity, hardness and equivalence results.

## 1 Introduction

*Motivating Examples.* Consider the classical *Tiling* problem. Given a set of tiles $T$ and integer $t$, decide whether it is possible to tile the $t \times t$ area with tiles from $T$. The general problem is NP-complete. Parameterizing by $t$ it is W[1]-complete. But what if we parameterize by $|T|$, is the problem FPT? where does it fall in the fixed parameter hierarchy? The naive algorithm takes $O(|T|^{t^2})$ steps which does not even constitute an XP algorithm for the parameter $|T|$. Is the problem in XP? (I.e. is there an $O(t^{f(|T|)})$ algorithm?

Next, consider the following generalization of the Hamiltonian cycle problem. Given a graph $G$ (on $n$ vertices) and integers $\boldsymbol{m} = (m_1, \ldots, m_n)$, determine if there is a cycle that visits node $v_i$ exactly $m_i$ times. Clearly, this problem is NP-complete. If parameterized by $\boldsymbol{m}$ it is para-NP complete. But what if we parameterize by $|G|$? Does the problem then become FPT? Is it XP? Para-NP-complete?

This paper aims at developing the theory and tools to answer questions such as the above.

*Fixed Structure Parametrization.* In general, many problems can naturally be viewed as composed of two parts: (i) a combinatorial structure (e.g. a set of tiles, a graph), and (ii) a number(s) (size of area to be tiled, number of visits). The standard parametrization most often takes the "number part" as the parameter, and asks whether the problem is polynomial in the combinatorial structure.

M. Grohe and R. Niedermeier (Eds.): IWPEC 2008, LNCS 5018, pp. 30–42, 2008.
© Springer-Verlag Berlin Heidelberg 2008

Here, we reverse the question, taking the combinatorial structure as the parameter, and asking whether the problem is polynomial in the "number part".[1] Accordingly, we call our parametrization *fixed structure parametrization*, producing *fixed structure problems*. Interestingly, despite the extended literature on parameterized complexity, *fixed structure* parameterizations have enjoyed little consideration, and in no systematic way.

As it turns out, *fixed structure* problems not only provide a new set of parameterized problems, but also result in problems that do not seem to fall within the classical parameterized complexity classes. Rather they seem to form complexity classes of their own. In this paper we aim at taking the first steps in understanding this *fixed structure complexity*.

*Summary of Results.* Some fixed structure problems, such as the generalized Hamiltonian cycle problem mentioned above, we could show to be FPT. For others, we identified three different equivalence classes of fixed structure problems, all of which seem to consist of problems that are not FPT. The first of these classes is defined by the following fixed structure variant of short-NSTM-Halt:

---

FIXED-STRUCTURE-SHORT-NTM-EXACT-HALT (*FS-Exact-Halt*)
   Instance: Non-deterministic Turing machine $M$; integer $t$ (in unary)
Parameter: $|M|$
   Problem: Does $M$ accept the empty string in exactly $t$ steps?

---

The naive algorithm for this problems runs in time $\Theta(|M|^t)$. Thus, the question is not only if the problem is FPT, but also if it is altogether in XP. We prove:

**Theorem 1.** *If FS-Exact-Halt $\in XP$ then NEXP=EXP.*

Accordingly, in this work we also consider XP reductions, not only FPT ones (when the reductions are also FPT, we note so). We prove:

**Theorem 2.** *The following are equivalent to FS-Exact-Halt:*
- *Fixed-structure tiling of the $t \times t$ torus (under FPT reductions).*
- *Fixed-structure clique, independent set and vertex-cover (under XP reductions).*

The exact definitions of the fixed-structure versions of the graph problems are provided in Section 3. We note that the classical reductions, used for proving the NP-completeness of the above problems, often fail in our fixed structure setting, as they "hardwire" the number part into combinatorial structure (for more details see the proof of Theorem 2 in Section 4).

We define two additional fixed-structure equivalence classes, using similar machine characterizations:

---

[1] Clearly, there is no formal distinction between the "number part" of a problem and the "combinatorial structure"; both are strings over some alphabet. In practice, however, the two elements are frequently clearly distinct.

FIXED-STRUCTURE-SHORT-NTM-HALT (*FS-Halt*)
Instance: Non-deterministic Turing machine $M$; integer $t$ (in unary)
Parameter: $|M|$
Problem: Does $M$ accept the empty string in at most $t$ steps?

FIXED-STRUCTURE-SHORT-NTM-NOT-HALT (*FS-Not-Halt*)
Instance: Non-deterministic Turing machine $M$; integer $t$ (in unary)
Parameter: $|M|$
Problem: Does $M$ have a computation on the empty string not halting
for at least $t$ steps?

Note that in all three problems we did not specify the number of tapes. The reason is that we prove that any of them is equivalent for any number of tapes, even under FPT reductions (unlike the case for the classical parametrization [1]).

Under classical complexity, as well under the standard parametrization, the three problems *FS-Halt*, *FS-Not-Halt* and *FS-Exact-Halt* are equivalent. Interestingly, under the fixed-structure parametrization this is not the case, even under XP reductions (hence clearly also under FPT reductions). We prove:

**Theorem 3**

1. *FS-Halt* $\leq^{fpt}$ *FS-Exact-Halt* and *FS-Not-Halt* $\leq^{fpt}$ *FS-Exact-Halt, but*
2. *If* *FS-Halt* $\equiv^{XP}$ *FS-Exact-Halt or* *FS-Not-Halt* $\equiv^{XP}$ *FS-Exact-Halt then*
   *NEXP=EXP.*

Some problems we show are equivalent to *FS-Halt* or to *FS-Not-Halt*:

**Theorem 4**

- *Fixed-Structure-Short-Post-Correspondence and Fixed-Structure-Short-Grammar-Derivation are equivalent to FS-Halt (under FPT reductions).*
- *Fixed-Structure-Restricted-Tiling (i.e. tiling with a specified origin tile) of the $t \times t$ plane is equivalent to FS-Not-Halt (under FPT reductions).*

Thus, in the fixed structure setting, tiling of the plane and tiling of the torus are *not* equivalent.

*Related Work.* To the best of our knowledge, there has been no systematic analysis of fixed-structure parametrization. Of the 376 parameterized problems described in the *Compendium of Parameterized Problems* [2], we identified few problems that we would classify as fixed-structure, in the sense we consider here. Notably, the following two: the parameterizations of *Bounded-DFA-Intersection* by $k$, $|\Sigma|$, and $q$, which is reported open in [3], and the parametrization of the *Rush-Hour-Puzzle* by $C$ (the set of cars), which is shown to be FPT in [4], by exhaustive search.

Related to fixed-structure problems are those problems where instances are composed of *two* combinatorial structures, and one of the structures is taken as the parameter. Many of the database query problems are of this type [5,6]. In this case there are two combinatorial structures – the query $Q$ and the database $d$, and the parameter is either $|Q|$ or the number of variables within. The full characterization of the complexity of several of these problems is still open [5]. It is interesting to note that Vardi's initial work [7] considers both parameterizations – both by $|Q|$ and by $|d|$ (though, naturally, without explicit use of parameterized complexity terminology).

Other problems with two combinatorial structures are ordering problems on graphs, lattices and the like, e.g. given graphs $H$ and $G$ decide whether $H$ is a minor of $G$ (see [8,9]). Some of these problems have been resolved, but the complexity of others is still open. While these problems are not strictly fixed-structure in the sense we consider here, it would be interesting to see if the theory developed here may be relevant to these problems as well.

*Organization of the Paper.* Unfortunately, the space limitations of this extended abstract allow us to provide only a small fraction of the results and proofs. In particular, all the positive results showing inclusion in FPT are omitted. Here, we focus on hardness results alone. The full set of results and proofs will appear in the full version. The rest of the paper is organized as follows. In the next section we prove Theorems 1 and 3. Section 3 introduces the exact definition of the fixed-structure versions of the graph problems. Section 4 provides the highlights of the proof of Theorem 2. We conclude with open problems in Section 5.

## 2   Complexity and Hardness

We now prove Theorems 1 and 3, as well as an additional theorem.

**Theorem 1 (repeated)** *If FS-Exact-Halt $\in$ XP then EXP = NEXP.*

*Proof.* Let $L \in$ NEXP. Then there exists a nondeterministic Turing machine $M_L$, which for some constant $c$ decides on every input $x$ whether $x \in L$ in time $2^{|x|^c}$. We construct a new Turing machine $M_L'$ as follows. On the empty tape, $M_L'$ first nondeterministically chooses some $x \in \Sigma^*$, and then runs $M_L$ on this $x$. If $M_L$ accepts $x$, then $M_L'$ idles until *exactly* $2^{|x|^c} + x$ steps have elapsed since the beginning of its run, and then accepts.[2] Otherwise, $M_L'$ rejects.

Assume that *FS-Exact-Halt* $\in$ XP. Then there exists an algorithm $A$ and an arbitrary function $f(\cdot)$, such that $A(M, t)$ decides in $|(M, t)|^{f(|M|)} + f(|M|)$ steps whether $M$ has a computation that accepts the empty string in exactly $t$ steps. Given an input $x$, for which we want to decide whether $x \in L$, we simply run $A$ on the input $(M_L', t)$, with $t = 2^{|x|^c} + x$. Note that the function $x \hookrightarrow 2^{|x|^c} + x$ is a bijection, and therefore $M_L'$ accepts in exactly $t$ steps iff $x \in L$. In addition, $A$ runs in time $|(M_L', t)|^{f(|M_L'|)} + f(|M_L'|) = O(2^{x^{c'}})$, for some $c'$ depending only on $M_L'$. Thus, $L \in$ EXP and EXP = NEXP.                                   $\square$

Next, we show an interesting, albeit easy, result, which will also serve us in our next proof:

**Theorem 5.** *FS-Halt and FS-Not-Halt are non-uniform FPT.*

---

[2] Note that it must be shown that this counting can be performed in the fixed structure setting. The proof is omitted here and provided in the full version.

*Proof.* Consider the *FS-Halt* problem (the proof for *FS-Not-Halt* is analogous). To prove that the problem is non-uniform FPT we need to construct, for every size $k$, an algorithm $A_k$, such that for every $M$, with $|M| = k$, and every $t$, decides whether $(M, t) \in$ *FS-Halt* in time $O(t^\alpha)$ for some constant $\alpha$. We do so by simply creating a table exhaustively listing, for each Turing machine $M$ with $|M| = k$, the minimum number of steps in which $M$ can accept on the empty string. Given an input $(M, t)$ the algorithm consults this table, comparing $t$ to this minimum. Clearly, the algorithm is correct and runs in polynomial time. Note, however, that constructing this table is, in general, undecidable.    □

**Theorem 3 (repeated)**

1. *FS-Halt $\leq^{fpt}$ FS-Exact-Halt and FS-Not-Halt $\leq^{fpt}$ FS-Exact-Halt, but*
2. *If FS-Halt $\equiv^{XP}$ FS-Exact-Halt or FS-Not-Halt $\equiv^{XP}$ FS-Exact-Halt then NEXP=EXP.*

*Proof.* Due to lack of space, the proof of (1) is omitted. The proof of (2) combines the techniques of Theorems 1 and 5. Suppose that $R$ is an XP reduction from *FS-Exact-Halt* to *FS-Halt*. Let $L$, $M_L$ and $M'_L$ be as defined in the proof of Theorem 1. Then, for any $x$, $x \in L$ iff $M'_L$ accepts the empty string in exactly $g(x) = 2^{|x|^c} + x$ steps. Denote $R(M'_L, t) = (N, s)$. Then, since $R$ is an XP reduction, $|N|$ must be bounded by a function of $M'_L$ alone. Thus, there is only a finite number such $N$'s (that are the result of applying $R$ to $M'_L$ for some $t$). Thus, after the reduction we need only consider a finite number of slices of *FS-Halt*. By Theorem 5 there is a polynomial algorithm for each of these slices. Hence, combining $R$ with the union of these algorithms we obtain an EXP algorithm for $L$. Note however, that this proof is non-constructive, as are the algorithms provided by Theorem 5.    □

# 3    Defining Fixed-Structure Graph Problems

We are interested in defining fixed-structure versions of common graph problems. This seems easy: many graph problems are naturally composed of a graph and a number. Thus, to obtain a fixed structure version simply parameterize by the graph structure. However, this approach results in non-interesting problems. The reason is that the size of the graph necessarily bounds "the number" (e.g. the number of colors is at most the number of nodes), and the resulting problems are trivially FPT. Thus, in order to obtain meaningful fixed structure graph problems, we must be able to define *families* of graphs (of increasing sizes), all of which share a *common* underlying structure. In a way, the grid is an example of such a graph family; it comes in many sizes, but all share a common core structure. Cliques, hypercubes and cycles are other examples of such graph families.

We now give a general definition of such graph families, which we call *parametric graphs*. The basic idea is to define the graphs using *expressions* that accept *parameters*. The expression defines a graph by applying standard *graph operations* to a set of *base graphs* and *parameters*. The *base graphs* can be any explicitly represented graphs. The *operations* combine these graphs to obtain larger and more complex ones. The operations we consider are:

- Union: for graphs $G_1 = (V_1, E_1)$ and $G_2 = (V_2, E_2)$, the union graph $G = G_1 \cup G_2$ is the graph $G = (V_1 \cup V_2, E_1 \cup E_2)$.
- Multiplication (by a scalar): for a graph $G$ and integer $i$, the $i$-multiplicity of $G$, denoted by $i \cdot G$, is the union of $i$ separate copies of $G$.
- Sum: defined on graphs *over the same set of vertices*, $G_1 = (V, E_1)$ and $G_2 = (V, E_2)$. The sum graph $G = G_1 + G_2$ has the union of the edges from both graphs, $G = (V, E_1 \cup E_2)$.
- Direct product (also known as *tensor product*): for graphs $G_1 = (V_1, E_1)$ and $G_2 = (V_2, E_2)$, their direct product is the graph $G = G_1 \times G_2 = (V_1 \times V_2, E)$ such that $((v_1, v_2), (w_1, w_2)) \in E$ iff $(v_1, w_1) \in E_1$ and $(v_2, w_2) \in E_2$.

Using these operations it is possible to construct large and complex graphs from smaller ones. For example, the cycle with 14 vertices can be constructed as:

$$C_{14} = \left( \left[ \begin{smallmatrix} \bullet \\ \bullet \end{smallmatrix} \right] \cup 3 \cdot \left[ \begin{smallmatrix} \bullet\!-\!\bullet \\ \bullet\!-\!\bullet \end{smallmatrix} \right] \right) + \left( 3 \cdot \left[ \begin{smallmatrix} \bullet\!-\!\bullet \\ \bullet\!-\!\bullet \end{smallmatrix} \right] \cup \left[ \begin{smallmatrix} \bullet \\ \bullet \end{smallmatrix} \right] \right) = \left[ \begin{smallmatrix} \bullet\!-\!\bullet\!-\!\bullet\!\cdots\!\bullet\!-\!\bullet \end{smallmatrix} \right]$$

Since the sum operation requires that both graphs share the same set of vertices, w.l.o.g. we assume that the vertex set of any graph we consider is simply the integers, $\{1, \ldots, |V|\}$. Thus, following a union, multiplication or product operation, the vertices must be renamed. We do so systematically "lexicographically", as follows. For the the product and multiplication operations, the new order is simply the lexicographic order on the new vertices. For $G = G_1 \cup G_2$, vertexes of each graphs retain their original order, and all those of $G_1$ are come before those of $G_2$.

The multiplication operation allows us to define expressions that accept parameters. This way a single expression can define graphs of varying sizes, all sharing a common, underlying combinatorial structure. Thus, we define a *parametric graph* as an expression of the above format that accepts parameters. Using this notion we can define fixed-structure versions of classical graph problems. For example, the fixed-structure version of Independent-Set is the following:

---

FIXED STRUCTURE INDEPENDENT SET $(\cup, \cdot, +, \times)$
  Instance: a *parametric graph* expression $G$ (using the operations $\cup, \cdot, +$, and $\times$); a vector $t$ of integer parameter values to $G$ (in unary); integer $\psi$.
Parameter: $|G|$
  Problem: Does $G(t)$ have an independent set of size $\psi$?

---

Note the complexity of the problem may depend on the set of operations used in the graph expressions. Hence, the definition of the problem explicitly lists these operations $((\cup, \cdot, +, \times)$ in our case). Also, note that the problem is not necessarily in NP, as the size of the resultant graph is not polynomially bounded in the input size.

Fixed-structure versions for other graph problems are defined similarly.

## 4    Problems Equivalent to *FS-Exact-Halt*

We now provide an outline for the proof of Theorem 2. Unfortunately, we cannot provide all the details, but do hope that our exposition provides a flavor of the problems one encounters in fixed structure reductions, and some of the methods we use to overcome these problems.

**Lemma 1.** *FS-Exact-Halt* $\leq^{fpt}$ *Fixed-structure-Torus-Tiling.*

*Proof.* Let $M$ be a Turing machine and $t$ an integer. We construct a tile set $T = T(M)$ and an integer $s = s(t)$, such that $T$ has a valid tiling of the $s \times s$ torus iff $M$ accepts the empty string in exactly $t$ steps. The core of the proof follows the reduction used for proving the undecidability of tiling of the infinite first-quadrant. The problem arises, however, when trying to convert this construction to the bounded case, as discussed below.

The basic idea of the reduction from Turing machine computation to tiling is to make each valid tiling represent a run of the Turing machine: every row in the tiling corresponds to a configuration, and one row can be placed on top of another only if the configuration corresponding to the bottom row yields the one corresponding to the top row. Provided that the first row represents the initial configuration of $M$, we obtain that the infinite first-quadrant can be tiled iff $M$ has a non-halting computation (see, for example, [10] for more details).

In order to use this reduction for proving hardness of torus-tiling, it is not difficult to augment the construction, so that: (i) the row corresponding to the initial configuration can only be placed directly above a row corresponding to an accepting configuration, and (ii) the leftmost end of each row can be placed directly to the right of the rightmost end. Thus, if $M$ accepts in $t$ steps then the $t \times t$ torus can be tiled with $T(M)$. Unfortunately, the converse is not necessarily true. The problem is that the $t \times t$ torus can be split into smaller regions, each corresponding to a shorter accepting computations. For example, the $10 \times 10$ torus can be tiled with four copies of a tiling of $5 \times 5$ tori. Another subtle point is guaranteeing that the first row indeed corresponds to the machine's initial configuration. For the unbounded case, this is provided by a careful and complex construction, provided in [11]. Unfortunately, this construction does not seem to carry over to the bounded case.

If we were to prove standard NP-completeness, the following simple and standard construction solves both of the above problems. Let $T = T(M)$ be the tile set obtained by the unbounded reduction. We create a new tile set $T'$, such that for each tile $z \in T$, we have (essentially) $t^2$ copies, $z^{(1,1)}, \ldots, z^{(t,t)}$, one for each torus location. It is now easy to configure the tiles such that tile $z^{(i,j)}$ can appear only at location $(i,j)$ (for all $z, i, j$). In this way we have eliminated the possibility to cover the torus by copies of smaller tori. In addition, we can force the first row to whatever we wish, by eliminating all but the appropriate tiles for this row. This construction fails, however, in the fixed structure setting. The reason is that the reduction "hardwires" the number $t$ into $T'$. By doing so, however, we have moved $t$ into the "parameter" part of the instance, which is forbidden in parameterized reductions.

Thus, we provide a solution that works independently of $t$, as follows. We construct a set of "base tiles" upon which the original "machine simulation" tiles are then superimposed. The "base tiles" are constructed such that they only admit a specific tiling, which forces the tiling to "behave" as desired.

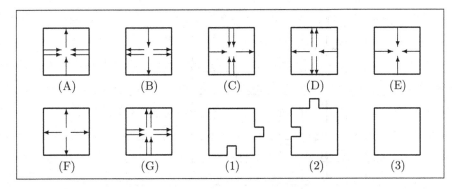

**Fig. 1.** The base tiles

*The Base Tiling.* We start by creating the tiles (A)-(G) depicted in Figure 1. One can observe that any torus can be tiled using this set of tiles, and that any tiling of an *odd* sized torus can be decomposed into rectangular regions such that (see Figure 2):

1. Tile (G) is placed at the bottom left corner of the region.
2. The rest of the bottom row is composed of tiles (A) and (B) only.
3. The rest of the leftmost column is composed of tiles (C) and (D) only.

We call such a region a *core-region*. Note that tile (G) can only be placed above tiles (D) or (G), and to the right of tiles (B) or (G). Thus, we obtain that if a core-region $R$ is directly above another core-region $S$, then $R$ and $S$ have the same width. Likewise, if $R$ is directly to the right of $S$, then $R$ and $S$ have the same height. Therefore, all core-regions must have the same size.

In order to force the core-regions to be *square*, we use the numbered tiles ((1)-(3)) of Figure 1, which will be superimposed on the tiles (A)-(G). It is immediate that tile (1) can only appear above tile (2), and that tile (2) can only appear to the right to tile (1). Thus, tiles (1) and (2) necessarily form a diagonal. We superimpose tile (1) on tiles (E) and (G); tile (2) on tiles (A), (D) and (F); and tile (3) on tiles (A) though (F). This forces the core-regions to be square, because the diagonal starting with the (G) on the bottom left corner must hit a (G) tile at its other end. In all, we obtain that any tiling of an odd-sized torus using this set of tiles, which we call the *base tiles*, can be decomposed into square core-regions, all of identical sizes.

*Machine Simulation Tiling.* The original reduction's "machine simulation" tiles are superimposed onto the "base tiles" as follows:

- Onto tile (G) we superimpose the tile representing the beginning of the first configuration (representing the machine's head, its start-state and a blank tape).
- Onto tiles (A) and (B) we superimpose the tiles representing the rest of the first configuration (a blank tape).

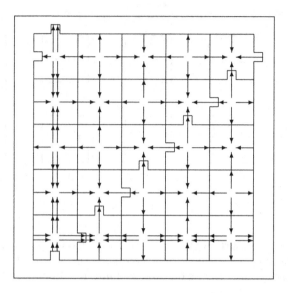

**Fig. 2.** The base tiling of a 5 × 5 region

- The rest of the machine simulation tiles are superimposed on all of the remaining base tiles.
- Tiles (A), (B) and (G) are configured such that only they can appear above a tile representing an accepting state.

Using this set of superimposed tiles, every tiling of an $i \times i$ core-region must correspond to a computation of $M$ accepting in exactly $i$ steps.

*Guaranteeing a Single Core-Region.* We now want to guarantee that the tiling of the torus consists of a *single* core-region, providing that it corresponds to a single computation of the full length (rather than copies of shorter computations). Note that the tiling of a prime-sized torus necessarily consists of a single core-region - covering the entire torus. This is because core-regions must be of the same size, and hence this size must divide the size of the torus. For technical reasons we cannot use prime numbers, but use their third-power instead. Given the Turing machine $M$, we construct a new machine $M'$ such that for every $t$, $M'$ accepts in exactly $P_t^3$ steps if and only if $M$ accepts in exactly $t$ steps (where $P_t$ is the $t$-th prime number). We do so using nondeterminism, by which $M'$ first "guesses" the number of steps $t$, and simulates $M$ to see if it accepts in exactly $t$ steps. In parallel, $M'$ computes $P_t^3$ on a second tape, and counts the number of steps on a third tape. If the simulation of $M$ on the first tape accepts in exactly $t$ steps, then $M'$ waits until exactly $P_t^3$ steps have elapsed and then accepts. We now build the tile set $T(M')$ corresponding to $M'$. If $M$ has a computation accepting in $t$ steps, then $T(M')$ has a tiling of the $P_t^3 \times P_t^3$ torus. Conversably, suppose that $T(M')$ has a tiling of the $P_t^3 \times P_t^3$ torus. Then, this tiling is either composed of a single core-region, or core-regions of size $P_t$ or $P_t^2$. However, core-regions of sizes $P_t$ or $P_t^2$ would correspond to computations of $M'$ accepting in $P_t$ or $P_t^2$ steps, respectively. But $M'$ only accepts in a number of steps which is a cube of a prime, which neither $P_t$ nor $P_t{}^2$ are. Hence, the tiling of the $P_t{}^3 \times P_t{}^3$ torus necessarily consists of a single core-region, corresponding to a $t$-step accepting computation of $M$. $\qquad\square$

**Lemma 2.** *Fixed-structure-Torus-Tiling $\leq^{fpt}$ Fixed-structure-Independent-Set.*

*Proof.* Given a tile set $T$ and a number $s$, we construct a parametric graph expression $G$, a vector $t$ of parameter values, and integer $\psi$, such that $G(t)$ has an independent set of size $\psi$ iff the $s \times s$ torus can be tiled using the tiles in $T$. Here we provide the proof for directed graphs. The proof for undirected graphs is considerably more complex and is provided in the full version.

The basic idea is to create a graph $G = G(t)$ that "represents" the torus. For each of the $s^2$ locations of the torus we create a "super-node" that is a $|T|$-clique. Each vertex in the super-node represents a different tile of $T$. The $s^2$ super-nodes are organized in $s$ rows and $s$ columns (as in the torus). Note that in each super-node at most one vertex can be chosen for the independent set. This chosen vertex will represent the tile chosen for this location in the torus tiling. Edges are placed between vertices in adjacent super-nodes (vertical and horizontal), to correspond to the adjacency constraints of the tiling. Specifically, let $H_T$ be the bipartite graph with $|T|$ vertices at each side, such that there is an edge $i \rightarrow j$ iff tile $t_i$ *cannot* be placed to the left of tile $t_j$. Similarly, let $V_T$ be the bipartite graph with $|T|$ vertices in each part (this time viewed as one part above the other), such that there is an edge $i \rightarrow j$ iff tile $t_j$ cannot be placed on top of tile $t_i$. Each super-node is connected with its right-neighbor with $H_T$ and with its neighbor on top by $V_T$. With this construction, $G$ has an independent set of size $s^2$ iff the $s \times s$ torus can be tiled by $T$. We now show how to construct the graph expression for $G$.

*The Directed $s$-Cycle.* The basic building block of our torus-graph is the directed $s$-cycle. For $s = 3 \bmod 4$ the directed $s$-cycle (denoted $C(s)$) is created using the following expression:

$$C(s) = \left( \left[\!\begin{smallmatrix}\bullet\\\bullet\end{smallmatrix}\!\right] \cup \left( \left( \left\lfloor \tfrac{s}{4} \right\rfloor - 1 \right) \cdot \left[\begin{smallmatrix}\bullet\!\!\!\to\!\!\!\bullet\\\bullet\!\!\!\leftarrow\!\!\!\bullet\end{smallmatrix}\right] \right) \cup \left[\begin{smallmatrix}\bullet\!\to\!\bullet\\\bullet\!\to\!\bullet\end{smallmatrix}\right]\right) + \left( \left( \left\lfloor \tfrac{s}{4} \right\rfloor \cdot \left[\begin{smallmatrix}\bullet\!\!\!\to\!\!\!\bullet\\\bullet\!\!\!\leftarrow\!\!\!\bullet\end{smallmatrix}\right] \right) \cup \left(\begin{smallmatrix}\bullet&\bullet\end{smallmatrix}\right) \right)$$

For $s = 0, 1$, and $2 \bmod 4$ the construction is similar (placing less nodes at the right-end of the expression).

Using the $s$-cycle, we construct two graphs, the sum of which is the $s \times s$ torus. The graphs, denoted $\mathrm{Tr}^H$ and $\mathrm{Tr}^V$, consist of the horizontal and vertical edges of the torus, respectively. Let $e_1$ be the graph with a single vertex with a self loop. Then, $\mathrm{Tr}^H$ and $\mathrm{Tr}^V$ are obtained by multiplying $C(s)$ by $s$ self-loops from the left and from the right, respectively:

$$\mathrm{Tr}^H(s) = (s \cdot e_1) \times C(s) \;,\quad \mathrm{Tr}^V(s) = C(s) \times (s \cdot e_1)$$

Next, we "blow-up" the graphs $\mathrm{Tr}^H$ and $\mathrm{Tr}^V$, substituting each vertex with a "super-node" consisting of $|T|$ vertices, and connecting the "super-nodes" by $H_T$ and $V_T$, respectively:

$$G^H(s) = \mathrm{Tr}^H(s) \times H_T \;,\quad G^V(s) = \mathrm{Tr}^V(s) \times V_T$$

(Note that this is where the directness of the graph comes to play, allowing to keep the directions of $H_T$ and $V_T$.) Together, these graphs have all the vertices and most of the edges, except for the clique edges within each super-node. These are obtained by adding $s^2$ copies of the fixed clique $K_{|T|}$. The complete graph expression is:

$$G(s) = G^H(s) + G^V(s) + (s^2 \cdot K_{|T|})$$

This concludes the construction of the expression $G$. The parameter for this graph is $s$. By construction, $G$ has an independent set of size $\psi = s^2$ iff the $s \times s$ torus can be tiled with $T$.  □

**Lemma 3.** *Fixed-Structure-Independent-Set $\leq^{XP}$ FS-Exact-Halt.*

*Proof.* Given a graph expression $G$, parameter vector $\boldsymbol{t}$, and integer $\psi$, we construct a Turing machine $M_G$ and integer $r$, such that $M_G$ accepts the empty string in exactly $r$ steps iff $G(\boldsymbol{t})$ has an independent set of size $\psi$. Assume that $\boldsymbol{t}$ has $m$ entries, and denote $t = \sum_{i=1}^{m} t_i$. First note that it is possible to construct $G(\boldsymbol{t})$ in at most $(|G| + t)^{3|G|}$ steps. This is true since there are at most $|G|$ operations, and the result graph of the $i$-th operation has at most $(|G| + t)^i$ vertices and $(|G| + t)^{2i}$ edges. Once the graph $G(\boldsymbol{t})$ is constructed, one can guess a subset of the vertices, and check if they are an independent set of size $\psi$. The only problem is that the machine $M_G$ operates on the empty input. Thus, we cannot explicitly provide it with the parameters $\boldsymbol{t}$ and $\psi$. Rather, we let the machine "guess" these values, and encode them into the number of steps. Specifically, let $code(\psi, \boldsymbol{t}) = \left( P_{|G|} \cdot P_{\psi}^2 \cdot \prod_{i=1}^{m} P_{t_i}^{i+2} \right)^{3|G|}$, where $P_j$ is the $j$-th prime number. Note that $code(\cdot, \cdot)$ is a bijection. Accordingly, given $G$ we construct the Turing machine $M_G$ to operate as follows:

1. Nondeterministically "guess" a vector $\boldsymbol{t}' = (t_1', \ldots, t_m')$ and integer $\psi'$.
2. Create the graph $G = G(\boldsymbol{t}')$.
3. Nondeterministically "guess" a subset of the vertices of $G$ and check if they are an independent set of size $\psi'$. If not, reject.
4. In parallel to the above, compute $code(\psi', \boldsymbol{t}')$. Run for a total of $code(\psi', \boldsymbol{t}')$ steps and accept.

It can be verified that $code(\psi', \boldsymbol{t}')$ steps suffice for steps (1)-(3). We obtain that $M_G$ accepts in exactly $r = code(\psi, \boldsymbol{t})$ steps iff $G(\boldsymbol{t})$ has an independent set of size $\psi$. Note that $code(\psi, \boldsymbol{t}) \leq (|G| + \psi + \sum_{i=1}^{m} t_i)^{4|G|}$, providing that the reduction is an XP one.  □

The equivalence of *Fixed-Structure-Clique* and *Fixed-Structure-Vertex-Cover* follows from the standard reductions between Independent-Set, Clique and vertex-Cover.

## 5   Open Problems

This work takes the first steps in understanding fixed-structure problems. Many important and interesting problems remain open. Here we list just a few:

- The results presented in this paper are hardness results. We were also able to show that some other fixed-structure problems are FPT. These results are omitted due to lack of space. However, we believe we are still lacking in tools for the design of FPT algorithms for fixed structure problems.
- We identified three core fixed-parameter problems, which we believe define three separate complexity classes. Are these "the right" complexity classes? Are there other important/interesting classes? Is there a hierarchy? Are *FS-Halt* and *FS-Not-Halt* indeed non-equivalent? Are they in FPT? What other problems are equivalent to these problems?
- We showed an XP equivalence between *Fixed-Structure-Independent-Set (FS-IS)* and *FS-Exact-Halt*. With our definition of FS-IS this is all but unavoidable, since the size of the graph may be exponential in its representation, and hence FS-IS need not be in NP. If we add the size of the graph (in unary) to the input, FS-IS becomes NP. Is this problem equivalent to *FS-Exact-Halt* under FPT reductions?
- In this paper, we only covered few fixed structure problems. A whole line of research is to analyze the complexity of the fixed-structure versions of the numerous problems for which the classical parametrization has been studied.
- The notion of graph products provides the basis for many interesting fixed-structure graph problems. For example, what is the fixed structure complexity of Independent-Set on $G \times K_t$ graphs? Similarly, for other graph problems, and other graph structures (e.g. tori, trees, butterflies, etc., instead of $K_t$). In addition, one may consider other types of graph products, i.e. cartesian, lexicographic and strong products (see [12]).
- We proved that fixed-structure tiling of the plane is equivalent to *FS-Not-Halt* for the version of the problem in which the origin tile is specified (the proof is not provided here). What is the complexity of the general problem when the origin tile is not specified?
- We already noted that problems with two combinatorial-structures, such as the database query problems and graph ordering problems, though different, are somewhat related to fixed-structure problems. Some of these problems are still open. It would be interesting to see if the directions developed here can shed some light on these problems.

**Acknowledgements.** We are grateful to Mike Fellows for helpful comments on an early version of this work.

# References

1. Cesati, M., Di Ianni, M.: Computational models for parameterized complexity. Mathematical Logic Quarterly 43, 179–202 (1997)
2. Cesati, M.: Compendium of parameterized problems (2006), http://bravo.ce. uniroma2.it/home/cesati/research/compendium/compendium.pdf
3. Wareham, T.: The parameterized complexity of intersection and composition operations on sets of finite-state automata. In: Yu, S., Păun, A. (eds.) CIAA 2000. LNCS, vol. 2088, pp. 302–310. Springer, Heidelberg (2001)
4. Fernau, H., Hagerup, T., Nishimura, N., Ragde, P., Reinhardt, K.: On the parameterized complexity of a generalized rush hour puzzle. In: Proceedings of Canadian Conference on Computational Geometry, CCCG, pp. 6–9 (2003)

5. Papadimitriou, C.H., Yannakakis, M.: On the complexity of database queries. In: Proceedings of the Sixteenth ACM SIGACT-SIGMOD-SIGART Symposium on Principles of Database Systems, pp. 12–14 (1997)
6. Downey, R.G., Fellows, M.R., Taylor, U.: On the parameteric complexity of relational database queries and a sharper characterization of w[1]. In: Combinatorics, Complexity and Logic, Proceedings of DMTCS 1996 (1996)
7. Vardi, M.: The complexity of relational query languages. In: Proceedings of the 14th ACM Symposium on Theory of Computing, pp. 137–146 (1982)
8. Downey, R.G., Fellows, M.R.: Parameterized Complexity. Springer, Heidelberg (1999)
9. Grohe, M., Schwentick, T., Segoufin, L.: When is the evaluation of conjunctive queries tractable? In: Proceedings of 33rd annual ACM Symposium on Theory of Computing, pp. 657–666 (2001)
10. Lewis, H.R., Papadimitriou, C.H.: Elements of the Theory of Computation. Prentice Hall, Englewood Cliffs (1981)
11. Berger, R.: The undecidability of the domino problem. Mem. AMS 66 (1966)
12. Imrich, W., Klavzer, S.: Product Graphs: Structure and Recognition. Wiley, Chichester (2000)

# An Improved Fixed-Parameter Algorithm for Minimum-Flip Consensus Trees

Sebastian Böcker, Quang Bao Anh Bui, and Anke Truss

Lehrstuhl für Bioinformatik, Friedrich-Schiller-Universität Jena,
Ernst-Abbe-Platz 2, 07743 Jena, Germany
{boecker,bui,truss}@minet.uni-jena.de

**Abstract.** In computational phylogenetics, the problem of constructing a consensus tree for a given set of input trees has frequently been addressed. In this paper we study the MINIMUM-FLIP PROBLEM: the input trees are transformed into a binary matrix, and we want to find a perfect phylogeny for this matrix using a minimum number of flips, that is, corrections of single entries in the matrix. In its graph-theoretical formulation, the problem is as follows: Given a bipartite graph $G = (V_t \cup V_c, E)$, the problem is to find a minimum set of edge modifications such that the resulting graph has no induced path with four edges which starts and ends in $V_t$.

We present a fixed-parameter algorithm for the MINIMUM-FLIP PROBLEM with running time $O(4.83^k (m + n) + mn)$ for $n$ taxa, $m$ characters, and $k$ flips. Additionally, we discuss several heuristic improvements. We also report computational results on phylogenetic data.

## 1  Introduction

When studying the relationship and ancestry of current organisms, discovered relations are usually represented as phylogenetic trees, that is, rooted trees where each leaf corresponds to a group of organisms, called *taxon*, and inner vertices represent hypothetical last common ancestors of the organisms located at the leaves of its subtree.

Supertree methods assemble phylogenetic trees with shared but overlapping taxon sets into a larger supertree which contains all taxa of every input tree and describes the evolutionary relationship of these taxa [2]. Constructing a supertree is easy for compatible input trees [12, 3], that is, in case there is no contradictory information encoded in the input trees. The major problem of supertree methods is dealing with incompatible data in a reasonable way [14]. The most popular supertree method is matrix representation with parsimony (MRP) [2]: MRP performs a maximum parsimony analysis on a binary matrix representation of the set of input trees. Problem is NP-complete [8], and so is MRP. The matrix representation with flipping (MRF) supertree method also uses a binary matrix representation of the input trees [5]. Unlike MRP, MRF seeks the minimum number of "flips" (corrections) in the binary matrix that make the matrix representation consistent with a phylogenetic tree. Evaluations

M. Grohe and R. Niedermeier (Eds.): IWPEC 2008, LNCS 5018, pp. 43–54, 2008.

by Chen et al. [6] indicate that MRF is superior to MRP and other common approaches for supertree construction, such as MinCut supertrees [14].

If all input trees share the same set of taxa, the supertree is called a consensus tree [1, 11]. As in the case of supertrees, we can encode the input trees in a binary matrix: Ideally, the input trees match nicely and a consensus tree can be constructed without changing the relations between taxa. In this case, we can construct the corresponding *perfect phylogeny* in $O(mn)$ time for $n$ taxa and $m$ characters [10]. Again, the more challenging problem is how to deal with incompatible input trees. Many methods for constructing consensus trees have been established, such as majority consensus or Adams consensus [1]. One method for constructing consensus trees is the *Minimum-flip* method [6]: Flip as few entries as possible in the binary matrix representation of the input trees such that the matrix admits a perfect phylogeny. Unfortunately, the MINIMUM-FLIP PROBLEM of finding the minimum set of flips which make a matrix compatible, is NP-hard [6, 7]. Based on a graph-theoretical interpretation of the problem and the forbidden subgraph paradigm of Cai [4], Chen et al. introduce a simple fixed-parameter algorithm with running time $O(6^k mn)$, where $k$ is the minimum number of flips [6, 7]. Furthermore, the problem can be approximated with approximation ratio $2d$ where $d$ is the maximum number of ones in a column [6,7].

*Our contributions.* We introduce a refined fixed-parameter algorithm for the MINIMUM-FLIP PROBLEM with $O(4.83^k (m+n)+mn)$ running time, and discuss some heuristic improvements to reduce the practical running time of our algorithm. To evaluate the performance and to compare it to the fixed-parameter algorithm of Chen et al., we have implemented both algorithms and evaluate them on perturbed matrix representations of phylogenetic trees. Our algorithm turns out to be significantly faster than the $O(6^k mn)$ strategy, and also much faster than worst-case running times suggest. We believe that our work is a first step towards exact computation of minimum-flip supertrees.

## 2  Preliminaries

Throughout this paper, let $n$ be the number of taxa characterized by $m$ characters. Let $M$ be an $n \times m$ binary matrix that represents the characteristics of our taxa: Each cell $M[i,j]$ takes a value of "1" if the taxon $t_i$ has character $c_j$, and "0" otherwise.

We say that $M$ admits a *perfect phylogeny* if there is a rooted tree such that each of the $n$ leaves corresponds to one of the $n$ taxa and, for each character $c_j$, there is an inner vertex of the tree such that for all taxa $t_i$, $M[i,j] = 1$ if and only if $t_i$ is a leaf of the subtree below $c_j$. The PERFECT PHYLOGENY problem is to recognize if a given binary matrix $M$ admits a perfect phylogeny. Gusfield [10] introduces an algorithm which checks if a matrix $M$ admits a perfect phylogeny and, if possible, constructs the corresponding phylogenetic tree in total running time $O(mn)$.

The MINIMUM-FLIP PROBLEM [6] asks for the minimum number of matrix entries to be flipped from "0" to "1" or from "1" to "0" in order to transform

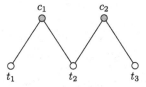

**Fig. 1.** An M-graph. Grey vertices denote characters, white vertices denote taxa.

$M$ into a matrix which admits a perfect phylogeny. The corresponding decision problem is to check whether there exists a solution with at most $k$ flips. Our fixed-parameter algorithm requires a maximum number of flips $k$ to be known in advance: To find an optimal solution we call this algorithm repeatedly, increasing $k$.

In this article we use a graph-theoretical model to analyze the MINIMUM-FLIP PROBLEM. First, we define a graph model of the binary matrix representation. The *character graph* $G = (V_t \cup V_c, E)$ of a $n \times m$ binary matrix $M$ is an undirected and unweighted bipartite graph with $n + m$ vertices $t_1, \ldots, t_n, c_1, \ldots, c_m$ where $\{c_i, t_j\} \in E$ if and only if $M[i,j] = 1$. The vertices in $V_c$ represent characters and those in $V_t$ represent taxa. We call the vertices c- or t-vertices, respectively.

An *M-Graph* is a path of length four where the end vertices and the center vertex are t-vertices and the remaining two vertices are c-vertices, see Fig. 1. We call a graph *M-free* if is does not have an M-graph as an induced subgraph. The following theorem provides an essential characterization with regard to the graph-theoretical modeling of the MINIMUM-FLIP PROBLEM.

**Theorem 1 (Chen et al. [7]).** *A binary matrix $M$ admits a perfect phylogeny if and only if the corresponding character graph $G$ does not contain an induced M-graph.*

With Theorem 1 the MINIMUM-FLIP PROBLEM is equivalent to the following graph-theoretical problem: Find a minimum set of edge modifications, that is, edge deletions and edge insertions, which transform the character graph of the input matrix into an M-free bipartite graph. Using this characterization, Chen et al. [6, 7] introduce a simple fixed-parameter algorithm with running time $O(6^k mn)$ where $k$ is the minimum number of flips. This algorithm follows a search-tree technique from [4]: It identifies an M-graph in the character graph and branches into all six possibilities of deleting or inserting one edge of the character graph such that the M-graph is eliminated (four cases of deleting one existing edge and two cases of adding a new edge).

The following notation will be used frequently throughout this article: Let $N(v)$ be the set of neighbors of a vertex $v$. For two c-vertices $c_i, c_j \in V_c$, let $X(c_i, c_j) := N(c_i) \setminus N(c_j)$, $Y(c_i, c_j) := N(c_i) \cap N(c_j)$, and $Z(c_i, c_j) := N(c_j) \setminus N(c_i)$. We call $c_i$ and $c_j$ *c-neighbors* if and only if $Y(c_i, c_j)$ is not empty.

# 3   The Algorithm

We present a search tree algorithm largely based on observations on the structure of intersecting M-graphs. First, we use a set of reduction rules to cut down the size of $G$, see Sect. 3.1 below. As long as there are c-vertices of degree three or higher in $G$, we use the branching strategy described in Sect. 3.2. If there are no such vertices left, we can use a simplified branching strategy described in Sect. 3.3. In the beginning of every recursion call of our search tree algorithm we execute the data reduction described in the following section.

## 3.1   Data Reduction

When the algorithm receives a character graph $G$ as input, it is reduced with respect to the following simple reduction rules:

*Rule 1.* Delete all c-vertices $v \in V_c$ of degree $|V_t|$ from the graph.

*Rule 2.* Delete all c-vertices $v \in V_c$ of degree one from the graph.

We verify the correctness of these reduction rules, starting with Rule 1. Let $c$ be a c-vertex of degree $|V_t|$ in the input graph $G$. Then $c$ is connected to all t-vertices in $G$ and there cannot be an M-graph containing $c$. Furthermore, it is not possible to insert any new edge incident to $c$. Assume there is an optimal solution for $G$ which deletes edges incident to $c$ in $G$. If we execute all edit operations of the optimal solution except deletions of edges incident to $c$, we also obtain an M-free graph, since every M-graph which does not contain $c$ is destroyed by edit operations of the optimal solution and there is no M-graph containing $c$. This is a contradiction to the assumption that the solution is optimal, so edges incident to c-vertices of degree $|V_t|$ are never deleted in an optimal solution. Thus the corresponding c-vertices need not be observed and can be removed safely.

The correctness of Rule 2 is obvious.

After computing and saving the degree of every vertex in $G$ in time $O(mn)$, Rules 1 and 2 of the data reduction can be done in time $O(m + n)$.

## 3.2   Solving Instances with c-Vertices of Degree at Least Three

In this section we describe the branching strategy we use as long as there are c-vertices of degree three or higher. The efficiency of this branching is based on the following observation:

**Lemma 1.** *A character graph $G$ reduced with respect to the abovementioned data reduction rules has a c-vertex of degree at least three if and only if $G$ contains $F_1$ or $F_2$ (see Fig. 2) as induced subgraph.*

*Proof.* Assume that there are c-vertices in $G$ with degree at least three. Let $c_i$ be a c-vertex with maximum degree in $G$. Then $c_i$ must have a c-neighbor $c_j$ which has at least one neighbor $t_j$ outside $N(c_i)$ because otherwise $c_i$ would be

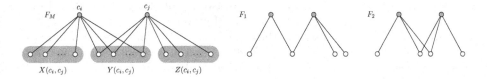

**Fig. 2.** The big M-graph $F_M$ and its special cases $F_1$ and $F_2$

removed from $G$ by the data reduction. Let $t_k$ be a common neighbor of $c_i$ and $c_j$ which has to exist since $c_i$ and $c_j$ are c-neighbors. There also exists a neighbor $t_i$ of $c_i$ which is not a neighbor of $c_j$. If $t_i$ is the only one t-vertex, which is neighbor of $t_i$ but not of $t_j$, then $t_i$ and $t_j$ must share an another common neighbor $t'_k$ besides $t_k$ (since $t_i$ has degree of at least three) and the M-graph $t_i c_i t_k c_j t_j$ and edges $\{c_i, t'_k\}, \{c_j, t'_k\}$ form an $F_2$ graph. Otherwise let $t'_i$ be an another t-vertex, which is neighbor of $t_i$ but not of $t_j$, the M-graph $t_i c_i t_k c_j t_j$ and the edge $\{t'_i, c_i\}$ form an $F_1$ graph. We conclude that if $G$ has t-vertices with degree at least three, then $G$ contains $F_1$ or $F_2$ as induced subgraph.

If $G$ contains $F_1$ or $F_2$ as induced subgraph, it is obvious that $G$ has c-vertices of degree at least three.                                                                 □

We now consider the following structure of intersecting M-graphs, called *big M-graph*: this graph is a subgraph of the character graph and consists of two c-vertices $c_i, c_j$ and t-vertices in the nonempty sets $X(c_i, c_j)$, $Y(c_i, c_j)$, $Z(c_i, c_j)$ where at least one of these sets has to contain two or more t-vertices, see Fig. 2. The graphs $F_1$ and $F_2$ are big M-graphs of minimum size. In view of Lemma 1, any character graph with at least one c-vertex of degree three or higher has to contain big M-graphs as induced subgraphs. Furthermore, it should be clear that a big M-graph contains many M-graphs as induced subgraphs. Therefore, if there are big M-graphs in the character graph, our algorithm first branches into subcases to eliminate all M-graphs contained in those big M-graphs. The branching strategy to eliminate all M-graphs contained in a big M-graph is based on the following lemma:

**Lemma 2.** *If a character graph $G$ is M-free, then for every two distinct c-vertices $c_i$, $c_j$ of $G$ it holds that at least one of the sets $X(c_i, c_j)$, $Y(c_i, c_j)$, $Z(c_i, c_j)$ must be empty.*

Since there is at least one M-graph containing $c_i$ and $c_j$ if $X(c_i, c_j)$, $Y(c_i, c_j)$, $Z(c_i, c_j)$ are simultaneously non-empty, the correctness of Lemma 2 is obvious.

Lemma 2 leads us to the following branching strategy for big M-graphs. Given a character graph $G$ with at least one c-vertex of degree three or higher, our algorithm chooses a big M-graph $F_M$ in $G$ and branches into subcases to eliminate all M-graphs in $F_M$. Let $c_i, c_j$ be the c-vertices of $F_M$. According to Lemma 2, one of the sets $X(c_i, c_j)$, $Y(c_i, c_j)$, $Z(c_i, c_j)$ must be "emptied" in each subcase. Let $x$, $y$, $z$ denote the cardinalities of sets $X(c_i, c_j), Y(c_i, c_j)$ and $Z(c_i, c_j)$. We now describe how to empty $X(c_i, c_j)$, $Y(c_i, c_j)$, and $Z(c_i, c_j)$.

For each t-vertex $t$ in $X(c_i, c_j)$ there are two possibilities to remove it from $X(c_i, c_j)$: we either disconnect $t$ from the big M-graph by deleting the edge $\{c_i, t\}$, or move $t$ to $Y(c_i, c_j)$ by inserting the edge $\{c_j, t\}$. Therefore the algorithm has to branch into $2^x$ subcases to empty $X(c_i, c_j)$ and in each subcase, it executes $x$ edit-operations. The set $Z(c_i, c_j)$ is emptied analogously.

To empty the set $Y(c_i, c_j)$ there are also two possibilities for each t-vertex $t$ in $Y(c_i, c_j)$, namely moving it to $X(c_i, c_j)$ by deleting the edge $\{c_j, t\}$ or moving it to $Z(c_i, c_j)$ by deleting the edge $\{c_i, t\}$. This also leads to $2^y$ subcases and in each subcase, $y$ edit operations are executed.

Altogether, the algorithm branches into $2^x + 2^y + 2^z$ subcases when dealing with a big M-graph $F_M$. In view of Lemma 2, at least $\min\{x, y, z\}$ edit-operations must be executed to eliminate all M-graphs in $F_M$. Our branching strategy has branching vector

$$(\underbrace{x, \ldots, x}_{2^x}, \underbrace{y, \ldots, y}_{2^y}, \underbrace{z, \ldots, z}_{2^z})$$

which leads to a branching number of 4.83 as shown in the following lemma (see [13] for details on branching vectors and branching numbers).

**Lemma 3.** *The worst-case branching number of the above branching strategy is 4.83.*

*Proof.* The branching number $b$ of the above branching strategy is the single positive root of the equation

$$2^x \frac{1}{b^x} + 2^y \frac{1}{b^y} + 2^z \frac{1}{b^z} = 1 \iff \frac{1}{(b/2)^x} + \frac{1}{(b/2)^y} + \frac{1}{(b/2)^z} = 1.$$

Considering $\frac{b}{2}$ a variable, the single positive root of the second equation is the branching number corresponding to the branching vector $(x, y, z)$. The smaller values $x$, $y$, $z$ take, the higher $\frac{b}{2}$ and, hence, $b$. Due to the definition of a big M-graph, $x$, $y$, and $z$ cannot equal one simultaneously, so $b$ is maximal if one of the variables $x$, $y$, $z$ equals two and the other two equal one. Without loss of generality, assume that $x = 2$ and $y = z = 1$. Then the single positive root of the equation $(\frac{2}{b})^2 + \frac{2}{b} + \frac{2}{b} = 1$ is $\frac{b}{2} = 2.414214$. Therefore, $b$ is at most 4.83. $\square$

From Lemma 2 we infer an interesting property. When we take into consideration that we need an edit operation for each vertex we remove from set $X$, $Y$, or $Z$, the following corollary is a straightforward observation.

**Corollary 1.** *There is no solution with at most $k$ flips if there exist two c-vertices $c_i, c_j \in V_c$ satisfying $\min\{|X(c_i, c_j)|, |Y(c_i, c_j)|, |Z(c_i, c_j)|\} > k$.*

We use this property for pruning the search tree in the implementation of our algorithm, see Sect. 4.

Corollary 1 implies that we can abort a program call whenever we find a big M-graph where $x$, $y$, $z$ simultaneously exceed $k$. Furthermore, if one or two of the values $x$, $y$, $z$ are greater than $k$, we do not branch into subcases deleting the respective sets. Anyway, the number of subroutine calls in this step of the algorithm is fairly large, up to $3 \cdot 2^k$. But large numbers of program calls at this point are a result of large numbers of simultaneous edit operations which lower $k$ to a greater extent. Therefore the branching number of our strategy goes to 2 for large $x$, $y$, $z$, and this is confirmed by the growth of running times in our computational experiments (see Sect. 5).

### 3.3   Solving Instances with c-Vertices of Degree at Most Two

In this section we assume that there is no big M-graph in the character graph. As we proved in Lemma 1, if a character graph $G$ reduced with respect to our data reduction does not contain any big M-graph, every c-vertex in $G$ has degree two. In this case we use the branching strategy based on the following lemma to transform $G$ into an M-free character graph.

**Lemma 4.** *If every c-vertex in a character graph has degree two, there is an optimal solution for the* MINIMUM-FLIP PROBLEM *without inserting any edge into the character graph.*

*Proof.* Let $G$ be a character graph where all c-vertices have degree two and $G'$ be the resulting graph of an optimal solution for $G$ where we did add new edges to $G$ and $\{c, t\}$ be such a new edge.

In the following we show that there is another optimal solution for $G$ without any edge insertion. Let $t_i, t_j$ be the t-vertices connected with $c$. Since all M-graphs eliminated by inserting $\{c, t\}$ into $G$ contain edges $\{c, t_i\}$ and $\{c, t_j\}$, we can delete $\{c, t_i\}$ or $\{c, t_j\}$ from $G$ to eliminate these M-graphs instead of adding $\{c, t\}$ to $G$. Deleting an edge can only cause new M-graphs containing the c-vertex incident to this edge. But after removing one of the edges $\{c, t_i\}$ or $\{c, t_j\}$, vertex $c$ has degree one and cannot be vertex of any M-graph. Therefore the resulting graph is still M-free and the number of edit operations does not increase since we swap an insertion for only one deletion. Hence, edge insertions are not necessary for optimal solutions when all c-vertices of the character graph have degree two.                                                                            □

Now that all c-vertices in our graph $G$ have degree two, we define a weighted graph $G_w$ as follows: We adopt the set $V_t$ of t-vertices in $G$ as vertex set for $G_w$. Two vertices $t_1, t_2$ are connected if and only if they possess a common neighbor in $G$. The weight of an edge $\{t_1, t_2\}$ is the number of common neighbors of $t_1, t_2$ in $G$, see Fig. 3.

On the weighted graph $G_w$ and the number $k$ of remaining edit operations, the MINIMUM-FLIP problem turns out to be the problem of deleting a set of edges with minimum total weight such that there are no paths of length two in $G_w$, that is, the graph is split into connected components of size one or two.

Since it is unknown if the abovementioned problem can be solved in polynomial time, we used the fixed-parameter algorithm described in the following text to deal with this problem. Deleting a weighted edge in $G_w$ corresponds to deleting one of the edges incident to each of the respective c-vertices in $G$. That

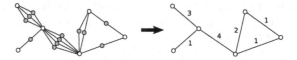

**Fig. 3.** Left: When all c-vertices (gray) have degree two, we can regard each c-vertex with its incident edges as a single edge in a multigraph whose vertex set is the set of t-vertices (white). Right: We merge all those "edges" between two t-vertices into a single weighted edge whose weight equals the number of c-vertices adjacent to the t-vertices and we obtain a simplified model of our graph.

is, whenever we delete an edge $\{t_i, t_j\}$ of weight $m$ from $G_w$, we include, for each vertex $c$ of the $m$ original c-vertices which were used for $e$ (see Fig. 3), one of the edges $\{c, t_i\}$ and $\{c, t_j\}$ from $G$ in our solution set and lower $k$ by $m$.

Let us describe our branching strategy for the weighted problem in detail. If we consider an edge $e$ in $G_w$, we observe that either $e$ has to be deleted or all other edges which are incident to a vertex in $e$.

Now we pick an edge $e$ which has weight greater than one, shares a vertex with an edge of weight greater than one, or is incident to a vertex of degree three or higher. Then we branch into two cases: Delete $e$ or keep $e$ but delete all edges which share vertices with $e$. In each case, lower $k$ by the weights of the deleted edges. If the graph decomposes, we treat each connected component separately. With this branching strategy we receive a branching vector of $(1, 2)$ or better which corresponds to a branching number of 1.62. We use this strategy as long as there are degree-three vertices or edges of weight greater than one.

As soon as all edges have weight one and all vertices have degree at most two, the remaining graph is either a path or a cycle. We solve each of these graphs by alternately keeping and deleting edges such that solving a path with $l$ edges costs $\lfloor \frac{l}{2} \rfloor$ and solving a cycle of length $l$ costs $\lceil \frac{l}{2} \rceil$. Clearly, this operation can be done in linear time.

We prove in the following theorem that our algorithm solves the MINIMUM-FLIP PROBLEM in time $O(4.83^k (m + n) + mn)$.

**Theorem 2.** *The above algorithm solves the* MINIMUM-FLIP PROBLEM *for a character graph with $n$ t-vertices and $m$ c-vertices in $O(4.83^k (m + n) + mn)$ time.*

*Proof.* At the beginning of the algorithm, the execution of the data reduction takes $O(mn)$ time. By saving the degree of each vertex in $G$, the algorithm needs $O(m + n)$ time to execute the data reduction in each recursion call.

The algorithm distinguishes two cases: there are c-vertices with degree at least three, or every c-vertex has degree two. In the first case it uses the branching strategy described in Sect. 3.2 with branching number 4.83, in the second case it executes the branching strategy in Sect. 3.3 with branching number 1.62. Therefore, the size of the search tree is $O(4.83^k)$. All in all, the running time of the algorithm is $O(4.83^k (m + n) + mn)$. □

# 4    Algorithm Engineering

In the course of algorithm design, we found some improvements which do not affect the theoretical worst-case running time or even increase the polynomial factor but as they manage to prune the search tree, they are highly advisable in practice. These are a few heuristic improvements we included in the course of implementation.

*Treat connected components separately.* If a given character graph is not connected or decomposes in the course of the algorithm, we compute the solutions for each of its connected components separately because connecting different connected components never deletes an M-graph.

*Avoid futile program calls.* If $\min\{X(c_i, c_j), Y(c_i, c_j), Z(c_i, c_j)\} > k$ holds for two c-vertices $c_i, c_j$, we know that it is impossible to solve the current instance (see Corollary 1). Therefore, whenever we find such an M-graph we abort the current search tree branch and call the algorithm with an appropriately increased parameter, thus skipping program runs which are doomed to failure.

*Avoid redundant search tree branches.* When executing an edit operation in a big M-graph, we fix the outcome of the operation, that is, whenever we insert an edge, this edge is set to "permanent" and when we delete an edge, it is set to "forbidden". With this technique we make sure that edit operations are not undone later in the search tree.

*Try promising search tree branches first.* In the first part of our branching strategy, branching on a big M-graph $F_M$ with c-vertices $c_i$, $c_j$ leads to $2^{|X(c_i,c_j)|} + 2^{|Y(c_i,c_j)|} + 2^{|Z(c_i,c_j)|}$ branches. It is likely that a minimum solution destroys the $F_M$ with as few edge modifications as possible. As we use depth-first search and stop when we find a solution, we branch on the edges incident to the smallest of sets $X, Y, Z$ first.

*Calculate branching numbers in advance.* When dealing with big M-graphs, we save, for each pair of c-vertices, the branching number corresponding to a branching at the $F_M$ associated with these vertices in a matrix. The minima of each row are saved in an extra column in order to allow faster searching for the overall minimum. We use a similar technique to deal with the weighted graph in the second part of the algorithm.

The polynomial factor in the running time proved in Theorem 2 cannot be hold when applying the abovementioned heuristic improvements, since initializing the matrix used to calculate the branching numbers takes time $O(m^2 n)$ and updating this matrix needs time $O(mn)$ in each recursion call. While initializing or updating this matrix, we also check if the data reduction rules can be applied. Testing for futile program calls and redundant search tree branches can be executed in the same time. Altogether the running time of our algorithm with the abovementioned heuristic improvements is $O(4.83^k mn + m^2 n)$. Despite that, the heuristic improvements lead to drastically reduced running times in practice.

**Table 1.** Comparison of running times of our $O(4.83^k)$ algorithm and the $O(6^k)$ algorithm. $|V_t|$ and $|V_c|$ denote the number of t- and c-vertices, respectively. # flips is the number of perturbances in the matrix whereas $k$ is the true number of flips needed to solve the instance. Each row corresponds to ten datasets. *Six out of ten computations were finished in under ten hours.

| Dataset | $|V_t|$ | $|V_c|$ | # flips | avg. $k$ | time $4.83^k$ | time $6^k$ |
|---|---|---|---|---|---|---|
| Marsupials | 21 | 20 | 10 | 9.6 | 9.5 s | 2 h |
| | | | 12 | 10.7 | 25.5 s | > 10 h* |
| | | | 14 | 13.2 | 3 min | > 10 h |
| | | | 16 | 15.4 | 12 min | > 10 h |
| | | | 18 | 17 | 47 min | > 10 h |
| | | | 20 | 18.9 | 3.3 h | > 10 h |
| Marsupials | 51 | 50 | 10 | 10 | 17 s | 19 h |
| | | | 12 | 10.5 | 30 s | > 10 h |
| | | | 14 | 12.5 | 2.3 min | > 10 h |
| | | | 16 | 15.5 | 50 min | > 10 h |
| | | | 18 | 17.5 | 3 h | > 10 h |
| | | | 20 | 19 | 8 h | > 10 h |
| Tex (Bacteria) | 97 | 96 | 10 | 9.7 | 17 s | 59.3 h |
| | | | 12 | 11.9 | 12.5 min | > 10 h |
| | | | 14 | 13.9 | 18 min | > 10 h |
| | | | 16 | 15.4 | 1.1 h | > 10 h |
| | | | 18 | 17.3 | 4 h | > 10 h |
| | | | 20 | 19.7 | 10.3 h* | > 10 h |

## 5    Experiments

To evaluate in how far our improved branching strategy affects running times in practice, we compared our algorithm against Chen et al.'s $O(6^k mn)$ algorithm [6, 7]. Both algorithms were implemented in Java. Computations were done on an AMD Opteron-275 2.2 GHz with 6 GB of memory running Solaris 10.

Each program receives a binary matrix as input and returns a minimum set of flips needed to solve the instance. The parameter need not be given as the program starts with calling the algorithm with $k = 0$ and repeatedly increases $k$ by one until a solution is found. As soon as it finds a solution with at most $k$ flips, the program call is aborted instantly and the solution is returned without searching further branches. All data reduction rules and heuristic improvements described in Sect. 3.1 and 4 were used for our algorithm and, if applicable, also for the $O(6^k mn)$ algorithm. For our experiments we used matrix representations [9] of real phylogenetic trees, namely two phylogenetic trees of marsupials with 21 and 51 taxa (data provided by Olaf Bininda-Emonds) and one tree of 97 bacteria computed using Tex protein sequences (data provided by Lydia Gramzow). Naturally, these matrices admit perfect phylogenies.

We perturbed each matrix by randomly flipping different numbers of entries, thus creating instances where the number of flips needed for resolving all M-graphs in the corresponding character graph is at most the number of perturbances. For each matrix representation and each number of perturbances we created ten different instances and compared the running times of both algorithms on all instances. In many datasets we created it was possible to solve the instance with a smaller number of flips.

Each dataset was allowed ten hours of computation. Running times for the $O(6^k mn)$ algorithm for ten flips on the two larger datasets were calculated despite this restriction to show the order of magnitude. The results of the computations are summed up in Table 1. When the average running time was below ten hours, all instances were finished in less then ten hours. When the average was more than ten hours, all ten instances took more than ten hours, except for the small Marsupial datasets with $k = 12$ for the $O(6^k mn)$ algorithm and the Tex datasets with $k = 20$ for our algorithm. In both cases, six of ten instances were solved.

Our experiments show that our method is constantly significantly faster than Chen et al.'s algorithm. In the course of computations we observed that, on average, increasing $k$ by one resulted in about 2.2-fold running time for a program call of our algorithm and 5-fold running time for the $O(6^k)$ search tree algorithm. The reason for the factor of 2.2 is probably that big M-graphs can be fairly large in practice such that the real branching number is close to two as analyzed in Sect. 3.2.

# 6   Conclusion

We have presented a new refined fixed-parameter algorithm for the MINIMUM-FLIP PROBLEM. This method improves the worst-case running time for the exact solution of this problem mainly by downsizing the search tree from $O(6^k)$ to $O(4.83^k)$. The experiments show that in practice the difference in running times is by far larger than one would expect from the worst-case analysis. Our algorithm outperformed the $O(6^k)$ algorithm dramatically. We believe that this a big step towards computing exact solutions efficiently.

Since the MINIMUM-FLIP PROBLEM is fixed-parameter tractable with respect to the minimum number of flips, a problem kernel must exist [13]. Finding a kernelization procedure is a natural next step in the theoretical analysis, and, to our expectation, may also greatly improve running times. Even if our program may never be fast enough to solve very large instances, it is certainly useful for tuning and evaluating heuristic algorithms such as Chen et al.'s heuristic and approximation algorithms [6, 7].

To create not only consensus trees but also arbitrary supertrees, we have to consider a version of the MINIMUM-FLIP PROBLEM where, besides zeros and ones, a considerable amount of matrix entries are '?' (unknown) and we have to create a matrix without question marks which admits a perfect phylogeny with as few flips of zeros and ones as possible. It is an interesting open question if this problem is fixed-parameter tractable with respect to the minimum number of flips. An algorithm which can handle this problem would make an interesting tool in computational biology.

# Acknowledgment

All programming and experiments were done by Patrick Seeber. We thank Olaf Bininda-Emonds and Lydia Gramzow for providing the datasets.

# References

1. Adams III, E.N.: Consensus techniques and the comparison of taxonomic trees. Syst. Zool. 21(4), 390–397 (1972)
2. Bininda-Emonds, O.R.: Phylogenetic Supertrees: Combining Information to Reveal the Tree of Life. Computational Biology Book Series, vol. 4. Kluwer Academic, Dordrecht (2004)
3. Bryant, D., Steel, M.A.: Extension operations on sets of leaf-labelled trees. Adv. Appl. Math. 16(4), 425–453 (1995)
4. Cai, L.: Fixed-parameter tractability of graph modification problems for hereditary properties. Inf. Process. Lett. 58(4), 171–176 (1996)
5. Chen, D., Diao, L., Eulenstein, O., Fernández-Baca, D., Sanderson, M.: Flipping: A supertree construction method. In: Bioconsensus. DIMACS Series in Discrete Mathematics and Theoretical Computer Science, vol. 61, pp. 135–160. American Mathematical Society, Providence, RI (2003)
6. Chen, D., Eulenstein, O., Fernández-Baca, D., Sanderson, M.: Supertrees by flipping. In: H. Ibarra, O., Zhang, L. (eds.) COCOON 2002. LNCS, vol. 2387, pp. 391–400. Springer, Heidelberg (2002)
7. Chen, D., Eulenstein, O., Fernandez-Baca, D., Sanderson, M.: Minimum-flip supertrees: Complexity and algorithms. IEEE/ACM Trans. Comput. Biol. Bioinform. 3(2), 165–173 (2006)
8. Day, W., Johnson, D., Sankoff, D.: The computational complexity of inferring rooted phylogenies by parsimony. Math. Biosci. 81, 33–42 (1986)
9. Farris, J., Kluge, A., Eckhardt, M.: A numerical approach to phylogenetic systemetics. Syst. Zool. 19, 172–189 (1970)
10. Gusfield, D.: Efficient algorithms for inferring evolutionary trees. Networks 21, 19–28 (1991)
11. Kannan, S., Warnow, T., Yooseph, S.: Computing the local consensus of trees. In: Proc. of Symposium on Discrete Algorithms (SODA 1995) (1995)
12. Ng, M.P., Wormald, N.C.: Reconstruction of rooted trees from subtrees. Discrete Appl. Math. 69(1–2), 19–31 (1996)
13. Niedermeier, R.: Invitation to Fixed-Parameter Algorithms. Oxford University Press, Oxford (2006)
14. Semple, C., Steel, M.: A supertree method for rooted trees. Discrete Appl. Math. 105(1–3), 147–158 (2000)

# An $O^*(1.0977^n)$ Exact Algorithm for MAX INDEPENDENT SET in Sparse Graphs

N. Bourgeois, B. Escoffier, and V. Th. Paschos

LAMSADE, CNRS UMR 7024 and Université Paris-Dauphine
Place du Maréchal De Lattre de Tassigny, 75775 Paris Cedex 16, France
{bourgeois,escoffier,paschos}@lamsade.dauphine.fr

**Abstract.** We present an $O^*(1.0977^n)$ search-tree based exact algorithm for MAX INDEPENDENT SET in graphs with maximum degree 3. It can be easily seen that this algorithm also works in graphs with average degree 3.

## 1 Introduction

Very active research has been recently conducted around the development of optimal algorithms for **NP**-hard problems with non-trivial worst-case complexity. In this paper we handle MAX INDEPENDENT SET-3, that is the MAX INDEPENDENT SET problem in graphs with maximum degree 3.

Given a graph $G(V, E)$, MAX INDEPENDENT SET consists of finding a maximum-size subset $V' \subseteq V$ such that for any $(v_i, v_j) \in V' \times V'$, $(v_i, v_j) \notin E$. MAX INDEPENDENT SET is a paradigmatic problem in theoretical computer science and numerous studies carry either on its approximation or on its solution by exact algorithms with non-trivial worst-case complexity. The best such complexity is, to our knowledge, the $O^*(1.1889^n)$ algorithm claimed by [1].

One of the most studied versions of MAX INDEPENDENT SET is its restriction in graphs with maximum degree 3, denoted by MAX INDEPENDENT SET-3 in what follows. Dealing with exact computation of MAX INDEPENDENT SET-3, several algorithms have been devised successively improving worst case complexity of its solution. Let us quote here the $O^*(1.1259^n)$ algorithm by [2], the $O^*(1.1254)$ algorithm by [3], the $O^*(1.1225^n)$ algorithm by [4], the $O^*(1.1120)$ algorithm by [5] and the $O^*(1.1034^n)$ algorithm by [6]. In this paper, based upon a refined branching with respect to [5], we devise an exact algorithm for MAX INDEPENDENT SET-3 with worst-case running time of $O^*(1.0977^n)$. In fact, the main difference of our analysis with respect to [5] lies in a more careful examination of the cases where all the vertices have degree 3. Also, as it hopefully will be understood from the analysis, our result remains valid also for graphs where the maximum degree is higher but the average degree is bounded by 3.

Let $T(\cdot)$ be a super-polynomial and $p(\cdot)$ be a polynomial, both on integers. In what follows, using notations in [7], for an integer $n$, we express running-time bounds of the form $p(n) \cdot T(n)$ as $O^*(T(n))$, the star meaning that we ignore polynomial factors. We denote by $T(n)$ the worst-case time required to exactly

M. Grohe and R. Niedermeier (Eds.): IWPEC 2008, LNCS 5018, pp. 55–65, 2008.
© Springer-Verlag Berlin Heidelberg 2008

solve the considered combinatorial optimization problem on an instance of size $n$. We recall (see, for instance, [8]) that, if it is possible to bound above $T(n)$ by a recurrence expression of the type $T(n) \leq \sum T(n-r_i)+O(p(n))$, we have $\sum T(n-r_i) + O(p(n)) = O^*(\alpha(r_1, r_2, \ldots)^n)$ where $\alpha(r_1, r_2, \ldots)$ is the largest root of the function $f(x) = 1 - \sum x^{-r_i}$.

Consider a graph $G(V, E)$. Denote by $\alpha(G)$ the size of an optimal solution for MAX INDEPENDENT SET on $G$. For convenience, if $H \subseteq V$, we will denote by $\alpha(H)$ the cardinality of an optimal solution on the subgraph of $G$ induced by $H$. For any vertex $v$, we denote by $\Gamma(v)$ the set of its neighbors and by $deg(v) = |\Gamma(v)|$ its degree. We denote by $\alpha(G|v)$ (resp., $\alpha(G|\bar{v})$) the size of the optimal solution if we include $v$ (resp. if we do not include $v$).

## 2   Preprocessing

The MAX INDEPENDENT SET-3-instance tackled is parameterized by $d = m - n$, where $n = |V|$ and $m = |E|$; $T(d)$ will denote the maximum running time of our algorithm on a graph whose parameter is smaller than or equal to $d$.

Before running the algorithm, we perform a preprocessing of the graph, in order to first remove vertices of degree 1 and 2 as well as dominated vertices. Some of the properties this preprocessing is based upon are easy and already known, mainly from [5]. We keep them for legibility.

**Lemma 1.** *Assume that there exists $v \in V$ such that $deg(v) = 1$. Then, there exists a maximum independent set $S^*$ such that $v \in S^*$.*

*Proof.* Let $w$ be the only neighbor of $v$. If $v$ is not selected, then $w$ must be selected (else the solution would not even be maximal for inclusion). Then:

$$\alpha(G|v) = \alpha(V - \{v, w\}) + 1 \geq \alpha(V - \{v, w\} - \Gamma(w)) + 1 = \alpha(G|\bar{v})$$

**Lemma 2.** *Let $v, w \in V$ be such that $v \in \Gamma(w)$ and $\Gamma(w) - \{v\} \subseteq \Gamma(v) - \{w\}$. Then, there exists a maximum independent set $S^*$ such that $v \notin S^*$.*

*Proof.* Since $V - \{v\} - \Gamma(v) \subseteq V - \{w\} - \Gamma(w)$, $\alpha(G|v) \leq \alpha(G|w)$. We say that $v$ is dominated by $w$ and we can remove it from our graph. If $v$ both dominates and is dominated by $w$, we choose at random the only one we keep.

**Lemma 3.** *Let $v, w_1, w_2 \in V$ be such that $deg(v) = 2$, $\Gamma(v) = \{w_1, w_2\}$ and $w_1$ is not a neighbor of $w_2$. Then MAX INDEPENDENT SET on $G(V, E)$ is equivalent to the following problem:*

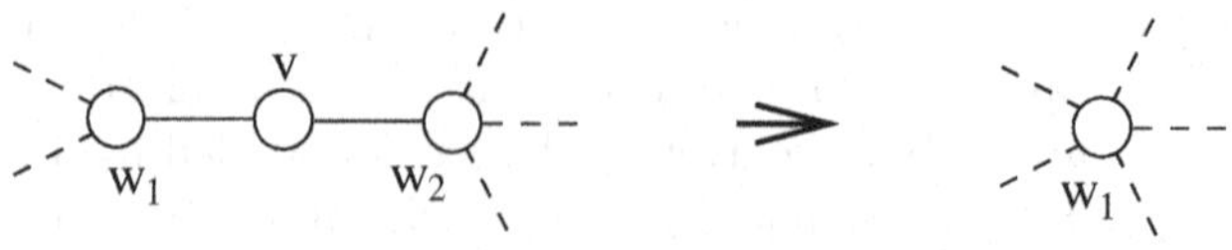

**Fig. 1.** Vertex folding

1. *form the subgraph $G'(V', E')$ induced by $V - \{v, w_2\}$. For any $x \neq v$ such that $\{x, w_2\} \in E$, add $\{x, w_1\}$ to $E'$ (see Figure 1);*
2. *compute a solution, say $S$, for MAX INDEPENDENT SET on $G'(V', E')$;*
3. *if $w_1 \in S$, $S^* = S \cup \{w_2\}$. Else $S^* = S \cup \{v\}$.*

*Proof.* Notice at first that, according to Lemma 1:

$$\alpha(G|\bar{v}) = \alpha(G|w_1, w_2) = \alpha(G'|w_1) + 1$$
$$\alpha(G|v) = \alpha(G|\bar{w}_1, \bar{w}_2) = \alpha(G'|\bar{w}_1) + 1$$

From what it holds $\alpha(G) = \alpha(G') + 1$.

This reduction is called *vertex folding* in [3].

Summing up the previous properties, we are now able to operate a reduction as soon as the graph has a vertex of degree 1 or 2.

**Proposition 1.** *Assume that there exists some vertex $v$ such that $1 \leq deg(v) \leq 2$. Then, there exists a graph $G'(V', E')$ with $|V'| < |V|$ and $d' = |E'| - |V'| \leq d$ such that it is equivalent to compute MAX INDEPENDENT SET on $G$ or on $G'$.*

*Proof.* The following holds:

- if $deg(v) = 1$, according to Lemma 1 we may add it to the solution and remove its neighbor $w$ from the graph. $d' = d - deg(w) + 2 \leq d$;
- if $deg(v) = 2$ and its neighbors $w_1, w_2$ are adjacent to each other, then $v$ dominates them. According to Lemma 2 we may remove $w_1$ from the graph; then Lemma 1 allows us to add $v$ to the solution and remove $w_2$. $d' = d - deg(w_1) - deg(w_2) + 4 \leq d$;
- finally, if $deg(v) = 2$ and its neighbors $w_1, w_2$ are not adjacent to each other, the equivalent graph we built in Lemma 3 verifies $d' \leq d$.

In other terms, we can always consider any vertex from our graph has at least degree 3.

We conclude this section by a remark that will be helpful later in the branching analysis.

*Remark 1.* Note that in the last case of Proposition 1 ($deg(v) = 2$ and its neighbors $w_1, w_2$ are not adjacent to each other) then (at least) one of the two following cases occurs when reducing the graph: either (i) $d' \leq d - 1$ (if $w_1$ and $w_2$ are both adjacent to a third vertex $x \neq v$), or (ii) a vertex of degree at least 4 is created in $G'$.

## 3   Branching

In this section, we consider that the whole preprocessing described in Section 2 has been computed as long as possible. That means no vertex has degree 2 or less, and no vertex is dominated.

**Lemma 4.** *Consider a graph that has no vertex of degree smaller than 2 and no dominated vertex and fix some vertex v. Then, the number $\mathcal{N}$ of edges that are incident to at least one of the neighbors of v is bounded below by:*

$$deg(v) + \frac{1}{2} \sum_{w \in \Gamma(v)} deg(w)$$

*Proof.* Let $I$ (resp. $\Omega$) be the inner (resp. outer) edges of $\Gamma(v)$, that means edges linking two vertices from $\Gamma(v)$ (resp. one vertex from $\Gamma(v)$ and one vertex from $V - \Gamma(v)$). Then:

$$\mathcal{N} = |\Omega| + |I|$$
$$\sum_{w \in \Gamma(v)} deg(w) = |\Omega| + 2|I|$$

From what we get:

$$\mathcal{N} = \sum_{w \in \Gamma(v)} deg(w)/2 + |\Omega|/2 \tag{1}$$

Notice that any $w \in \Gamma(v)$ has at least one neighbor in $V - \{v\} - \Gamma(v)$; else $w$ would dominate $v$. Moreover, $w$ is adjacent to $v$. Thus, $|\Omega| \geq 2deg(v)$.

**Proposition 2.** *Assume that there exists some vertex v whose degree is at least 5. Then, $T(d) \leq T(d-4) + T(d-7)$.*

*Proof.* We branch on $v$. If we choose not to add $v$ to the solution, then we can remove $v$ and any vertex adjacent to it, that means $d' = d - deg(v) + 1 \leq d - 4$. On the other hand, if $v$ belongs to the optimal solution, we remove from our graph $v$ and all its neighbors, that means (according to Lemma 4):

$$n' = n - 1 - deg(v)$$
$$m' \leq m - deg(v) - \frac{1}{2} \sum_{w \in \Gamma(v)} deg(w)$$
$$d' \leq d + 1 - \frac{1}{2} \sum_{w \in \Gamma(v)} deg(w)$$

Furthermore, $\sum_{w \in \Gamma(v)} deg(w) \geq 15$. Since $d'$ has to be an integer, this leads to the expected result.

**Proposition 3.** *Assume that there exists some vertex v whose degree is 4. Then, $T(d) \leq T(d-3) + T(d-5)$. Moreover, assume that one of the following cases holds:*

1. *one neighbor of v has degree 4;*
2. *any neighbor has degree 3 but the subgraph induced by $\Gamma(v)$ contains at most one edge.*

*Then,* $T(d) \leq T(d-3) + T(d-6)$.

*Proof.* We branch on $v$. If we choose not to add $v$ to the solution, then we can remove $v$ and any vertex adjacent to it, that means $d' = d - deg(v) + 1 \leq d - 3$. On the other hand, if $v$ belongs to the optimal solution, we remove from our graph $v$ and all its neighbors:

$$d' \leq d + 1 - \frac{1}{2} \sum_{w \in \Gamma(v)} deg(w)$$

Dealing with the general case, $\sum_{w \in \Gamma(v)} deg(w) \geq 12$. If *1.* holds, then

$$\sum_{w \in \Gamma(v)} deg(w) \geq 13$$

In case *2.*, we just have to notice that $|I| \leq 1$ means $|\Omega| \geq 10$; replacing this inequality in (1) we get $d' \leq d - 6$.

**Proposition 4.** *Assume that the degree of any vertex in our graph is exactly* 3. *Assume also that $G$ contains some 3-clique $\{a, b, c\}$. Then,* $T(d) \leq 2T(d-4)$.

*Proof.* Let $v$ be the third neighbor of $a$ and $u, w$ be the two other neighbors of $v$. Note that $u$ (and $w$) differs from $b$ and $c$, otherwise $a$ would dominate $b$ or $c$. We branch on $v$. If $v$ does not belong to the optimal solution, it is removed, and so are its incident edges. In the remaining graph, $a$ is dominated by $b$ and $c$ (see Figure 2). According to Lemma 2 we may add it to the optimal solution and remove the whole clique. Eventually,

$$d' = (m - 8) - (n - 4) = d - 4$$

On the other hand, if $v$ has to be added to the solution, we may remove it and its neighbors from the graph. That means, according to Lemma 4,

$$d' \leq (m - 3 - 9/2) - (n - 4) = d - 7/2$$

Since $d'$ has to be an integer, that leads to the expected inequality.

**Proposition 5.** *Assume that $G$ is a 3-clique-free graph where any vertex has degree exactly* 3. *Assume also that $G$ contains the subgraph described in Figure 3 (two pentagons sharing two incident edges). Then,* $T(d) \leq T(d-3) + T(d-5)$.

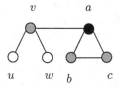

**Fig. 2.** Clique $\{a, b, c\}$

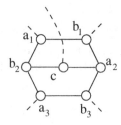

**Fig. 3.** Pentagons sharing two incident edges

*Proof.* Notice at first that in a 3-clique-free graph where any vertex has degree exactly 3, each time we add a vertex to the solution and remove its neighbors, we decrease $d$ by 5. Indeed, the set of neighbors of a vertex contains no inner edge, else it would form a clique. From what we get $\mathcal{N} = \sum_{w \in \Gamma(v)} deg(w) = 9$ and finally

$$d' = (m - 9) - (n - 4) = d - 5$$

We branch on $c$, and consider now the case it does not belong to our solution. Once $c$ and its incident edges have been removed, $deg(a_2) = deg(b_2) = 2$. We reduce them in any case. In the remaining graph,

$$d' = (m - 8) - (n - 5) = d - 3$$

**Proposition 6.** *Assume that $G$ is a 3-clique-free graph where any vertex has degree exactly 3. Assume also that $G$ does not contain the subgraph described in Figure 3. Then, $T(d) \leq T(d - 5) + 2T(d - 8) + T(d - 10)$.*

*Proof.* In this proposition, instead of branching on a single vertex as previously, we successively branch on several vertices. First of all, consider any $v \in V$. As we saw previously, since there is no 3-clique, if we add $v$ to the solution, we decrease $d$ by 5, else we decrease $d$ by 2. Thus, we can write $T(d) \leq T(d - 5) + Q(d - 2)$, for some function $Q \leq T$. If we had no further information about the remaining graph, we would have no choice but to write $T(d) \leq T(d - 5) + T(d - 2)$. But our graph is not any graph; in particular some higher than 3 degree vertices may have been created during our first branching. From now on, the proof will focus on refining analysis of $Q$ thanks to our graph properties.

Let $w_1, w_2, w_3$ be the neighbors of $v$. Once $v$ and its incident edges have been removed, they all have degree 2. Moreover, no couple of them can be adjacent, else they would form a clique with $v$. According to Lemma 3, we now reduce our graph. We distinguish some different cases.

Consider at first an easy case, where there exists a couple of vertices $u_1, u_2$ both adjacent to, say, $w_1$ and $w_2$ (see Figure 4). We can easily see that taking $u_1$ and/or $u_2$ is never interesting, it is never worse to take $w_1$ and $w_2$. So, we can add $\{w_1, w_2\}$ to the optimal solution and delete $\{u_1, u_2\}$. This operation decreases $d$ by 2: 4 vertices and 6 edges are removed. Indeed, $u_1$ and $u_2$ cannot be adjacent

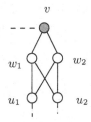

**Fig. 4.** $u_1, u_2$ both adjacent to $w_1$ and $w_2$

otherwise $u_1$ would dominate $u_2$ (and vice-versa) before branching on $v$. In other words, $Q(d) \le T(d-2)$. Consequently,

$$T(d) \le T(d-5) + T(d-4)$$

Assume now that there exists one single vertex $u$ that is adjacent to, say, $w_1$ and $w_2$. Let us denote $s$ and $t$ the third neighbor of respectively $w_1$ and $w_2$ (see Figure 5). Notice that in this case $u$ cannot be adjacent to $w_3$, else it would dominate $v$. When operating reductions of $w_1$ and $w_2$, $s$, $u$ and $t$ are merged together in a single vertex. Two cases may occur (see Remark 1): either 2 of them have another common neighbor and $d$ decreases by at least one, or our graph contains a vertex of degree 5. In the first case, we get $Q(d) \le T(d-1)$ and finally $T(d) \le T(d-5) + T(d-3)$. In the latter case, we now branch on the degree 5 vertex; according to Proposition 2, $Q(d) \le T(d-4) + T(d-7)$, that means:

$$T(d) \le T(d-5) + T(d-6) + T(d-9)$$

Let us now focus on the main case, where $\Gamma(w_i)$'s are disjoint. Let us denote by $u_i^1$ and $u_i^2$ the two other neighbors of $w_i$. When reducing vertex $w_i$, going back to Remark 1, either $d$ decreases by at least one, or a vertex - say $u_i$ - of degree 4 is created. If for at least one $w_i$ $d$ decreases, then $Q(d) \le T(d-1)$ and eventually $T(d) \le T(d-5) + T(d-3)$. Otherwise, when reductions of degree 2 vertices $w_i$'s have been proceeded, the remaining graph contains exactly three vertices whose degree is 4, namely $u_1, u_2, u_3$. Since our graph is not fully

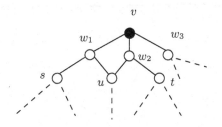

**Fig. 5.** $u$ adjacent to $w_1$ and $w_2$

reduced (branching may have disturbed it), it is possible that some neighbor of some $u_i$ dominates it. In this case we can remove it and its incident edges without branching; $Q(d) \leq T(d - 3)$, that means

$$T(d) \leq 2T(d - 5)$$

Thus, we may now assume that no $u_i$ is dominated. We have to consider whether they are adjacent or not.

If say $u_1$ and $u_2$ are adjacent, then we branch on $u_1$. According to Proposition 3 (in the specific case the vertex we branch has at least one neighbor of degree 4), either we add $u_1$ to the solution and we decrease $d$ by at least 6, or we remove $u_1$ and the four edges incident to it and we decrease $d$ by 3. Hence, $Q(d) \leq T(d - 6) + T(d - 3)$.

If no $u_i$ is adjacent to any other $u_j$, this time if we use Proposition 3, we cannot assert that two degree 4 vertices are adjacent, that would mean decreasing $d$ only by 5. Fortunately, it is possible to prove that $\Gamma(u_1)$ does not have more than one inner edge. Indeed, assume that it has two inner edges; since, before branching on $v$, we assumed that the graph does not contain any 3-clique, then the unique possibility is described in Figure 6. It means that, before branching on $v$, our graph contained two pentagons sharing two edges (edges $(w_1, u_1^1)$ and $(w_1, u_1^2)$). This is in contradiction with hypothesis of Proposition 6. Hence, in this case also $Q(d) \leq T(d - 6) + T(d - 3)$.

At this step, we get $T(d) \leq T(d - 5) + Q(d - 2) \leq 2T(d - 5) + T(d - 8)$. This sums up to $T(d) = O^*(1.2076^d)$.

We make a final remark that further improves the running time. Indeed, we will see that after performing the branching on $u_1$, then there exists another vertex of degree 4 in the remaining graph (or an even better case occurs). By branching on it, we decrease $d$ either by 3 or by 5 (Proposition 3). This leads to $Q(d) \leq T(d - 6) + T(d - 3 - 3) + T(d - 3 - 5) = 2T(d - 6) + T(d - 8)$, and finally $T(d) \leq T(d - 5) + 2T(d - 8) + T(d - 10)$, as claimed in the proposition. To see this, consider two cases. If say $u_1$ and $u_2$ are not adjacent, then when branching along the branch "take $u_1$", $u_2$ will still have degree at least 4 (or would have been deleted by domination which is even better).

The difficult case occurs when $u_1$, $u_2$ and $u_3$ form a clique. Let $s$ and $t$ be the two neighbors of $u_1$ with degree 3 (see Figure 7).

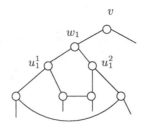

**Fig. 6.** Two pentagons sharing two consecutive edges

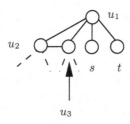

**Fig. 7.** $u_1, u_2$ and $u_3$ is a clique

- If $s$ and $t$ are adjacent, then in $\Gamma(u_1)$ there are only 2 edges (the other one being $(u_2, u_3)$), otherwise $s$ or $t$ would dominate $u_1$. Hence, when taking $u_1$, we delete 12 edges and 5 vertices, which means $Q(d) \leq T(d-3) + T(d-7)$, i.e., $T(d) \leq 2T(d-5) + T(d-9)$.
- If $s$ and $t$ are not adjacent, $Q(d) \leq T(d-3) + T(d-6)$ but when reducing $s$ and $t$ (after taking $u_1$), we are in one of the two cases of Remark 1, i.e., either $d$ decreases and $Q(d) \leq T(d-3) + T(d-7)$, or a degree 4 vertex is created.

The analysis of the solutions of the equations induced by the previous analysis shows that the worst case running time corresponds to the case $T(d) \leq T(d-5) + 2T(d-8) + T(d-10)$, which sums up to $T(d) = O^*(1.2048^d)$.

## 4   Dealing with Trees

As noticed in [5], we have to be careful with our previous analysis. Indeed, the adopted measure $d = m - n$ creates a somehow unexpected problematic situation, occurring when several connecting components are created when branching, one or several of them being tree(s). Indeed, applying preprocessing rules immediately reduces a tree to a single vertex, which can be added to the solution. However, when removing this vertex (i.e., when dealing with the tree), the measure $m - n$ *increases*. In other words, when creating a tree, if the measure globally decreases by say $x$, it decreases only by $x - 1$ in the remaining graph, tree excepted.

To deal with this situation, we study the situations when one or several tree(s) are created when branching. We show that in these particular cases the number of edges deleted is sufficient to compensate the loss induced by the tree(s).

To obtain this, we use a property shown by Fürer [5] that handles the case where there is a separator of size one or two in the graph. More precisely, if there exists one vertex $u$ or a couple of vertices $(u_1, u_2)$ the deletion of which disconnects the graph, creating one connecting component of say constant size, then this connected component can be eliminated before branching. This reduction, not increasing $d$, can be added for instance to the preprocessing step.

Let us consider that we branch on a vertex $v$. As pointed out by Fürer [5], if we don't take $v$, then no tree can be created (since every vertex of the graph has degree at least 3).

Let us now handle the case where we take $v$. Note that since each vertex in the graph has degree 3, when a tree with $r$ vertices is created, at least $r+2$ edges link one vertex of the tree to one vertex of $\Gamma(v)$. If $v$ has degree 3 (the graph is three regular), if a tree is created, this tree is a single vertex $w$ linked to the 3 vertices in $\Gamma(v)$. Indeed, each vertex in $\Gamma(v)$ has at least one edge linking it to the rest of the graph (else there would be a separator of size 2), hence at most one edge linking it to the tree. In this case, the graph reduces (before branching on $v$): we delete $v$ and $w$ and merge the vertices in $\Gamma(v)$ in a single vertex. We delete 6 edges and 4 vertices, hence $d$ decreases.

Assume now that $deg(v) \geq 4$.

Let us note by:

- $|I|$ (as previously) the number of edges linking two vertices in $\Gamma(v)$;
- $l$ the number of trees created when branching on $v$ (taking $v$);
- $t$ the number of edges linking a vertex in $\Gamma(v)$ to a vertex in one of the trees created;
- $e$ the number of edges linking one vertex of $\Gamma(v)$ to the rest of the graph;
- $D = \sum deg(w)$ for $w \in \Gamma(v)$.

When branching on $v$, either we don't take $v$ and $d' = d - deg(v) + 1 \leq d - 3$, or we take $v$. In this latter case: we delete $1 + deg(v)$ vertices $v$ and $\Gamma(v)$, $deg(v) + |I| + t + e$ edges incident to vertices in $\Gamma(v)$, and we loose 1 for each tree created. Hence, after handling the tree(s), $d$ has globally decreased by: $\delta = |I| + t + e - 1 - l$.

Note that $t \geq 3$ (or we have a separator of size 2 or less) and $e \geq 3l$ (at least 3 edges for each tree). Hence: $\delta \geq |I| + 3 + 3l - 1 - l \geq 2 + 2l$. In particular, if $l > 1$ then $\delta \geq 6$.

If $l = 1$, then $\delta = |I| + t + e - 2$. $D = 2|I| + t + e + 4$. Then $\delta = D/2 + t/2 + e/2 - 4$. Since $e \geq 3$ and $t \geq 3$, if $D \geq 13$ then $\delta \geq 13/2 - 1$. Since it is an integer, $\delta \geq 6$.

The only remaining case is $D = 12$ (hence $deg(v) = 4$ and each vertex in $\Gamma(v)$ has degree 3), $t = 3$ and $e = 3$ (see Figure 8). But in this case, the tree separated has only one vertex $w$. Then, the only inner edge in $\Gamma(v)$ is incident to the vertex $s$ in $\Gamma(v)$ non adjacent to $w$ (otherwise there would be a dominated vertex), as depicted in Figure 8. Hence, it is never interesting to take $s$ in the solution (take either $v$ or $t$ instead), and we don't need to branch.

Hence, in the cases where branching is needed, we decrease $d$ either by 3 (not taking $v$ in the solution) or by $\delta \geq 6$ (taking $v$ in the solution). This is in particular the recurrence relation $T(d) \leq T(d-3) + T(d-6)$ needed when operating branching in the analysis of Section 3.

## 5   The Concluding Theorem

The analysis provided in Sections 2, 3 and 4 allows to state the following result.

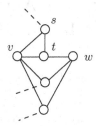

**Fig. 8.** Taking $s$ is never interesting

**Theorem 1.** *On any graph, it is possible to solve* MAX INDEPENDENT SET *with running time* $O^*(1.2048^{m-n})$. *In particular, a solution to* MAX INDEPENDENT SET-3 *may be computed with running time* $O^*(1.0977^n)$.

As the result of Theorem 1 depends only on the difference $m - n$, the time-bound claimed for MAX INDEPENDENT SET-3 remains valid also if only the *average* degree of the input-graph is bounded by 3 since, also in this case, $m \leq 3n/2$.

# References

1. Robson, J.M.: Finding a maximum independent set in time $O(2^{n/4})$. Technical Report 1251-01, LaBRI, Université de Bordeaux I (2001)
2. Beigel, R.: Finding maximum independent sets in sparse and general graphs. In: Proc. Symposium on Discrete Algorithms, SODA 1999, pp. 856–857 (1999)
3. Chen, J., Kanj, I.A., Xia, G.: Labeled search trees and amortized analysis: improved upper bounds for NP-hard problems. In: Ibaraki, T., Katoh, N., Ono, H. (eds.) ISAAC 2003. LNCS, vol. 2906, pp. 148–157. Springer, Heidelberg (2003)
4. Fomin, F.V., Høie, K.: Pathwidth of cubic graphs and exact algorithms. Inform. Process. Lett. 97, 191–196 (2006)
5. Fürer, M.: A faster algorithm for finding maximum independent sets in sparse graphs. In: Correa, J.R., Hevia, A., Kiwi, M. (eds.) LATIN 2006. LNCS, vol. 3887, pp. 491–501. Springer, Heidelberg (2006)
6. Razgon, I.: A faster solving of the maximum independent set problem for graphs with maximal degree 3. In: Proc. Algorithms and Complexity in Durham, ACiD 2006, pp. 131–142 (2006)
7. Wœginger, G.J.: Exact algorithms for NP-hard problems: A survey. In: Jünger, M., Reinelt, G., Rinaldi, G. (eds.) Combinatorial Optimization - Eureka, You Shrink! LNCS, vol. 2570, pp. 185–207. Springer, Heidelberg (2003)
8. Eppstein, D.: Improved algorithms for 3-coloring, 3-edge-coloring, and constraint satisfaction. In: Proc. Symposium on Discrete Algorithms, SODA 2001, pp. 329–337 (2001)

# New Fixed-Parameter Algorithms for the Minimum Quartet Inconsistency Problem⋆

Maw-Shang Chang[1], Chuang-Chieh Lin[1], and Peter Rossmanith[2]

[1] Department of Computer Science and Information Engineering,
National Chung Cheng University, Ming-Hsiung, Chiayi, Taiwan, R.O.C.
mschang@cs.ccu.edu.tw, lincc@cs.ccu.edu.tw
[2] Department of Computer Science, RWTH Aachen University, Germany
rossmani@informatik.rwth-aachen.de

**Abstract.** Given a set of $n$ taxa $S$, exactly one topology for every subset of four taxa, and a positive integer $k$ as the parameter, the parameterized Minimum Quartet Inconsistency (MQI) problem is to decide whether we can find an evolutionary tree inducing a set of quartet topologies that differs from the given set in at most $k$ quartet topologies. The best fixed-parameter algorithm devised so far for the parameterized MQI problem runs in time $O(4^k n + n^4)$. In this paper, first we present an $O(3.0446^k n + n^4)$ algorithm and an $O(2.0162^k n^3 + n^5)$ algorithm. Finally, we give an $O^*((1 + \epsilon)^k)$ algorithm with an arbitrarily small constant $\epsilon > 0$.

## 1 Introduction

Nowadays, to determine the evolutionary relationship of a set of taxa is very important in biological research, especially in computational biology. For this relationship, an *evolutionary tree* is a common model, which is widely considered. Let $S$ be a set of taxa and $|S| = n$. An evolutionary tree $T$ on $S$ is an *unrooted*, *leaf-labeled* tree such that the leaves of $T$ are bijectively labeled by the taxa in $S$, and each internal node of $T$ has degree three. In the past decade, *quartet methods* for building an evolutionary tree for a set of taxa have received much attention [1,3,4,6,8,9,10,11,14].

A *quartet* is a set of four taxa $\{a, b, c, d\}$ in the taxon set $S$. The *quartet topology* for $\{a, b, c, d\}$ induced by $T$ is the path structure connecting $a, b, c$, and $d$ in $T$ (see Fig. 1 for an illustration). A quartet $\{a, b, c, d\}$ has three possible topologies $[ab|cd], [ac|bd]$, and $[ad|bc]$, which are the bipartitions of $\{a, b, c, d\}$ (see Fig. 2 for an illustration).

A *quintet* is a set of five taxa in $S$, while a *sextet* is a set of six taxa in $S$. The *quintet topology* of a quintet $\{a, b, c, d, e\}$ induced by an evolutionary tree $T$ is the path structure connecting $a, b, c, d$, and $e$ in $T$. Similarly, we have the *sextet topology* of a sextet. Without loss of generality, assume that we have

---

⋆ This research was supported by NSC-DAAD Sandwich Program and partially supported by the National Science Council of Taiwan under grant no. NSC 96-2221-E-194-045-MY3, and was carried out at RWTH Aachen University, Germany.

M. Grohe and R. Niedermeier (Eds.): IWPEC 2008, LNCS 5018, pp. 66–77, 2008.

$[bc|de]$ induced by $T$ and another taxon $a$, then there are five possible quintet topologies since there are five positions for inserting $a$ into the tree structure of $[bc|de]$ (see Fig. 3 for an illustration).

There are 15 possible quintet topologies for a quintet $\{a, b, c, d, e\}$. A quintet has five quartets, and hence a quintet topology has 5 different induced quartet topologies. Two taxa $a, b$ are *siblings* on an evolutionary tree $T$ if $a$ and $b$ are both adjacent to the same internal vertex in $T$. Here we consider sextet topologies of the sextet $\{a, b, w, x, y, z\}$ where $a, b$ are siblings. It is clear that there are 15 possible sextet topologies with siblings $a, b$ (refer to Fig. 4).

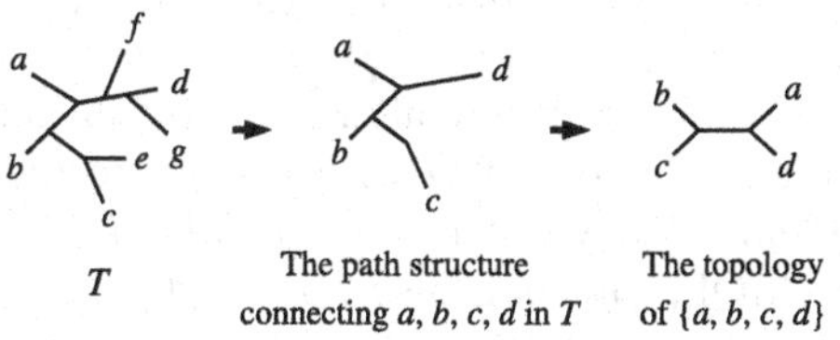

**Fig. 1.** The quartet topology of $\{a, b, c, d\}$ induced by evolutionary tree $T$

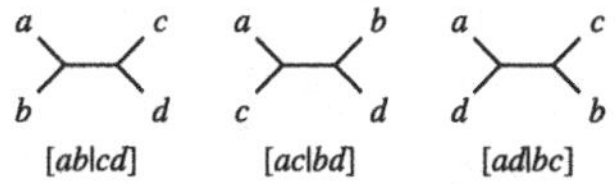

**Fig. 2.** Three topologies for the quartet $\{a, b, c, d\}$

Given a set of quartet topologies $Q$ over the taxon set $S$, we say that a quintet $\{a, b, c, d, e\} \subseteq S$ is *resolved* if there exists an evolutionary tree $T'$, on which $a, b, c, d, e$ are leaves, such that all the quartet topologies induced by $T'$ are in $Q$. Otherwise, we say that $\{a, b, c, d, e\}$ is *unresolved*. Similarly, we say that a sextet $\{a, b, w, x, y, z\} \subseteq S$ is $\{a, b\}$-*resolved* if there exists an evolutionary tree $T''$, on which $a, b, w, x, y, z$ are leaves and $a, b$ are siblings, such that all quartet topologies induced by $T''$ are in $Q$. Otherwise, we say that $\{a, b, w, x, y, z\}$ is $\{a, b\}$-*unresolved*.

Let $Q_T$ be the set of quartet topologies induced by $T$. If there exists an evolutionary tree $T$ such that $Q \subseteq Q_T$, we say that $Q$ is *tree-consistent* [2] (with $T$) or $T$ *satisfies* $Q$. If there exists a tree $T$ such that $Q = Q_T$, we say that $Q$ is *tree-like* [2]. $Q$ is called *complete* if $Q$ contains exactly one topology for

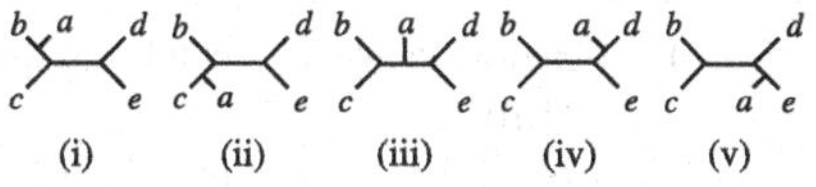

**Fig. 3.** Five possible topologies for the quintet $\{a, b, c, d, e\}$ when $[ab|cd]$ is given

every quartet, otherwise, *incomplete*. For two complete sets of quartet topologies $Q$ and $Q^*$ where $Q^*$ is tree-like but $Q$ is not, the *quartet errors* of $Q$ with respect to $Q^*$ are the quartet topologies in $Q$ that differ from those in $Q^*$. We denote the number of quartet errors of $Q$ with respect to $Q^*$ by $\Delta(Q, Q^*)$. The *number of quartet errors of $Q$* is defined to be $\min\{\Delta(Q, Q^*) : Q^* \text{ is tree-like}\}$.

## 1.1   Related Work

The *Quartet Compatibility Problem* (QCP) is to determine if there exists an evolutionary tree $T$ on $S$ satisfying all quartet topologies $Q$. The QCP problem can be solved in polynomial time if $Q$ is complete [6], but it becomes **NP**-complete when $Q$ is not necessarily complete [11]. The optimization problem, called the *Maximum Quartet Consistency Problem* (MQC), is to construct an evolutionary tree $T$ on $S$ to satisfy as many quartet topologies of $Q$ as possible. The *Minimum Quartet Inconsistency Problem* (MQI), which is a dual problem to the MQC problem, is to construct an evolutionary tree $T$ on $S$ such that the number of quartet errors of $Q$ with respect to $Q_T$ is minimized. The MQC problem is **NP**-hard [3], yet it has a *polynomial time approximation scheme* (PTAS) [10]. The MQI problem is also **NP**-hard [3], while the best approximation ratio found so far for the MQI problem is $O(n^2)$ [9]. For the case that the input set of quartet topologies $Q$ is not necessarily complete, Ben-Dor *et al.* gave an $O(3^n m)$ algorithm to solve the MQI problem by dynamic programming [1], where $m$ is the number of quartet topologies. When the number of quartet errors is smaller than $(n-3)/2$, Berry *et al.* [3] devised an $O(n^4)$ algorithm for the MQI problem. If the number of quartet errors is at most $cn$ for some positive constant $c$, Wu *et al.* [15] compute the optimal solution for the MQI problem in $O(n^5 + 2^{4c}n^{12c+2})$ time. While this is a polynomial time algorithm, the degree of the polynomial in the runtime grows quickly. Therefore parameterized algorithms are faster for practical values of $k$ and $n$.

Provided with a positive integer $k$ as an additional part of the input, the *parameterized MQI problem* is to determine whether there exists an evolutionary tree $T$ such that the number of quartet errors of $Q$ is at most $k$. Gramm and Niedermeier proved that the parameterized MQI problem is *fixed parameter tractable* in time $O(4^k n + n^4)$ [8]. In [14], Wu *et al.* presented a lookahead branch-and-bound algorithm for the MQC prob-

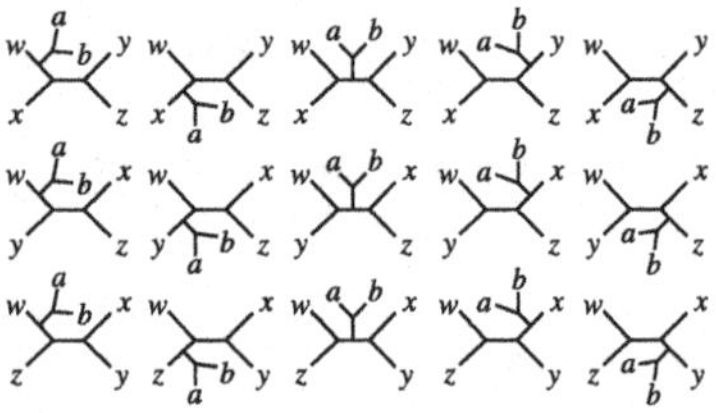

**Fig. 4.** The 15 possible sextet topologies for sextet $\{a, b, w, x, y, z\}$ with siblings $a, b$

lem which runs in time $O(4^{k'} n^2 k' + n^4)$, where $k'$ is an upper bound on the number of quartet errors of $Q$.

## 1.2  Our Result

In this paper, we focus on the parameterized MQI problem as follows. Given a complete set of quartet topologies $Q$ and a parameter $k$ as the input, determine whether there is a tree-like quartet topology set that differs from $Q$ in at most $k$ quartet topologies, that is, determine whether $Q$ has at most $k$ quartet errors. Using the *depth-bounded search tree* strategy, we propose an $O(3.0446^k n + n^4)$ fixed-parameter algorithm for this problem. With slight refinement, we obtain an $O(2.0162^k n^3 + n^5)$ algorithm and an $O^*((1+\epsilon)^k)$ time algorithm with arbitrarily small $\epsilon > 0$.

The paper is organized as follows. In Sect. 2, we will give additional theoretical background for the MQI problem. In Sect. 3, we present an $O(3.0446^k n + n^4)$

fixed-parameter algorithm for the parameterized MQI problem. In Sect. 4, we will introduce the *two-siblings-determined minimum quartet inconsistency problem* (2SDMQI), then an $O(2.0162^k n + n^4)$ fixed-parameter algorithm for this problem will be given. At the end of this section, we will present an $O(2.0162^k n^3 + n^5)$ fixed-parameter algorithm for the parameterized MQI problem by solving the 2SDMQI problem. Finally in Sect. 5, we will present an $O^*((1 + \epsilon)^k)$ fixed-parameter algorithm for the parameterized MQI problem, where $\epsilon > 0$ is an arbitrarily small constant. For the sake of brevity, many proofs are omitted in this extended abstract, but can be found in the full paper.

## 2   Preliminaries

Recall that $S$ is a set of taxa and $|S| = n$. Let $Q$ denote the complete set of quartet topologies over $S$. The set $Q$ is of size $\binom{n}{4}$. We say that a set of quartet topologies $Q'$ over $S$ *involves a taxon* $f$ if there exists at least one quartet topology $t = [v_1 v_2 | v_3 v_4] \in Q'$, where $v_1, v_2, v_3, v_4 \in S$, such that $f = v_i$ for some $i \in \{1, 2, 3, 4\}$. If a set of quartet topologies is not tree-consistent, we say that it has a *conflict*. We say that a set of three topologies has a *local conflict* if it is not tree-consistent.

**Lemma 1.** [8] *A set of three quartet topologies involving more than five taxa is tree-consistent.*

**Theorem 1.** [8] *Given a set of taxa $S$ and a complete set of quartet topologies $Q$ over $S$, and some taxon $f \in S$, then $Q$ is tree-like if and only if every set of three quartet topologies in $Q$ that involves $f$ has no local conflict.*

**Lemma 2.** *Assume that $\mathbf{q} \subseteq S$ is a quintet such that $f \in \mathbf{q}$ and let $Q_{\mathbf{q}} \subseteq Q$ denote the set of quartet topologies of quartets in $\mathbf{q}$. Then $\mathbf{q}$ is resolved if and only if every set of three quartet topologies in $Q_{\mathbf{q}}$ has no local conflict.*

**Corollary 1.** *Given a set of taxa $S$, a complete set of quartet topologies $Q$ over $S$, and some taxon $f \in S$, then $Q$ is tree-like if and only if every quintet containing $f$ is resolved.*

There are ten sets of three quartets with respect to a quintet $\{a, b, c, d, e\}$. Checking whether a set of three quartet topologies has a local conflict requires only constant time [8]. It is then clear that checking whether a quintet is resolved requires only constant time. Given a taxon $f \in S$ which is fixed, there are $\binom{n-1}{4}$ quintets containing $f$. Thus we have the following theorem.

**Theorem 2.** *Given a set $S$ of taxa, some taxon $f \in S$, and a complete set $Q$ of quartet topologies, then all unresolved quintets involving $f$ can be found in $O(n^4)$ time.*

Let $\prec$ be some order (e.g. lexicographic order) on the taxon set $S$. For the three possible topologies of a quartet, we denote them by *type* 0, 1, and 2 according to this order. Consider a quartet $\{a, b, c, d\} \subset S$ as an example. If $a \prec b \prec c \prec d$, we denote $[ab|cd]$ by 0, $[ac|bd]$ by 1, and $[ad|bc]$ by 2. For a quintet $\{s_1, s_2, s_3, s_4, s_5\}$ where $s_1 \prec s_2 \prec s_3 \prec s_4 \prec s_5$, we define its *topology vector* to be an ordered sequence of types of the quartet topologies over the quintet. For example, consider a quintet $\{a, b, c, d, e\} \subseteq S$ with $[ab|cd]$, $[ae|bc]$, $[ab|de]$, $[ae|cd]$ and $[bd|ce]$ in $Q$, then the topology vector of $\{a, b, c, d, e\}$ is $(0, 2, 0, 2, 1)$. Recall that there are 15 possible quintet topologies for a quintet $\{s_1, s_2, s_3, s_4, s_5\}$. We denote by $\mathcal{V}$ the set of topology vectors of all the possible quintet topologies of a quintet, then we have $\mathcal{V} = \{(0,0,0,0,0), (1,1,0,0,0), (2,2,0,0,0), (2,2,1,1,0),$ $(2,2,2,2,0)$, $(0,0,0,1,1)$, $(2,0,1,1,1)$, $(1,0,2,1,1)$, $(1,1,2,0,1)$, $(1,2,2,2,1)$, $(0,0,0,2,2), (0,2,2,2,2), (0,1,1,2,2), (1,1,1,0,2), (2,1,1,1,2)\}$.

Consider the sextet $\{s_1, s_2, s_3, s_4, s_5, s_6\} \subseteq S$ where $s_1, s_2$ are siblings in an evolutionary tree over $S$. There are six quartets in the sextet having fixed quartet topologies (for example, the quartet topology of $\{s_1, s_2, s_3, s_4\}$ must be $[s_1s_2|s_3s_4]$). Given two siblings $s_1, s_2$, the $\{s_1, s_2\}$-*reduced topology vector* of sextet $\{s_1, s_2, s_3, s_4, s_5, s_6\}$ is an ordered sequence of types of the quartet topologies which are not fixed. For example, consider a sextet $\{a, b, w, x, y, z\} \subseteq S$ with siblings $a, b$ such that $[aw|xy]$, $[ax|wz]$, $[az|wy]$, $[ay|xz]$, $[bw|xy]$, $[bx|wz]$, $[bz|wy]$, $[by|xz]$, and $[wx|yz]$ are in $Q$. The $\{a, b\}$-reduced topology vector of $\{a, b, w, x, y, z\}$ is $(0, 1, 2, 1, 0, 1, 2, 1, 0)$. Let us denote by $\mathcal{V}_2$ the set of $\{a, b\}$-reduced topology vectors of all possible sextet topologies of $\{a, b, w, x, y, z\}$. Then we have $\mathcal{V}_2 = \{(0,0,0,0,0,0,0,0,0), (1,1,0,0,1,1,0,0,0), (2,2,0,0,2,2,0,0,0),$ $(2,2,1,1,2,2,1,1,0), (2,2,2,2,2,2,2,2,0), (0,0,0,1,0,0,0,1,1), (2,0,1,1,2,0,1,$ $1,1), (1,0,2,1,1,0,2,1,1), (1,1,2,0,1,1,2,0,1), (1,2,2,2,1,2,2,2,1), (0,0,0,2,$ $0,0,0,2,2), (0,2,2,2,0,2,2,2,2), (0,1,1,2,0,1,1,2,2), (1,1,1,0,1,1,1,0,2), (2,$ $1,1,1,2,1,1,1,2)\}$.

## 3   An $O(3.0446^k n + n^4)$ Algorithm

*The Algorithm.* Our first fixed-parameter algorithm is called FPA1-MQI, which runs recursively. The concepts of the algorithm are as follows. We build a list of unresolved quintets $\mathcal{C}_f$ containing some fixed taxon $f$ and the list $\mathcal{V}$ of topologies vectors of possible quintet topologies for a quintet as preprocessing steps. In each recursion, the algorithm selects an unresolved quintet $\mathbf{q} = \{a, b, c, d, e\} \in \mathcal{C}_f$ arbitrarily and then tries to make $\mathbf{q}$ resolved by the procedure update according to all the possible 15 quintet topologies of $\mathbf{q}$.

Recall that each topology vector $\mu \in \mathcal{V}$ represents a quintet topology of a quintet. The procedure update changes quartet topologies according to the quartet topologies which $\mu$ stands for, and updates the set $\mathcal{C}_f$ and the parameter $k$ to be $\mathcal{C}'_f$ and $k'$ respectively. For example, assume that we have $[ab|cd]$, $[ae|bc]$, $[ab|de]$, $[ae|cd]$, and $[bd|ce]$ in $Q$ for the quintet $\{a, b, c, d, e\}$ (the corresponding topology vector is then $(0, 2, 0, 2, 1)$), and assume that $\mu = (2, 1, 1, 1, 2)$. According to $\mu$, the procedure update changes these quartet topologies to $[ad|bc]$, $[ac|be]$, $[ad|be]$, $[ad|ce]$, and $[be|cd]$ respectively, and these quartets are marked so that their topologies will not be changed again. However, if there is a branch node

---

**Algorithm 1: FPA1-MQI (a complete set of quartet topologies $Q$, an integer parameter $k$, a list $C_f$ of unresolved quintets)**

---

1: **if** $C_f$ is empty and $k \geq 0$ **then**
2:    return ACCEPT;
3: **else if** $k \leq 0$ **then**
4:    return;
5: **end if**
6: Extract an unresolved quintet $\mathbf{q}$ from $C_f$;
7: **for each** $\mu \in \mathcal{V}$ **do**
8:    $(Q', C_f', k') \leftarrow \mathsf{update}(Q, C_f, \mathbf{q}, \mu, k)$;
9:    FPA1-MQI $(Q', k', C_f')$;
10: **end for**

---

in the search tree such that some quartet, which has been marked, must be changed in all the possible 15 branches to make an unresolved quintet resolved, the algorithm stops branching here and just returns. Let $Q_\mu$ be the set of quartet topologies changed according to $\mu$. The procedure update obtains the updated set of unresolved quintets $C_f'$ by removing the newly resolved quintets and adding the newly unresolved quintets, and gets the updated parameter $k' = k - |Q_\mu|$.

By Corollary 1, we know $C_f$ is empty if and only if the set of quartet topologies is tree-like. Algorithm FPA1-MQI branches in all possible ways to eliminate each unresolved quintet in $C_f$ and it changes at most $k$ quartet topologies from the root to each branch node in the search tree. Thus it is easy to see that the algorithm is correct.

*The Time Complexity.* The algorithm works as a depth-bounded search tree. Each tree node has 15 branches and each branch corresponds to a quintet topology. Since there are 243 possible topology vectors of a quintet but 15 of them are in $\mathcal{V}$, we have 228 possible branching vectors and the corresponding branching numbers as well. Consider the case that the algorithm selects a quintet $\mathbf{q} = \{a, b, c, d, e\}$ which has induced quartet topologies $[ab|cd]$, $[ac|be]$, $[ae|bd]$, $[ad|ce]$, and $[bc|de]$ in $Q$. By comparing its corresponding topology vector $(0,1,2,1,0)$ with each topology vector $\mu \in \mathcal{V}$, we obtain the numbers of quartet topologies changed by Algorithm FPA1-MQI, and then we have a branching vector $(3, 3, 4, 3, 3, 3, 4, 3, 3, 4, 4, 3, 3, 4, 3)$ hence a branching number between 2.3004 and 2.3005 is obtained. It can be derived that the branching number in the worst case is between 3.0445 and 3.0446. Thus the size of the search tree is $O(3.0446^k)$. Then we obtain the following theorem by careful analysis.

**Theorem 3.** *There exists an $O(3.0446^k n + n^4)$ fixed-parameter algorithm for the parameterized minimum quartet inconsistency problem.*

## 4   An $O(2.0162^k n^3 + n^5)$ Algorithm

We define the *two-siblings-determined minimum quartet inconsistency problem* (2SDMQI) as follows. Given a complete quartet topology set $Q$ over a taxon set

---

**Algorithm 2:** FPA-2SDMQI (a complete set of quartet topologies $Q$, an integer parameter $k$, a list $C_a$ of unresolved quintets, two taxa $a, b$)

---

1: **if** $C_a$ is empty and $k \geq 0$ **then**
2:     return ACCEPT;
3: **else if** $k \leq 0$ **then**
4:     return;
5: **end if**
6: **for** every two taxa $u, v \in S \setminus \{a, b\}$ **do**
7:     **if** $k \leq 0$ **then** return;
8:     **else** change the quartet topology of $\{a, b, u, v\}$ to be $[ab|uv]$ if $[ab|uv] \notin Q$ and update $C_a$ and $k \leftarrow k - 1$;
9: **end for**
10: Resolve$(Q, k, C_a, a, b)$;

---

$S$, a parameter $k$ and two taxa $a, b \in S$ as the input, determine whether there exists an evolutionary tree $T$ on which $a$ and $b$ are siblings such that $Q_T$ differs from $Q$ in at most $k$ quartet topologies.

We present a fixed-parameter algorithm called FPA-2SDMQI for the 2SDMQI problem as follows. First, for every $u, v \in S \setminus \{a, b\}$ such that $[ab|uv] \notin Q$, we change the quartet topology of $\{a, b, u, v\}$ to be $[ab|uv]$ and decrease $k$ by 1. Second, we build two lists $C_a$ and $V_2$, where $C_a$ is a list of unresolved quintets containing $a$ while $V_2$ is a list of $\{a, b\}$-reduced topologies vectors of possible sextet topologies on which $a, b$ are siblings. Then the algorithm calls Algorithm Resolve to resolve all $\{a, b\}$-unresolved sextets by changing at most $k$ quartet topologies. In each recursion of Algorithm Resolve, we arbitrarily select an unresolved quintet $\mathbf{q}$ and try to make $\mathbf{q} \cup \{b\}$ be $\{a, b\}$-resolved by the procedure update$_2$ according to all possible 15 sextet topologies of $\mathbf{q} \cup \{b\}$ having $a, b$ as siblings. Similar to the procedure update in Sec. 3, we mark the quartets whose topologies are changed, and if there is a branch node in the search tree such that some quartet, which has been marked, must be changed in all the possible branches to make $\mathbf{q} \cup \{b\}$ be $\{a, b\}$-resolved, the algorithm stops branching here and just returns. The procedure update$_2$ updates the set of unresolved quintets $C_a$ by removing the newly resolved quintets and adding the newly unresolved quintets, and updates the parameter $k$. It is easy to see that Algorithm FPA-2SDMQI is also valid.

Similar to the analysis in Sec. 3, we obtain that the size of the search tree is $O(2.0162^k)$. Hence we have an $O(2.0162^k n + n^4)$ fixed-parameter algorithm for the 2SDMQI problem. Let us turn to consider the parameterized MQI problem. It is easy to see that every evolutionary tree with $|S| \geq 4$ leaves has at least two pairs of taxa which are siblings. Thus by building the list of unresolved quintets involving taxon $s$ for every $s \in S$ and running Algorithm FPA-2SDMQI for every two taxa in $S$, we obtain an $O(2.0162^k n^3 + n^5)$ fixed-parameter algorithm for the parameterized MQI problem.

**Theorem 4.** *There exists an $O(2.0162^k n^3 + n^5)$ fixed-parameter algorithm for the parameterized minimum quartet inconsistency problem.*

---

**Algorithm 3:** Resolve (a complete set of quartet topologies $Q$, an integer parameter $k$, a list $\mathcal{C}_a$ of unresolved quintets, two taxa $a, b$)

---

1: **if** $\mathcal{C}_a$ is empty and $k \geq 0$ **then**
2:     return ACCEPT;
3: **else if** $k \leq 0$ **then**
4:     return;
5: **end if**
6: Extract an unresolved quintet $\mathbf{q}$ from $\mathcal{C}_a$;
7: **if** $b \in \mathbf{q}$ **then** $\mathbf{q} \leftarrow \mathbf{q} \cup \{s\}$, for some arbitrary taxon $s \notin \mathbf{q}$; **else** $\mathbf{q} \leftarrow \mathbf{q} \cup \{b\}$;
8: **for** each $\nu \in \mathcal{V}_2$ **do**
9:     $(Q', \mathcal{C}_a', k') \leftarrow \mathsf{update}_2(Q, \mathcal{C}_a, \mathbf{q}, \nu, k)$;
10:     $\mathsf{Resolve}(Q', k', \mathcal{C}_a', a, b)$;
11: **end for**

---

# 5 An $O^*((1 + \epsilon)^k)$ Algorithm

*The Algorithm.* At the beginning of this section, let us consider some additional preliminaries. Let $T$ denote an evolutionary tree on $S$ such that $Q_T$ differs from $Q$ in at most $k$ quartet topologies. For an integer $m \geq 2$, we say that taxa $a_1, \ldots, a_m$ are *adjacent* if there exists an edge $\mathbf{e} = (w, v)$ on $T$ such that cutting $\mathbf{e}$ will produce a bipartition $(\{a_1, \ldots, a_m\}, S \setminus \{a_1, \ldots, a_m\})$ of $S$. In Fig. 5, cutting the edge $\mathbf{e}$ will derive four adjacent taxa $a_1, a_2, a_3$, and $a_4$. After $\mathbf{e} = (w, v)$ is cut, two binary trees will be produced which are rooted at $w$ and $v$ respectively. Note that two taxa on $T$ are adjacent if and only if they are siblings on $T$.

**Lemma 3.** *Given an evolutionary tree $T$ and an integer $2 \leq \omega \leq n/2$, there exists a set of $m$ adjacent taxa as leaves on $T$, where $\omega \leq m \leq 2\omega - 2$.*

By extending the idea of Algorithm FPA2-MQI to consider $m \geq 3$ adjacent taxa, we obtain another fixed-parameter algorithm, called FPA3-MQI, with two subroutines Algorithm MAKE-ADJ and Algorithm ADJ-Resolve. Assume that $A_m = \{a_1, \ldots, a_m\}$ is a set of adjacent taxa on $T$. We introduce the main concepts of Algorithm FPA3-MQI as follows.

$(2, 2)$-*cleaning*: For every two taxa $a_i, a_j \in A_m$ and every two taxa $u, v \in S \setminus A_m$, we modify the topology of $\{a_i, a_j, u, v\}$ to be $[a_i a_j | uv]$. We call this part of the algorithm $(2, 2)$-*cleaning*.

$(3, 1)$-*cleaning*: Assume the parameter is $k'$. For $a_h, a_i, a_j \in A_m$ and $s \in S \setminus A_m$, without loss of generality we denote the type of quartet topology $[a_h a_i | a_j s]$ by 0, $[a_h a_j | a_i s]$ by 1, and $[a_h s | a_i a_j]$ by 2. We construct a set of all possible evolutionary trees $\mathcal{T}_{m+1}$ on the taxa in $A_m \cup \{x\}$, where $x$ is an arbitrary taxon in $S \setminus A_m$, such that each $T' \in \mathcal{T}_{m+1}$ has at most $k'$ different induced quartet topologies from $Q$. Afterwards, for each $T' \in \mathcal{T}_{m+1}$, we change the type of topology of every quartet $\{a_h, a_i, a_j, s\}$ into the same type of topology as $\{a_h, a_i, a_j, x\}$ has on $T'$. We call this part of the algorithm $(3, 1)$-*cleaning*.

$(1, 3)$-*cleaning*: Without loss of generality, we denote the type of quartet topology $[a_i w | xy]$ by 0, $[a_i x | wy]$ by 1, and $[a_i y | wx]$ by 2 for $a_i \in A_m$ and

$w, x, y \in S \setminus A_m$. We build a list $B_m$ of sets of three taxa $\{w, x, y\} \subseteq S \setminus A_m$ such that the topologies of $\{a_i, w, x, y\}$ are not all the same for $i = 1, \ldots, m$. Then we make all these quartet topologies be the same type by Algorithm MAKE-ADJ, which recursively branches on three possible types of these quartet topologies. We call this part of the algorithm $(1, 3)$-*cleaning*.

*Quintet cleaning*: After $(2, 2)$-cleaning, $(3, 1)$-cleaning and $(1, 3)$-cleaning, assume that the parameter is $k''$ for the moment. We try to resolve all the unresolved quintets in $\mathcal{C}_{a_1}$ through Algorithm ADJ-Resolve, which changes at most $k''$ quartet topologies in $Q$. We call this part of the algorithm *quintet cleaning*.

**Lemma 4.** *Assume that $A_m = \{a_1, \ldots, a_m\}$ and the list of unresolved quintet is $\mathcal{C}_{a_1}$, then after $(2, 2)$-cleaning, $(3, 1)$-cleaning and $(1, 3)$-cleaning, $\mathbf{q} \cap A_m = \{a_1\}$ for every $\mathbf{q} \in \mathcal{C}_{a_1}$.*

Note that there does not always exist $\omega$ adjacent taxa in an evolutionary tree for an arbitrary integer $\omega$. By Lemma 3, we know there must be $m$ taxa which are adjacent in an evolutionary tree, where $\omega \leq m \leq 2\omega - 2$. Assume that we are given an integer $\omega$ as an additional input. Then to solve the parameterized MQI problem, first we build a list of unresolved quintet involving $s$ for each $s \in S$, then we run Algorithm FPA3-MQI for every $m \in \{\omega, \ldots, 2\omega - 2\}$. The procedure update$_m$ is similar to the procedure update in Sect. 3. Yet if a quartet topology of $\{a_1, w, x, y\}$, where $w, x, y \in \mathbf{q} \setminus a_1$, is changed, the procedure not only changes quartet topologies according to $\mu$, but also changes the topologies of $\{a_2, w, x, y\}, \ldots, \{a_m, w, x, y\}$ together into the same type as $\{a_1, w, x, y\}$ has. Then the procedure updates the set $\mathcal{C}_{a_1}$ and the parameter $k$.

Recall that we denote $T$ to be an evolutionary tree on $S$ such that $Q_T$ differs from $Q$ in at most $k$ quartelarget topologies. Given an arbitrary integer $2 \leq \omega \leq n/2$, there exists $m$ adjacent taxa in $T$, where $\omega \leq m \leq 2\omega - 2$. Assume that there is a set of adjacent taxa $A_m = \{a_1, \ldots, a_m\} \subseteq S$ on $T$. Since the taxa in $A_m$ are adjacent, the path connecting every two taxa $a_i, a_j \in A_m$ and the path connecting two taxa $u, v \in S \setminus A_m$ will be disjoint and hence the topology of $\{a_i, a_j, u, v\}$ must be $[a_i a_j | uv]$. So $(2, 2)$-cleaning is valid. In addition, once the topology of $\{a_h, a_i, a_j, x\}$ is

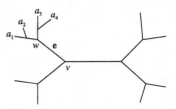

**Fig. 5.** An evolutionary tree with adjacent taxa $a_1, a_2, a_3, a_4$

fixed for $a_h, a_i, a_j \in A_m$ and some $x \in S \setminus A_m$, the quartets $\{a_h, a_i, a_j, s\}$ must have the same type of quartet topologies as $\{a_h, a_i, a_j, x\}$ has one $T$. Hence $(3, 1)$-cleaning is valid. Besides, the path structure connecting $a_i, w, x, y$ on $T$ must be the same for all $i \in \{1, \ldots, m\}$ and every three taxa $w, x, y \in S \setminus A_m$, so $(1, 3)$-cleaning is valid. After $(2, 2)$-cleaning, $(3, 1)$-cleaning, and $(1, 3)$-cleaning, there are only unresolved quintets involving $a_1$ by Lemma 4. Thus quintet cleaning, which is done by Algorithm ADJ-Resolve together with the procedure update$_m$, is valid. The number of unresolved quintets in $\mathcal{C}_{a_1}$ can be always decreased until $Q$ is tree-like. Hence it is easy to see that the algorithm is correct.

**Algorithm 4: FPA3-MQI (a complete set of quartet topologies $Q$, an integer parameter $k$, a list of unresolved quintets $C_{a_1}$, an arbitrary integer $m$)**

1: $Q^* \leftarrow Q$; $C_{a_1}^* \leftarrow C_{a_1}$; $k^* \leftarrow k$;
2: **for** every set of $m$ taxa $A_m = \{a_1, \ldots, a_m\} \subseteq S$ **do**
3:     **for** every two taxa $a_i, a_j \in A_m$ and every two taxa $u, v \in S \setminus A_m$ **do**
4:         **if** $k \leq 0$ **then** return; **else** Change the quartet topology of $\{a_i, a_j, u, v\}$ in $Q^*$ to be $[a_i a_j | uv]$ if $[a_i a_j | uv] \notin Q^*$, and then update $C_{a_1}^*$ and $k^* \leftarrow k^* - 1$;
5:     **end for**
6:     Build a set of all possible evolutionary trees $\mathcal{T}_{m+1}$ such that each $T' \in \mathcal{T}_{m+1}$ is an evolutionary tree on $A_m \cup \{x\}$, where $x$ is an arbitrary taxon in $S \setminus A_m$ and $|Q_{T'} \setminus Q^*| \leq k^*$;
7:     Build a list $B_m$ of sets of three taxa $w, x, y \in S \setminus A_m$ such that topologies of $\{a_i, w, x, y\}$ in $Q^*$ are not all the same for all $1 \leq i \leq m$;
8:     $Q^{**} \leftarrow Q^*$; $C_{a_1}^{**} \leftarrow C_{a_1}^*$; $k^{**} \leftarrow k^*$;
9:     **if** $\mathcal{T}_{m+1} = \emptyset$ **then**
10:         return;
11:     **else**
12:         **for** each $T' \in \mathcal{T}_{m+1}$ **do**
13:             $k^{**} \leftarrow k^{**} - |Q_{T'} \setminus Q^{**}|$;
14:             Change the quartet topologies in $Q^{**}$ over $A_m$ to those in $Q_{T'}$;
15:             For every taxon $s \in S \setminus A_m$ and every three taxa $a_h, a_i, a_j \in A_m$, change the topology of $\{a_h, a_i, a_j, s\}$ to the one of the same type as $\{a_h, a_i, a_j, x\}$ has; update $C_{a_1}^{**}$;
16:             **if** MAKE-ADJ$(Q^{**}, C_{a_1}^{**}, k^{**})$ returns ACCEPT **then**
17:                 return ACCEPT;
18:             **else**
19:                 Restore $(Q^{**}, C_{a_1}^{**})$ to $(Q^*, C_{a_1}^*)$, and $k^{**} \leftarrow k^*$;
20:             **end if**
21:         **end for**
22:     **end if**
23:     Restore $(Q^*, C_{a_1}^*)$ to $(Q, C_{a_1})$, delete $B_m$, and $k^* \leftarrow k$;
24: **end for**

---

**Algorithm 5: MAKE-ADJ (a complete set of quartet topologies $Q$, a list of unresolved quintets $C_{a_1}$, an integer parameter $k$)**

1: **if** $C_{a_1}$ is empty and $k \geq 0$ **then**
2:     return ACCEPT;
3: **else if** $k \leq 0$ **then**
4:     return;
5: **end if**
6: **while** $B_m \neq \emptyset$ **do**
7:     Extract $\{w, x, y\}$ from $B_m$;
8:     **for** each type $i \in \{0, 1, 2\}$ **do**
9:         Change all the topologies of $\{a_1, w, x, y\}, \ldots, \{a_m, w, x, y\}$ to topologies of type $i$; let $Q', C_{a_1}', k'$ be the changed $Q, C_{a_1}, k$ respectively;
10:         MAKE-ADJ$(Q', C_{a_1}', k')$;
11:     **end for**
12: **end while**
13: **if** ADJ-Resolve$(Q, k, C_{a_1})$ returns ACCEPT **then** return ACCEPT;

---

**Algorithm 6:** ADJ-Resolve (a complete set of quartet topologies $Q$, an integer parameter $k$, a list $C_{a_1}$ of unresolved quintets)

---

1: **if** $C_{a_1}$ is empty and $k \geq 0$ **then**
2:    return ACCEPT;
3: **else if** $k \leq 0$ **then**
4:    return;
5: **end if**
6: Extract an unresolved quintet $\mathbf{q}$ from $C_{a_1}$;
7: **for** each $\mu \in \mathcal{V}$ **do**
8:    $(Q', C'_{a_1}, k') \leftarrow \text{update}_m(Q, C_{a_1}, \mathbf{q}, \mu, k)$;
9:    ADJ-Resolve($Q', k', C'_{a_1}$);
10: **end for**

---

*The Time Complexity.* Let us consider the recursive structure of the algorithm, i.e., $(1,3)$-cleaning and quintet cleaning, as follows. Consider the quartets $\{a_1, w, x, y\}, \ldots, \{a_m, w, x, y\}$, where $w, x, y \in S \setminus A_m$. Without loss of generality, we denote the quartet topologies $[a_i w | xy]$, $[a_i x | wy]$, and $[a_i y | wx]$ to be type $0$, $1$, and $2$ respectively, for all $i = 1, \ldots, m$. Let $m_j$ be the number of quartets in $\{\{a_1, w, x, y\}, \ldots, \{a_m, w, x, y\}\}$ which have topologies of type $j$. $(1,3)$-cleaning branches on these three types to make every quartet $\{a_i, w, x, y\}$, where $a_i \in A_m$, have the same type of topology. It is clear that $m_0 + m_1 + m_2 = m$. Then the depth-bounded search tree for $(1,3)$-cleaning has a branching vector $(m_1 + m_2, m_0 + m_2, m_0 + m_1)$. By careful analysis we can deduce that the size of the depth-bounded search tree of $(1,3)$-cleaning is $O((1 + 2m^{-1/2})^k)$ for $m \geq 19$.

As to quintet cleaning, assume that the list of unresolved quintets is $C_{a_1}$. Let $\mathbf{q} = \{a_1, w, x, y, z\}$ be an unresolved quintet in $C_{a_1}$, and let $\mathbf{v_q} = (\mathbf{v_q}(1), \mathbf{v_q}(2), \mathbf{v_q}(3), \mathbf{v_q}(4), \mathbf{v_q}(5))$ denote the topology vector of $\mathbf{q}$, where $\mathbf{v_q}(1), \mathbf{v_q}(2), \mathbf{v_q}(3), \mathbf{v_q}(4)$, and $\mathbf{v_q}(5)$ are the types of topologies of quartets of $\mathbf{q}$ with respect to $Q$. Recall that $\mathcal{V} = \{\mu_1, \ldots, \mu_{15}\}$ is a set of topology vectors of 15 possible quintet topologies for a quintet, where $\mu_i = (\mu_i(1), \mu_i(2), \mu_i(3), \mu_i(4), \mu_i(5))$ stands for the $i$th topology vector in $\mathcal{V}$. For $1 \leq j \leq 5$, we denote $\mathbf{v_q}(j) \oplus \mu_i(j) = 1$ if $\mathbf{v_q}(j) \neq \mu_i(j)$, and $\mathbf{v_q}(j) \oplus \mu_i(j) = 0$ otherwise. For an unresolved quintet $\mathbf{q}$, let $\mathbf{b}(\mathbf{q})$ denote the branching vector of the recurrence of the quintet cleaning for $\mathbf{q}$. By the descriptions of quintet cleaning and the procedure $\text{update}_m$ of the algorithm, we derive that $\mathbf{b}(\mathbf{q}) = (\mathbf{b}_1(\mathbf{q}), \ldots, \mathbf{b}_{15}(\mathbf{q}))$, where $\mathbf{b}_i(\mathbf{q}) = m \left( \sum_{j=1}^{4} \mathbf{v_q}(j) \oplus \mu_i(j) \right) + \mathbf{v_q}(5) \oplus \mu_i(5)$. Then it can be derived that the depth-bounded search tree of quintet cleaning has size of $O((1 + 2m^{-1/2})^k)$ for $m \geq 17$. Therefore by detailed analysis, we have the following concluding theorem.

**Theorem 5.** *There exists an $O^*((1 + \epsilon)^k)$ time fixed-parameter algorithm for the parameterized minimum quartet inconsistency problem, where $\epsilon > 0$ is an arbitrarily small constant.*

**Acknowledgments.** We thank the anonymous referees for their helpful comments.

# References

1. Ben-Dor, A., Chor, B., Graur, D., Ophir, R., Pelleg, D.: From four-taxon trees to phylogenies: The case of mammalian evolution. In: Proceedings of the RECOMB, pp. 9–19 (1998)
2. Bandelt, H.J., Dress, A.: Reconstructing the shape of a tree from observed dissimilarity date. Adv. Appl. Math. 7, 309–343 (1986)
3. Berry, V., Jiang, T., Kearney, P.E., Li, M., Wareham, H.T.: Quartet cleaning: Improved algorithms and simulations. In: Nešetřil, J. (ed.) ESA 1999. LNCS, vol. 1643, pp. 313–324. Springer, Heidelberg (1999)
4. Cho, B.: From quartets to phylogenetic trees. In: Rovan, B. (ed.) SOFSEM 1998. LNCS, vol. 1521, pp. 36–53. Springer, Heidelberg (1998)
5. Downey, R.G., Fellows, M.R.: Parameterized Complexity. Springer, Heidelberg (1999)
6. Erdős, P., Steel, M., Székely, L., Warnow, T.: A few logs suffice to build (almost) all trees (Part 1). Random Struct. Alg. 14, 153–184 (1999)
7. Greene, D.H., Knuth, D.E.: Mathematics for the Analysis of Algorithms, 2nd edn. Progress in Computer Science. Birkhauser, Boston (1982)
8. Gramm, J., Niedermeier, R.: A fixed-parameter algorithm for minimum quartet inconsistency. J. Comput. System Sci. 67, 723–741 (2003)
9. Jiang, T., Kearney, P.E., Li, M.: Some open problems in computational molecular biology. J. Algorithms 34, 194–201 (2000)
10. Jiang, T., Kearney, P.E., Li, M.: A polynomial time approximation scheme for inferring evolutionary tree from quartet topologies and its application. SIAM J. Comput. 30, 1942–1961 (2001)
11. Steel, M.: The complexity of reconstructing trees from qualitative characters and subtrees. J. Classification 9, 91–116 (1992)
12. Niedermeier, R.: Invitation to Fixed-Parameter Algorithms. Oxford University Press, Oxford (2006)
13. Niedermeier, R., Rossmanith, P.: A general method to speed up fixed-parameter algorithms. Inform. Process. Lett. 73, 125–129 (2000)
14. Wu, G., You, J.-H., Lin, G.: A lookahead branch-and-bound algorithm for the maximum quartet consistency problem. In: Casadio, R., Myers, G. (eds.) WABI 2005. LNCS (LNBI), vol. 3692, pp. 65–76. Springer, Heidelberg (2005)
15. Wu, G., You, J.-H., Lin, G.: A polynomial time algorithm for the minimum quartet inconsistency problem with $O(n)$ quartet errors. Inform. Process. Lett. 100, 167–171 (2006)

# Capacitated Domination and Covering: A Parameterized Perspective

Michael Dom[1], Daniel Lokshtanov[2], Saket Saurabh[2], and Yngve Villanger[2]

[1] Institut für Informatik, Friedrich-Schiller-Universität Jena, Ernst-Abbe-Platz 2,
D-07743 Jena, Germany
dom@minet.uni-jena.de
[2] Department of Informatics, University of Bergen, N-5020 Bergen, Norway
{daniello,saket,yngvev}@ii.uib.no

**Abstract.** Capacitated versions of VERTEX COVER and DOMINATING SET have been studied intensively in terms of polynomial time approximation algorithms. Although the problems DOMINATING SET and VERTEX COVER have been subjected to considerable scrutiny in the parameterized complexity world, this is not true for their capacitated versions. Here we make an attempt to understand the behavior of the problems CAPACITATED DOMINATING SET and CAPACITATED VERTEX COVER from the perspective of parameterized complexity.

The original, uncapacitated versions of these problems, VERTEX COVER and DOMINATING SET, are known to be fixed parameter tractable when parameterized by a structure of the graph called the *treewidth (tw)*. In this paper we show that the capacitated versions of these problems behave differently. Our results are:

– CAPACITATED DOMINATING SET is W[1]-hard when parameterized by treewidth. In fact, CAPACITATED DOMINATING SET is W[1]-hard when parameterized by both treewidth and solution size $k$ of the capacitated dominating set.
– CAPACITATED VERTEX COVER is W[1]-hard when parameterized by treewidth.
– CAPACITATED VERTEX COVER can be solved in time $2^{O(\text{tw} \log k)} n^{O(1)}$ where tw is the treewidth of the input graph and $k$ is the solution size. As a corollary, we show that the weighted version of CAPACITATED VERTEX COVER in general graphs can be solved in time $2^{O(k \log k)} n^{O(1)}$. This improves the earlier algorithm of Guo et al. [15] running in time $O(1.2^{k^2} + n^2)$. CAPACITATED VERTEX COVER is, therefore, to our knowledge the first known "subset problem" which has turned out to be fixed parameter tractable when parameterized by solution size but W[1]-hard when parameterized by treewidth.

## 1 Introduction

DOMINATING SET and VERTEX COVER are problems representative for domination and covering, respectively. Given a graph $G$ and an integer $k$, VERTEX COVER asks for a size-$k$ set of vertices that cover all edges of the graph, while

M. Grohe and R. Niedermeier (Eds.): IWPEC 2008, LNCS 5018, pp. 78–90, 2008.
© Springer-Verlag Berlin Heidelberg 2008

DOMINATING SET asks for a size-$k$ set of vertices such that every vertex in the graph either belongs to this set or has a neighbor which does. These fundamental problems in algorithms and complexity have been studied extensively and find applications in various domains [3,4,5,8,9,12,13,15,17,21].

VERTEX COVER and DOMINATING SET have a special place in parameterized complexity [7,10,20]. VERTEX COVER was one of the earliest problems that was shown to be fixed parameter tractable (FPT) [7]. On the other hand, DOMINATING SET turned out to be intractable in the realm of parameterized complexity—specifically, it was shown to be W[2]-complete [7]. VERTEX COVER has been put to intense scrutiny, and many papers have been written on the problem. After a long race, the currently best algorithm for VERTEX COVER runs in time $O(1.2738^k + kn)$ [4]. VERTEX COVER has also been used as a testbed for developing new techniques for showing that a problem is FPT [7,10,20]. Though DOMINATING SET is a fundamentally hard problem in the parameterized W-hierarchy, it has been used as a benchmark problem for developing *sub-exponential time* parameterized algorithms [1,6,11] and also for obtaining a *linear kernels* in planar graphs [2,14,10,20], and more generally, in graphs that exclude a fixed graph $H$ as a minor.

Different applications of VERTEX COVER and DOMINATING SET have initiated studies of different generalizations and variations of these problems. These include CONNECTED VERTEX COVER, CONNECTED DOMINATING SET, PARTIAL VERTEX COVER, PARTIAL SET COVER , CAPACITATED VERTEX COVER and CAPACITATED DOMINATING SET, to name a few. All these problems have been investigated extensively and are well understood in the context of polynomial time approximation [5,12,13]. However, these problems hold a lot of promise and remain hitherto unexplored in the light of parameterized complexity; with exceptions that are few and far between [3,15,18,21,22].

*Problems Considered:* Here we consider two problems, CAPACITATED VERTEX COVER (CVC) and CAPACITATED DOMINATING SET (CDS). To define these problems, we need to introduce the notions of *capacitated graphs, vertex covers,* and *dominating sets.* A capacitated graph is a graph $G = (V, E)$ together with a capacity function $c : V \to \mathbb{N}$ such that $1 \le c(v) \le d(v)$, where $d(v)$ is the degree of the vertex $v$. Now let $G = (V, E)$ be a capacitated graph, $C$ be a vertex cover of $G$ and $D$ be a dominating set of $G$.

**Definition 1.** *We call $C \subseteq V$ a capacitated vertex cover if there exists a mapping $f : E \to C$ which maps every edge in $E$ to one of its two endpoints such that the total number of edges mapped by $f$ to any vertex $v \in C$ does not exceed $c(v)$.*

**Definition 2.** *We call $D \subseteq V$ a capacitated dominating set if there exists a mapping $f : (V \setminus D) \to D$ which maps every vertex in $(V \setminus D)$ to one of its neighbors such that the total number of vertices mapped by $f$ to any vertex $v \in D$ does not exceed $c(v)$.*

Now we are ready to define CAPACITATED VERTEX COVER and CAPACITATED DOMINATING SET.

CAPACITATED VERTEX COVER (CVC): Given a capacitated graph $G = (V, E)$ and a positive integer $k$, determine whether there exists a capacitated vertex cover $C$ for $G$ containing at most $k$ vertices.

CAPACITATED DOMINATING SET (CDS): Given a capacitated graph $G = (V, E)$ and a positive integer $k$, determine whether there exists a capacitated dominating set $D$ for $G$ containing at most $k$ vertices.

*Our Results:* To describe our results we first need to define the *treewidth* (tw) of a graph. Let $V(U)$ be the set of vertices of a graph $U$. A tree decomposition of an (undirected) graph $G = (V, E)$ is a pair $(X, U)$ where $U$ is a tree whose vertices we will call *nodes* and $X = \{X_i \mid i \in V(U)\}$ is a collection of subsets of $V$ such that **(1)** $\bigcup_{i \in V(U)} X_i = V$, **(2)** for each edge $\{v, w\} \in E$, there is an $i \in V(U)$ such that $v, w \in X_i$, and **(3)** for each $v \in V$ the set of nodes $\{i \mid v \in X_i\}$ forms a subtree of $U$. The *width* of a tree decomposition $(\{X_i | i \in V(U)\}, U)$ equals $\max_{i \in V(U)} \{|X_i| - 1\}$. The *treewidth* of a graph $G$ is the minimum width over all tree decompositions of $G$.

There is a tendency to think that most combinatorial problems, especially "subset problems", are tractable for graphs of bounded treewidth (tw) when parameterized by tw. In fact, the non-capacitated versions of the problems considered here, namely VERTEX COVER and DOMINATING SET, are known to be fixed parameter tractable when parameterized by the treewidth of the input graph. The algorithms for VERTEX COVER and DOMINATING SET run in time $O(2^{\text{tw}} n)$ [20] and time $O(4^{\text{tw}} n)$ [1], respectively. In contrast, the capacitated versions of these problems behave differently. More precisely, we show the following:

– CAPACITATED DOMINATING SET is W[1]-hard when parameterized by the treewidth. In fact, CDS is W[1]-hard when parameterized by both treewidth and solution size $k$ of the capacitated dominating set.

– CAPACITATED VERTEX COVER is W[1]-hard when parameterized by treewidth.

– CAPACITATED VERTEX COVER can be solved in time $2^{O(\text{tw} \log k)} n^{O(1)}$ where tw is the treewidth of the input graph and $k$ is the solution size. As a corollary of the last result we obtain an improved algorithm for the weighted version of CAPACITATED VERTEX COVER in general graphs. Here, every vertex of the input graph has, in addition to the capacity, a weight, and the question is if there is a capacitated vertex cover whose weight is at most $k$. Our algorithm running in time $O(2^{O(k \log k)} n^{O(1)})$ improves the earlier algorithm of Guo et al. [15] running in time $O(1.2^{k^2} + n^2)$.

The so-called "subset problems" are known to go either way, that is, FPT or W[i]-hard ($i \geq 1$) when parameterized by solution size. However, when parameterized by treewidth they have invariably been FPT. Examples favoring this claim include, but are not limited to, INDEPENDENT SET, DOMINATING SET, PARTIAL VERTEX COVER. Contrary to these observed patterns, our hardness result for CVC when parameterized by treewidth makes it possibly the first known "subset problem" which has turned out to be FPT when parameterized by solution size, but W[1]-hard when parameterized by treewidth.

## 2  Preliminaries

We assume that all our graphs are simple and undirected. Given a graph $G = (V, E)$, the number of its vertices is represented by $n$ and the number of its edges by $m$. For a subset $V' \subseteq V$, by $G[V']$ we mean the subgraph of $G$ induced by $V'$. With $N(u)$ we denote all vertices that are adjacent to $u$, and with $N[u]$, we refer to $N(u) \cup \{u\}$. Similarly, for a subset $D \subseteq V$, we define $N[D] = \cup_{v \in D} N[v]$ and $N(D) = N[D] \setminus D$. Let $f$ be the function associated with a capacitated dominating set $D$. Given $u \in D$ and $v \in V \setminus D$, we say that $u$ *dominates* $v$ if $f(v) = u$; moreover, every vertex $u \in D$ dominates itself. Note that the capacity of a vertex $v$ only limits the number of neighbors that $v$ can dominate, that is, a vertex $v \in D$ can dominate $c(v)$ of its neighbors plus $v$ itself.

Parameterized complexity is a two-dimensional framework for studying the computational complexity of problems [7,10,20]. One dimension is the input size $n$ and the other one a *parameter* $k$. A problem is called *fixed-parameter tractable (FPT)* if it can be solved in time $f(k) \cdot n^{O(1)}$, where $f$ is a computable function only depending on $k$. The basic complexity class for fixed-parameter intractability is W[1]. To show that a problem is W[1]-hard, one needs to exhibit a parameterized reduction from a known W[1]-hard problem: We say that a parameterized problem $A$ is *(uniformly many:1) reducible* to a parameterized problem $B$ if there is an algorithm $\Phi$ which transforms $(x, k)$ into $(x', g(k))$ in time $f(k) \cdot |x|^\alpha$, where $f, g : \mathbb{N} \to \mathbb{N}$ are arbitrary functions and $\alpha$ is a constant independent of $|x|$ and $k$, so that $(x, k) \in A$ if and only if $(x', g(k)) \in B$.

## 3  Parameterized Intractability – Hardness Results

### 3.1  CDS Is W[1]-Hard Parameterized by Treewidth and Solution Size

In this section we show that CAPACITATED DOMINATING SET is W[1]-hard when parameterized by treewidth and solution size. We reduce from the W[1]-hard problem MULTICOLOR CLIQUE, a restriction of the CLIQUE problem:

MULTICOLOR CLIQUE: Given an integer $k$ and a connected undirected graph $G = (V[1] \cup V[2] \cdots \cup C[k], E)$ such that for every $i$ the vertices of $V[i]$ induce an independent set, is there a $k$-clique $C$ in $G$?

In fact, we will reduce to a slightly modified version of CAPACITATED DOMINATING SET called MARKED CAPACITATED DOMINATING SET where we *mark* some vertices and demand that all marked vertices must be in the dominating set. We can then reduce from MARKED CAPACITATED DOMINATING SET to CAPACITATED DOMINATING SET by attaching $k + 1$ leaves to each marked vertex and increasing the capacity of each marked vertex by $k + 1$. It is easy to see that the new instance has a size-$k$ capacitated dominating set if and only if the original one had a size-$k$ capacitated dominating set that contained all marked vertices, and that this operation does not increase the treewidth of the graph. Thus, to prove that CAPACITATED DOMINATING SET is W[1]-hard when parameterized

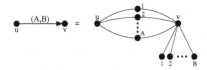

**Fig. 1.** Adding an $(A, B)$-arrow from $u$ to $v$

by treewidth and solution size, it is sufficient to prove that MARKED CAPACITATED DOMINATING SET is.

We will now show how, given an instance $(G, k)$ of MULTICOLOR CLIQUE, we can build an instance $(H, c, k')$ of MARKED CAPACITATED DOMINATING SET such that

- $k' = 7k(k-1) + 2k$,
- $G$ has a clique of size $k$ if and only if $H$ has a capacitated dominating set of size $k'$, and
- the treewidth of $H$ is $O(k^4)$.

For a pair of distinct integers $i, j$, let $E[i, j]$ be the set of edges with one endpoint in $V[i]$ and the other in $V[j]$. Without loss of generality, we will assume that $|V[i]| = N$ and $|E[i, j]| = M$ for all $i$, $j$, $i \neq j$. To each vertex $v$ we assign a unique identification number $v^{up}$ between $N + 1$ and $2N$, and we set $v^{down} = 2N - v^{up}$. For two vertices $u$ and $v$, by adding an $(A, B)$-*arrow* from $u$ to $v$ we will mean adding $A$ subdivided edges between $u$ and $v$ and attaching $B$ leaves to $v$ (see Fig. 1). Now we describe how to build the graph $H$ for a given instance $(G = (V[1] \cup V[2] \cdots \cup V[k], E), k)$ of MULTICOLOR CLIQUE.

For every integer $i$ between 1 and $k$ we add a marked vertex $\hat{x}_i$ that has a neighbor $\bar{v}$ for every vertex $v$ in $V[i]$. For every $j \neq i$, we add a marked vertex $\hat{y}_{ij}$ and a marked vertex $\hat{z}_{ij}$. Now, for every vertex $v \in V[i]$ and every integer $j \neq i$ we add a $(v^{up}, v^{down})$-arrow from $\bar{v}$ to $\hat{y}_{ij}$ and a $(v^{down}, v^{up})$-arrow from $\bar{v}$ to $\hat{z}_{ij}$. Finally we add a set $S_i$ of $k' + 1$ vertices and make every vertex in $S_i$ adjacent to every vertex $\bar{v}$ with $v \in V[i]$. See left part of Fig. 2 for an illustration.

Similarly, for every pair of integers $i$, $j$ with $i < j$, we add a marked vertex $\hat{x}_{ij}$ with a neighbor $\bar{e}$ for every edge $e$ in $E[i, j]$. Moreover, we add four new marked vertices $\hat{p}_{ij}, \hat{p}_{ji}, \hat{q}_{ij}$, and $\hat{q}_{ji}$. For every edge $e = \{u, v\}$ in $E[i, j]$ with $u \in V[i]$ and $v \in V[j]$, we add a $(u^{down}, u^{up})$-arrow from $\bar{e}$ to $\hat{p}_{ij}$, a $(u^{up}, u^{down})$-arrow from $\bar{e}$ to $\hat{q}_{ij}$, a $(v^{down}, v^{up})$-arrow from $\bar{e}$ to $\hat{p}_{ji}$ and a $(v^{up}, v^{down})$-arrow from $\bar{e}$ to $\hat{p}_{ji}$. We also add a set $S_{ij}$ of $k' + 1$ vertices and make every vertex in $S_{ij}$ adjacent to every vertex $\bar{e}$ with $e \in E[i, j]$. See right part of Fig. 2 for an illustration.

Finally, we add a marked vertex $\hat{r}_{ij}$ and a marked vertex $\hat{s}_{ij}$ for every $i \neq j$. For every $i \neq j$, we add $(2N, 0)$-arrows from $\hat{y}_{ij}$ to $\hat{r}_{ij}$, from $\hat{p}_{ij}$ to $\hat{r}_{ij}$, from $\hat{z}_{ij}$ to $\hat{s}_{ij}$, and from $\hat{q}_{ij}$ to $\hat{s}_{ij}$ (see Fig. 3). This concludes the description of the graph $H$.

We now describe the capacities of the vertices. For every $i \neq j$, the vertex $\hat{x}_i$ has capacity $N - 1$, the vertex $\hat{x}_{ij}$ has capacity $M - 1$, the vertices $\hat{y}_{ij}$ and $\hat{z}_{ij}$ both have capacity $2N^2$, the vertices $\hat{p}_{ij}$ and $\hat{q}_{ij}$ have capacity $2NM$, and both $\hat{r}_{ij}$

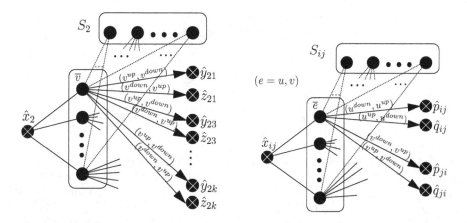

**Fig. 2.** Left part: Gadget constructed for the vertices in the color class $V[2]$. Right part: Gadget constructed for the edges between $V[i]$ and $V[j]$.

and $\hat{s}_{ij}$ have capacity $2N$. For all other vertices, their capacity is equal to their degree in $H$.

**Observation 1** *The treewidth of $H$ is $O(k^4)$.*

*Proof.* If we remove all marked vertices ($\bigcup_{i=1}^{k} S_i$ and $\bigcup_{i \neq j} S_{ij}$), a total of $O(k^4)$ vertices, from $H$, we obtain a forest. As deleting a vertex reduces the treewidth by at most one, this concludes the proof. □

**Lemma 1.** *If $G$ has a multicolor clique $C = \{v_1, v_2, \ldots, v_k\}$ then $H$ has a capacitated dominating set $D$ of size $k'$ containing all marked vertices.*

*Proof.* For every $i < j$ let $e_{ij}$ be the edge from $v_i$ to $v_j$ in $G$. In addition to all the marked vertices, let $D$ contain $\overline{v_i}$ and $\overline{e_{ij}}$ for every $i < j$. Clearly $D$ contains exactly $k'$ vertices, so it remains to prove that $D$ is indeed a capacitated dominating set.

For every $i < j$, let $\hat{x}_i$ and $\hat{x}_{ij}$ dominate all their neighbors except for $\overline{v_i}$ and $\overline{e_{ij}}$ respectively. The vertices $\overline{v_i}$ and $\overline{e_{ij}}$ can dominate all their neighbors, since their capacity is equal to their degree. Let $\hat{r}_{ij}$ dominate $v_i^{down}$ of the vertices in the $(2N, 0)$-arrow from $\hat{y}_{ij}$, and $v_i^{up}$ of the vertices of the $(2N, 0)$-arrow from $\hat{p}_{ij}$. Similarly let $\hat{s}_{ij}$ dominate $v_i^{up}$ of the vertices in the $(2N, 0)$-arrow from $\hat{z}_{ij}$, and $v_i^{down}$ of the vertices of the $(2N, 0)$-arrow from $\hat{q}_{ij}$. Finally, for every $i \neq j$ we let $\hat{y}_{ij}$, $\hat{z}_{ij}$, $\hat{p}_{ij}$ and $\hat{q}_{ij}$ dominate all their neighbors that have not been dominated yet. One can easily check that every vertex of $H$ will either be a dominator or dominated in this manner, and that no dominator dominates more vertices than its capacity. □

**Lemma 2.** *If $H$ has a capacitated dominating set $D$ of size $k'$ containing all marked vertices, then $G$ has a multicolor clique of size $k$.*

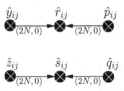

**Fig. 3.** Vertex-Edge incidence gadget

*Proof.* Observe that for every integer $1 \leq i \leq k$, there must be a $v_i \in V[i]$ such that $\overline{v}_i \in D$. Otherwise we have that $S_i \subset D$ and, since $|S_i| > k'$, we obtain a contradiction. Similarly, for every pair of integers $i, j$ with $i < j$ there must be an edge $e_{ij} \in E[i, j]$ such that $\overline{e}_{ij} \in D$. We let $e_{ji} = e_{ij}$. Since $|D| \leq k'$ it follows that these are the only unmarked vertices in $D$. Since all the unmarked vertices in $D$ have capacity equal to their degree, we can assume that each such vertex dominates all its neighbors. We now proceed with proving that for every pair of integers $i, j$ with $i \neq j$, the edge $e_{ij} = uv$ is incident to $v_i$. We prove this by showing that if $u \in V[i]$ then $v_i^{up} + u^{down} = 2N$.

Suppose for a contradiction that $v_i^{up} + u^{down} < 2N$. Observe that each vertex of $T = (N(\hat{y}_{ij}) \cup N(\hat{r}_{ij}) \cup N(\hat{p}_{ij})) \setminus (N(\overline{v}_i) \cup N(\overline{e}_{ij}))$ must be dominated by either $\hat{y}_{ij}$, $\hat{r}_{ij}$, or $\hat{p}_{ij}$. However, by our assumption that $v_i^{up} + u^{down} < 2N$, it follows that $|T| = 2N^2 + 4N + 2MN - (v_i^{up} + u^{down}) > 2N^2 + 2N + 2MN$. The sum of the capacities of $\hat{y}_{ij}$, $\hat{r}_{ij}$, and $\hat{p}_{ij}$ is exactly $2N^2 + 2N + 2MN$. Thus it is impossible that every vertex of $T$ is dominated by one of $\hat{y}_{ij}$, $\hat{r}_{ij}$, and $\hat{p}_{ij}$, a contradiction. If $v_i^{up} + u^{down} > 2N$ then $v_i^{down} + u^{up} < 2N$, and we can apply an identical argument for $\hat{z}_{ij}$, $\hat{s}_{ij}$, and $\hat{q}_{ij}$.

Thus, it follows that for every $i \neq j$ there is an edge $e_{ij}$ incident both to $v_i$ and to $v_j$. Thus $\{v_1, v_2, \ldots, v_k\}$ forms a clique in $G$. As any $k$-clique in $G$ is a multicolor clique this completes the proof.                                    $\square$

**Theorem 1.** CDS *parameterized by treewidth and solution size is* W[1]-*hard.*

### 3.2   CVC Parameterized by Treewidth Is W[1]-Hard

Usually vertex cover problems can be seen as restrictions of domination problems, and therefore it is natural to expect CAPACITATED VERTEX COVER to be somewhat easier than CAPACITATED DOMINATING SET. In this section, we give a result similar to the hardness result for CAPACITATED DOMINATING SET, but weaker in the sense that we only show that CAPACITATED VERTEX COVER is hard when parameterized by the treewidth, while we have seen in the previous section that CAPACITATED DOMINATING SET is hard when parameterized by the treewidth and the solution size.

To obtain our result we reduce from MULTICOLOR CLIQUE, as in the previous section. Again, we reduce to a marked version of CAPACITATED VERTEX COVER, where we search for a size $k'$ capacitated vertex cover that contains all the marked vertices. The reduction from MARKED CAPACITATED VERTEX COVER to CAPACITATED VERTEX COVER is almost identical to the reduction from MARKED CAPACITATED DOMINATING SET to CAPACITATED DOMINATING SET described in the previous section. Notice also that in MARKED CAPACITATED VERTEX COVER it makes sense to have marked vertices with capacity zero, as they will get non-zero capacity after the reduction to CAPACITATED VERTEX COVER.

We reduce by building for an instance $(G, k)$ of MULTICOLOR CLIQUE an instance $(H, c, k')$ of MARKED CAPACITATED VERTEX COVER in almost the same manner as in the reduction to MARKED CAPACITATED DOMINATING SET. The only differences are:

- We do not add the vertex sets $S_i$ and $S_{ij}$ for every $i, j$.
- When we add an $(A, B)$-arrow from $u$ to $v$, the $A$ vertices on the subdivided edges are marked and have capacity 1, while the $B$ leaves attached to $v$ are also marked but have capacity 0.
- We set $k'$ to $k' = 7k(k - 1) + 2k + (2k^2 N + (2M + 4) \cdot k \cdot (k - 1)) \cdot 2N$.

The new term in the value of $k'$ is simply a correction for all the extra marked vertices in the $(A, B)$-arrows. Notice that in this case the value of $k'$ is not a function of $k$ alone, and that therefore this reduction does not imply that CAPACITATED VERTEX COVER is W[1]-hard parameterized by treewidth and solution size. However, by applying arguments almost identical to the ones in the previous section, we can prove the following claims; the details are omitted.

- The treewidth of $H$ is $O(k^3)$.
- If $G$ has a multicolor clique $C = \{v_1, v_2, \ldots, v_k\}$ then $H$ has a capacitated vertex cover $S$ of size $k'$ containing all marked vertices.
- If $H$ has a capacitated vertex cover $S$ of size $k'$ containing all marked vertices, then $G$ has a multicolor clique of size $k$.

Together, the claims imply the following theorem.

**Theorem 2.** CVC *parameterized by treewidth is* W[1]-*hard.*

## 4 FPT Algorithm for CVC on Graphs of Bounded Treewidth

In the last sections we have shown that CAPACITATED VERTEX COVER, when parameterized only by the treewidth tw of the input graph, is W[1]-hard, while CAPACITATED DOMINATING SET remains W[1]-hard even when parameterized by both tw and the solution size $k$. We complement these hardness results by giving a time $2^{O(\text{tw} \log k)} n^{O(1)}$ algorithm for graphs of bounded treewidth, a result which was sketched independently by Hannes Moser [19]. Furthermore, using this algorithm, we give an improved algorithm for the weighted version of CAPACITATED VERTEX COVER in general graphs: Our algorithm, running in time $O(2^{O(k \log k)} n^{O(1)})$, improves the earlier algorithm of Guo et al. [15], which runs in time $O(1.2^{k^2} + n^2)$.

To solve CVC on graphs of bounded treewidth, we give a dynamic programming algorithm working on a so-called *nice tree decomposition* of the input graph $G$: A tree decomposition $(X, U)$ is a nice tree decomposition if one can root $U$ in such a way that the root and every inner node of $U$ is either an *insert node*, a *forget node*, or a *join node*. Thereby, a node $i$ of $U$ is an insert node if $i$ has exactly one child $j$, and $X_i$ consists of all vertices of $X_j$ plus one additional vertex; it is a forget node if $i$ has exactly one child $j$, and $X_i$ consists of all but one vertices of $X_j$; and it is a join node if $i$ has exactly two children $j_1, j_2$, and $X_i = X_{j_1} = X_{j_2}$. Given a tree decomposition of width tw, a nice tree decomposition of the same width can be found in linear time [16]. In what follows, we assume that the nice tree decomposition $(X, U)$ that we are using has the additional property that the bag associated with the root of $U$ is empty (such a decomposition can easily be constructed by taking an arbitrary nice tree decomposition and adding some forget nodes "above" the original root). Similarly, we assume that every bag associated with a leaf node different from the root of $U$ contains exactly one vertex. For a node $i$ in the tree $U$ of a tree decomposition $(X, U)$, let

$$Y_i := \bigcup \{v \in X_j \mid j \text{ is a node in the subtree of } U \text{ whose root is } i\},$$
$$Z_i := Y_i \setminus X_i, \quad \text{and} \quad E_i := \{\{v, w\} \in E \mid v \in Z_i \vee w \in Z_i\}.$$

Starting at the leaf nodes of $U$ that are different from the root, our dynamic programming algorithm assigns to every node $i$ of $U$ a table $A_i$ that has

- a column $\ell$ with $\ell \le k$,
- for every vertex $v \in X_i$ a column $\text{vc}(v)$ with $\text{vc}(v) \in \{true, false\}$, and
- for every vertex $v \in X_i$ a column $\text{s}(v)$ with $\text{s}(v) \in \{null, 0, 1, \ldots, k-1\}$.

Every row of such a table $A_i$ corresponds to a solution $(f, C)$ for CVC on the subgraph of $G$ that consists of all vertices in $Y_i$ and all edges in $E$ having at least one endpoint in $Z_i$. More exactly, for every row of a table $A_i$ there is a vertex set $C \subseteq Y_i$ and mapping $f : E_i \to C$ with the following properties:

- $C$ is a capacitated vertex cover for $G_i = (Y_i, E_i)$.
- $|C| \le \ell$.
- $C$ contains all vertices $v \in X_i$ with $\text{vc}(v) = true$ and no vertex $v \in X_i$ with $\text{vc}(v) = false$.
- For every vertex $v \in X_i \cap C$, we have $|\{\{v, w\} \in E_i \mid f(\{v, w\}) = w\}| = \text{s}(v)$, and for every vertex $v \in X_i \setminus C$, we have $\text{s}(v) = null$.

Intuitively speaking, for a vertex $v \in C$, the value $\text{s}(v)$ contains the number of edges incident to $v$ that are covered by vertices in $Z_i$ and, therefore, do not have to be covered by $v$. The simple observation that $\text{s}(v)$ can be at most $k-1$ (because $C$ can contain at most $k-1$ neighbors of $v$) is crucial for the running time of the algorithm.

Clearly, if the table associated with the root of $U$ is nonempty, the given instance of CVC is a *yes*-instance.

We will now describe the computation of the table $A_i$ for a node $i$ in $U$, depending on if $i$ is a leaf node different from the root, an insert node, a forget node, or a join node. If necessary, we write $\ell_i$, $\text{vc}_i(v)$, and $\text{s}_i(v)$ in order to make clear that a value $\ell$, $\text{vc}(v)$, and $\text{s}(v)$, respectively, stems from a row of a table $A_i$.

**The node $i$ is a leaf node different from the root.** Let $X_i = \{v\}$. Then we add one row to the table $A_i$ for the case that $v$ is not part of $C$ and one row for the case that $v$ is part of $C$, provided that $k > 0$. Because $i$ has no child and, hence, no neighbor of $v$ belongs to $Z_i$, the value s$(v)$ is set to 0 in the case that $v$ is part of $C$:

```
1  if k > 0: {add a new row to A_i and set  vc(v) := true;  s(v) := 0;  ℓ := 1  in this row; }
2  add a new row to A_i and set  vc(v) := false;  s(v) := null;  ℓ := 0  in this row;
```

**The node $i$ is an insert node.** Let $j$ be the child of $i$ in $U$, and let $X_i = X_j \cup \{v\}$. Here we extend the table $A_j$ by adding the values vc$(v)$ and s$(v)$. For every row of $A_j$, we add one row to the table $A_i$ for the case that $v$ is not part of $C$ and one row for the case that $v$ is part of $C$, provided that $\ell_j < k$. Because no neighbor of $v$ can belong to $Z_i$, the value s$(v)$ is set to 0 in the case that $v$ is part of $C$:

```
1  for every row r of A_j: {
2      if ℓ_j < k: {
3          copy the row r from A_j into A_i and set  vc(v) := true;  s(v) := 0;  ℓ := ℓ + 1  in this row; }
4      copy the row r from A_j into A_i and set  vc(v) := false;  s(v) := null  in this row; }
```

**The node $i$ is a forget node.** Let $j$ be the child of $i$ in $U$, and let $X_i = X_j \setminus \{v\}$. Clearly, all neighbors of $v$ belong to $Y_j$ due to the definition of a tree decomposition. What has to be done is to consider the edges $\{v, w\}$ with $w \in X_i$, to decide which of them shall be covered by $v$, and to set the value of $s_j(v)$ accordingly. Note that this approach ensures that for all edges $\{v, w\}$ with $w \in Z_j$ we have already decided in a previous step which of these edges are covered by $v$. More exactly, for every row of $A_j$, we perform the following steps. If vc$_j(v) = $ *true*, then we try all possibilities for which edges between $v$ and vertices $w \in X_i$ can be covered by $v$ and add rows to $A_i$ accordingly. If vc$_j(v) = $ *false*, then, of course, no edge between $v$ and vertices $w \in X_j$ can be covered by $v$, and we add one row to $A_i$. In both cases, we have to check that for every edge $\{v, w\}$ with $w \in X_i$ that is not covered by $v$ it holds that vc$_j(w) = $ *true* and the remaining capacity of $w$, which can be computed from s$(w)$ and the number of $w$'s neighbors in $Z_j$, is big enough to cover $\{v, w\}$:

```
1   N' := N(v) ∩ X_i;
2   for every row r of A_j: {
3       if vc_j(v) = true: {
4           for every subset N'' of N' with |N''| = min{|N'|, cap(v) − (|N(v) ∩ Z_j| − s_j(v))}: {
5               if ∀w ∈ N' \ N'' : vc_j(w) = true ∧ cap(w) > |N(w) ∩ Z_j| − s_j(w): {
6                   copy the row r from A_j into A_i;
7                   for every vertex w ∈ N'' with vc(w) = true: {
8                       update the new row in A_i and set s(w) := s(w) + 1; }}}
9           else: { if ∀w ∈ N' : vc(w) = true ∧ cap(w) > |N(w) ∩ Z_j| − s_j(w): {
10                  copy the row r from A_j into A_i; }}}
```

**The node $i$ is a join node.** Let $j_1$ and $j_2$ be the children of $i$ in $U$. Here we consider every pair $r_1, r_2$ of rows where $r_1$ is from $A_{j_1}$ and $r_2$ is from $A_{j_2}$. We say that two rows $r_1$ and $r_2$ are *compatible* if for every vertex $v$ in $X_i$ it holds that vc$_{j_1}(v) = $ vc$_{j_2}(v)$. If two rows are compatible, then we check whether for every vertex $v \in X_i$ with vc$_{j_1}(v) = $ vc$_{j_2}(v) = $ *true* the number of edges $\{v, w\}$

covered by $v$ with $w \in Z_{j_1}$ plus the number of edges $\{v, w\}$ covered by $v$ with $w \in Z_{j_2}$ is at most cap($v$). If this is the case, a new row is added to $A_i$:

```
1  for every compatible pair r₁, r₂ of rows where r₁ is from A_{j₁} and r₂ is from A_{j₂}: {
2      if ∀v ∈ Xᵢ : vc_{j₁}(v) = false ∨ cap(v) ≥ |N(v) ∩ Z_{j₁}| − s_{j₁}(w) + |N(v) ∩ Z_{j₂}| − s_{j₂}(w): {
3          add a new row to Aᵢ;
4          update the new row in Aᵢ and set ℓ := ℓ_{j₁} + ℓ_{j₂} − |{v ∈ Xᵢ | vc_{j₁}(v) = true}|;
5          for every vertex v ∈ Xᵢ: {
6              update the new row in Aᵢ and set vc(v) = vc_{j₁}(v);  s(v) = s_{j₁}(v) + s_{j₂}(v); }}}
```

In all four cases ($i$ is a leaf node different from the root, an insert node, a forget node, or a join node), after inserting a row to $A_i$, we delete *dominated* rows from $A_i$. A row $r_1$ is dominated by a row $r_2$ if $r_1$ and $r_2$ are compatible, the value of $\ell$ in $r_1$ is equal or greater than the value of $\ell$ in $r_2$, and for every vertex $v \in X_i$ with vc($v$) = *true* the value of s($v$) in $r_1$ is equal or less than the value of s($v$) in $r_2$. The correctness of this data reduction is obvious: If the solution corresponding to $r_1$ can be extended to a solution for the whole graph, then this is also possible with the solution corresponding to $r_2$ instead. Clearly, due to this data reduction, the table can never contain more than $k^{\text{tw}}$ rows, which leads to the following theorem.

**Theorem 3.** CVC *on graphs of treewidth* tw *can be solved in* $k^{3 \cdot \text{tw}} \cdot n^{O(1)}$ *time.*

*Proof.* The correctness of the algorithm follows from the above description. The running time for computing one table $A_i$ associated with a tree node $i$ is bounded from above by $k^{3 \cdot \text{tw}} \cdot n^{O(1)}$, due to the fact that every table contains at most $k^{\text{tw}}$ rows and that the tree decomposition has $O(n)$ tree nodes [16].     □

We mention in passing that with usual backtracking techniques it is possible to construct the mapping $f$ and the set $C$ after running the dynamic programming algorithm.

**CVC in General Undirected Graphs:** The algorithm described above can also be used for solving CVC on general graphs with the following two observations. Firstly, the treewidth of graphs that have a vertex cover of size $k$ is bounded above by $k$, and a corresponding tree decomposition of width $k$ can be found in $O(1.2738^k + kn)$ time [4]. (For a graph $G = (V, E)$ that has a vertex cover $C$ with $|C| = k$, a tree decomposition of width $k$ can be constructed as follows: Let $U$ be a path of length $|V \setminus C|$, and assign to every node $i$ of $U$ a bag $X_i$ that contains $C$ and one vertex from $V \setminus C$. The vertex cover of size $k$ can be found in time $O(1.2738^k + kn)$ [4].) Secondly, Theorem 3 can easily be adapted to the *weighted* version of CVC, where every vertex of the input graph has, in addition to the capacity, a weight, and the question is if there is a capacitated vertex cover whose weight is at most $k$. With these observations, we get the following corollary.

**Corollary 1.** *The weighted version of* CVC *on general graphs can be solved in* $k^{3k} \cdot n^{O(1)} = 2^{O(k \log k)} \cdot n^{O(1)}$ *time.*

# 5   Conclusion

We conclude with an open question. It is easy to observe that if a planar graph has a CDS of size at most $k$ then the treewidth of the input graph is at most $O(\sqrt{k})$ [1,6,11]. Hence, in order to show that CDS is FPT for planar graphs, it is sufficient to obtain a dynamic programming algorithm for it on *planar graphs* of bounded treewidth. The following question in this direction remains unanswered: Is CDS in planar graphs parameterized by solution size fixed parameter tractable?

# References

1. Alber, J., Bodlaender, H.L., Fernau, H., Kloks, T., Niedermeier, R.: Fixed parameter algorithms for DOMINATING SET and related problems on planar graphs. Algorithmica 33(4), 461–493 (2002)
2. Alber, J., Fellows, M.R., Niedermeier, R.: Polynomial-time data reduction for dominating set. J. ACM 51(3), 363–384 (2004)
3. Bläser, M.: Computing small partial coverings. Inf. Process. Lett. 85(6), 327–331 (2003)
4. Chen, J., Kanj, I.A., Xia, G.: Improved parameterized upper bounds for Vertex Cover. In: Královič, R., Urzyczyn, P. (eds.) MFCS 2006. LNCS, vol. 4162, pp. 238–249. Springer, Heidelberg (2006)
5. Chuzhoy, J., Naor, J.: Covering problems with hard capacities. SIAM J. Comput. 36(2), 498–515 (2006)
6. Demaine, E.D., Fomin, F.V., Hajiaghayi, M.T., Thilikos, D.M.: Subexponential parameterized algorithms on bounded-genus graphs and H-minor-free graphs. J. ACM 52(6), 866–893 (2005)
7. Downey, R.G., Fellows, M.R.: Parameterized Complexity. Springer, Heidelberg (1999)
8. Duh, R., Fürer, M.: Approximation of k-Set Cover by semi-local optimization. In: Proc. 29th STOC, pp. 256–264. ACM Press, New York (1997)
9. Feige, U.: A threshold of ln n for approximating Set Cover. J. ACM 45(4), 634–652 (1998)
10. Flum, J., Grohe, M.: Parameterized Complexity Theory. Springer, Heidelberg (2006)
11. Fomin, F.V., Thilikos, D.M.: Dominating sets in planar graphs: Branch-width and exponential speed-up. SIAM J. Comput. 36(2), 281–309 (2006)
12. Gandhi, R., Halperin, E., Khuller, S., Kortsarz, G., Srinivasan, A.: An improved approximation algorithm for Vertex Cover with hard capacities. J. Comput. System Sci. 72(1), 16–33 (2006)
13. Guha, S., Hassin, R., Khuller, S., Or, E.: Capacitated vertex covering. J. Algorithms 48(1), 257–270 (2003)
14. Guo, J., Niedermeier, R.: Linear problem kernels for NP-hard problems on planar graphs. In: Arge, L., Cachin, C., Jurdziński, T., Tarlecki, A. (eds.) ICALP 2007. LNCS, vol. 4596, pp. 375–386. Springer, Heidelberg (2007)
15. Guo, J., Niedermeier, R., Wernicke, S.: Parameterized complexity of Vertex Cover variants. Theory Comput. Syst. 41(3), 501–520 (2007)
16. Kloks, T.: Treewidth. LNCS, vol. 842. Springer, Heidelberg (1994)

17. Lovàsz, L.: On the ratio of optimal fractional and integral covers. Discrete Math. 13, 383–390 (1975)
18. Mölle, D., Richter, S., Rossmanith, P.: Enumerate and expand: Improved algorithms for Connected Vertex Cover and Tree Cover. In: Grigoriev, D., Harrison, J., Hirsch, E.A. (eds.) CSR 2006. LNCS, vol. 3967, pp. 270–280. Springer, Heidelberg (2006)
19. Moser, H.: Exact algorithms for generalizations of Vertex Cover. Diploma thesis, Institut für Informatik, Friedrich-Schiller Universität Jena (2005)
20. Niedermeier, R.: Invitation to Fixed-Parameter Algorithms. Oxford University Press, Oxford (2006)
21. Nishimura, N., Ragde, P., Thilikos, D.M.: Fast fixed-parameter tractable algorithms for nontrivial generalizations of Vertex Cover. Discrete Appl. Math. 152(1–3), 229–245 (2005)
22. Raman, V., Saurabh, S.: Short cycles make W-hard problems hard: FPT algorithms for W-hard problems in graphs with no short cycles. Algorithmica (to appear, 2008)

# Some Fixed-Parameter Tractable Classes of Hypergraph Duality and Related Problems

Khaled Elbassioni[1], Matthias Hagen[2,3], and Imran Rauf[1]

[1] Max-Planck-Institut für Informatik, D–66123 Saarbrücken
{elbassio,irauf}@mpi-inf.mpg.de
[2] Friedrich-Schiller-Universität Jena, Institut für Informatik, D–07737 Jena
hagen@cs.uni-jena.de
[3] University of Kassel, Research Group Programming Languages / Methodologies,
Wilhelmshöher Allee 73, D–34121 Kassel, Germany
hagen@uni-kassel.de

**Abstract.** In this paper we present fixed-parameter algorithms for the problem DUAL—given two hypergraphs, decide if one is the transversal hypergraph of the other—and related problems. In the first part, we consider the number of edges of the hypergraphs, the maximum degree of a vertex, and a vertex complementary degree as our parameters.

In the second part, we use an Apriori approach to obtain FPT results for generating all maximal independent sets of a hypergraph, all minimal transversals of a hypergraph, and all maximal frequent sets where parameters bound the intersections or unions of edges.

## 1 Introduction

In many situations, one might be interested in finding all objects or configurations satisfying a certain *monotone property*. Consider, for instance, the problem of finding all (inclusion-wise) maximal/minimal collections of items that are frequently/infrequently bought together by customers in a supermarket. More precisely, let $\mathcal{D} \in \{0,1\}^{m \times n}$ be a binary matrix whose rows represent the subsets of items purchased by different customers in a supermarket. For a given integer $t \geq 0$, a subset of items is said to be *t-frequent* if at least $t$ rows (transactions) of $\mathcal{D}$ contain it, and otherwise is said to be *t-infrequent*. Finding frequent itemsets is an essential problem in finding the so-called *association rules* in data mining applications [AIS93]. By monotonicity, it is enough to find the *border* which is defined by the minimal $t$-infrequent and maximal $t$-frequent sets. While it was shown in [BGKM02] that finding maximal frequent sets is an NP-hard problem, finding the minimal $t$-infrequent sets, as well as many other enumeration problems in different areas (see e.g. [BEGK03, EG95]), turn out to be polynomially equivalent with the *hypergraph transversal problem*, defined as follows.

Let $V$ be a finite set and $\mathcal{H} \subseteq 2^V$ be a hypergraph on $V$. A *transversal* of $\mathcal{H}$ is a subset of $V$ that intersects every hyperedge of $\mathcal{H}$. Let $\mathcal{H}^d \subseteq 2^V$ be the hypergraph of all inclusion-wise minimal transversals of $\mathcal{H}$, also called the *dual* of $\mathcal{H}$. For hypergraphs $\mathcal{F}$ and $\mathcal{G}$ on vertex set $V$, the hypergraph transversal

M. Grohe and R. Niedermeier (Eds.): IWPEC 2008, LNCS 5018, pp. 91–102, 2008.
© Springer-Verlag Berlin Heidelberg 2008

problem, denoted $\text{DUAL}(\mathcal{F}, \mathcal{G})$, asks to decide whether they are dual to each other, i.e., $\mathcal{F}^d = \mathcal{G}$. Equivalently, the problem is to check if two monotone Boolean functions $f, g : \{0,1\}^n \mapsto \{0,1\}$ are dual to each other, i.e., $f(x) = \bar{g}(\bar{x})$ for all $x \in \{0,1\}^n$.

Let $\mathcal{F}, \mathcal{G} \subseteq 2^V$ be Sperner hypergraphs (i.e. no hyperedge of which contains another), and let $(\mathcal{F}, \mathcal{G})$ be an instance of $\text{DUAL}$. By definition of dual hypergraphs we may assume throughout that

$$F \cap G \neq \emptyset \text{ for all } F \in \mathcal{F} \text{ and } G \in \mathcal{G}. \tag{1}$$

A *witness* for the non-duality of the pair $(\mathcal{F}, \mathcal{G})$ satisfying (1) is a subset $X \subseteq V$, such that

$$X \cap F \neq \emptyset \text{ for all } F \in \mathcal{F}, \text{ and } X \not\supseteq G \text{ for all } G \in \mathcal{G}. \tag{2}$$

We shall say that the hypergraphs $\mathcal{F}$ and $\mathcal{G}$ satisfying (1) are dual if no such witness exists. Intuitively, a witness of non duality of $(\mathcal{F}, \mathcal{G})$ is a transversal of $\mathcal{F}$ (not necessarily minimal) that does not include any hyperedge of $\mathcal{G}$. Also, by definition, the pair $(\emptyset, \{\emptyset\})$ is dual. Note that the condition (2) is symmetric in $\mathcal{F}$ and $\mathcal{G}$: $X \subseteq V$ satisfies (2) for the pair $(\mathcal{F}, \mathcal{G})$ if and only if $\bar{X}$ satisfies (2) for $(\mathcal{G}, \mathcal{F})$.

In the following we present some fixed-parameter algorithms for this problem. Briefly, a parameterized problem with parameter $k$ is *fixed-parameter tractable* if it can be solved by an algorithm running in time $O(f(k) \cdot poly(n))$, where $f$ is a function depending on $k$ only, $n$ is the size of the input, and $poly(n)$ is any polynomial in $n$. The class FPT contains all fixed-parameter tractable problems. For more general surveys on parameterized complexity and fixed-parameter tractability we refer to the monographs of Downey and Fellows, and Niedermeier [DF99, Nie06].

A related problem $\text{DUALIZATION}(\mathcal{F})$ is to generate $\mathcal{F}^d$ given $\mathcal{F}$. Given an algorithm for $\text{DUALIZATION}$ we can decide if $\mathcal{F}$ and $\mathcal{G}$ are dual by generating the dual hypergraph of one explicitly and compare it with the other (actually, $\text{DUAL}$ and $\text{DUALIZATION}$ are even equivalent in the sense of solvability in appropriate terms of polynomial time [BI95]). Due to the fact that the size of $\mathcal{F}^d$ may be exponentially larger than the the size of $\mathcal{F}$, we consider *output-sensitive* fixed-parameter algorithms for $\text{DUALIZATION}$, i.e., which are polynomial in both the input and output size $|\mathcal{F}| + |\mathcal{F}^d|$. In this sense, the time required to produce each new output is usually called the *delay* of the algorithm.

Both, $\text{DUAL}$ and $\text{DUALIZATION}$ have many applications in such different fields like artificial intelligence and logic [EG95, EG02], database theory [MR92], data mining and machine learning [GKM+03], computational biology [Dam06, Dam07], mobile communication systems [SS98], distributed systems [GB85], and graph theory [JPY88, LLK80]. The currently best known algorithms for $\text{DUAL}$ run in quasi-polynomial time or use $\mathcal{O}(\log^2 n)$ nondeterministic bits [EGM03, FK96, KS03]. Thus, on the one hand, $\text{DUAL}$ is probably not coNP-complete, but on the other hand a polynomial time algorithm is not yet known.

In this paper, we show that $\text{DUAL}(\mathcal{F}, \mathcal{G})$ is fixed parameter-tractable with respect to the following parameters:

- the numbers of edges $m = |\mathcal{F}|$ and $m' = |\mathcal{G}|$ (cf. Section 2),
- the maximum degrees of vertices in $\mathcal{F}$ and $\mathcal{G}$, i.e., $p = \max_{v \in V} |\{F \in \mathcal{F} : v \in F\}|$, $p' = \max_{v \in V} |\{G \in \mathcal{G} : v \in G\}|$ (cf. Section 3),
- the maximum complementary degrees $q = \max_{v \in V} |\{F \in \mathcal{F} : v \notin F\}|$ and $q' = \max_{v \in V} |\{G \in \mathcal{G} : v \notin G\}|$ (cf. Section 4), and
- the maximum $c$ such that $|F_1 \cup F_2 \cup \cdots \cup F_k| \geq n - c$, where $F_1, \ldots, F_k \in \mathcal{F}$ and $k$ is a constant—and the symmetric parameter $c'$ with respect to $\mathcal{G}$ (cf. Section 5.2).

We shall prove the bounds with respect to the parameters $m, p, q, c$; the other symmetric bounds follow by exchanging the roles of $\mathcal{F}$ and $\mathcal{G}$. Our results for the parameters $m$ and $q$ improve the respective results from [Hag07].

Other related FPT results were obtained by Damaschke who studied counting and generating minimal transversals of size up to $k$ and showed both problems to be FPT if hyperedges have constantly bounded size [Dam06, Dam07].

In Section 5.3 we consider the related problem of finding all maximal frequent sets, and show that it is fixed parameter-tractable with respect to the maximum size of intersection of $k$ rows of the database $\mathcal{D}$ for a constant $k$, thus generalizing the well-known Apriori algorithm, which is fixed-parameter with respect to the size of the largest transaction.

Let $V$ be a finite set of size $|V| = n$. For a hypergraph $\mathcal{F} \subseteq 2^V$ and a subset $S \subseteq V$, we use the following notations: $\bar{S} = V \setminus S$, $\mathcal{F}_S = \{F \in \mathcal{F} \mid F \subseteq S\}$ and $\mathcal{F}^S = \text{minimal}(\{F \cap S \mid F \in \mathcal{F}\})$, where for a hypergraph $\mathcal{H}$, minimal($\mathcal{H}$) denotes the Sperner hypergraph resulting from $\mathcal{H}$ by removing hyperedges that contain any other hyperedge of $\mathcal{H}$.

## 2  Number of Edges as Parameter

Let $(\mathcal{F}, \mathcal{G})$ be an instance of DUAL and let $m = |\mathcal{F}|$. We show that the problem is fixed-parameter tractable with parameter $m$ and improve the running time of [Hag07].

Given a subset $S \subseteq V$ of vertices, [BGH98] gave a criterion to decide if $S$ is a *sub-transversal* of $\mathcal{F}$, i.e., if there is a minimal transversal $T \in \mathcal{F}^d$ such that $T \supseteq S$. In general, testing if $S$ is a sub-transversal is an NP-hard problem even if $\mathcal{F}$ is a graph (see [BEGK00]). However, if $|S|$ is bounded by a constant, then such a check can be done in polynomial time. This observation was used to solve the hypergraph transversal problem in polynomial time for hypergraphs of bounded edge size in [BEGK00], or more generally of bounded conformality [BEGK04]. To describe this criterion, we need a few more definitions. For a subset $S \subseteq V$, and a vertex $v \in S$, let $\mathcal{F}_v(S) = \{H \in \mathcal{F} \mid H \cap S = \{v\}\}$. A selection of $|S|$ hyperedges $\{H_v \in \mathcal{F}_v(S) \mid v \in S\}$ is called *covering* if there exists a hyperedge $H \in \mathcal{F}_{V \setminus S}$ such that $H \subseteq \bigcup_{v \in S} H_v$.

**Proposition 2.1 (cf. [BGH98]).** *A non-empty subset $S \subseteq V$ is a sub-transversal for $\mathcal{F} \subseteq 2^V$ if and only if there exists a non-covering selection $\{H_v \in \mathcal{F}_v(S) \mid v \in S\}$ for $S$.*

If the size of $S$ is bounded we have the following.

**Lemma 2.2.** *Given a hypergraph $\mathcal{F} \subseteq 2^V$ of size $|\mathcal{F}| = m$ and a subset $S \subseteq V$, of size $|S| = s$, checking whether $S$ is a sub-transversal of $\mathcal{F}$ can be done in time $\mathcal{O}(nm(m/s)^s)$.*

*Proof.* For every possible selection $\mathcal{F} = \{H_v \in \mathcal{F}_v(S) \mid v \in S\}$, we can check if $\mathcal{F}$ is non-covering in $\mathcal{O}(n|\mathcal{F}_{\bar{S}}|)$ time. Since the families $\mathcal{F}_v(S)$ are disjoint, we have $\sum_{v \in S} |\mathcal{F}_v(S)| \leq m$, and thus the arithmetic-geometric mean inequality gives for the total number of selections

$$\prod_{v \in S} |\mathcal{F}_v(S)| \leq \left( \frac{\sum_{v \in S} |\mathcal{F}_v(S)|}{s} \right)^s \leq \left( \frac{m}{s} \right)^s.$$

$\square$

**Procedure DUALIZE1($\mathcal{F}, S, V$):**
   Input: A hypergraph $\mathcal{F} \subseteq 2^V$, and a subset $S \subseteq V$
   Output: The set $\{T \in \mathcal{F}^d \; : \; T \supseteq S\}$

1. **if** $S$ is not a sub-transversal for $\mathcal{F}$ **then return**
2. **if** $S \in \mathcal{F}^d$ **then** output $S$ and **return**
3. Find $e \in V \setminus S$, such that $S \cup \{e\}$ is a sub-transversal for $\mathcal{F}$
4. DUALIZE1($\mathcal{F}, S \cup \{e\}, V$)
5. DUALIZE1($\mathcal{F}^{V \setminus \{e\}}, S, V \setminus \{e\}$)

**Fig. 1.** The backtracking method for finding minimal transversals

The algorithm is given in Figure 1, and is based on the standard backtracking technique for enumeration (see e.g. [RT75, Eit94]). The procedure is called initially with $S = \emptyset$. It is easy to verify that the algorithm outputs all elements of the dual hypergraph $\mathcal{F}^d$, without repetition, and in lexicographic ordering (assuming some order on the vertex set $V$). Since the algorithm essentially builds a backtracking tree whose leaves are the minimal transversals of $\mathcal{F}$, the time required to produce each new minimal transversal is bounded by the depth of the tree (at most $\min\{n, m\}$) times the maximum time required at each node. By Lemma 2.2, the latter time is at most $n \cdot \mathcal{O}(nm) \cdot \max\{(m/s)^s \; : \; 1 \leq s \leq m\} = \mathcal{O}(n^2 m \cdot e^{m/e})$.

**Lemma 2.3.** *Let $\mathcal{F} \subseteq 2^V$ be a hypergraph with $|\mathcal{F}| = m$ edges on $|V| = n$ vertices. Then all minimal transversals of $\mathcal{F}$ can be found with $\mathcal{O}(n^2 m^2 e^{m/e})$ delay.*

**Theorem 2.4.** *Let $\mathcal{F}, \mathcal{G} \subseteq 2^V$ be two hypergraphs with $|\mathcal{F}| = m, |\mathcal{G}| = m'$ and $|V| = n$. Then $\mathcal{F}^d = \mathcal{G}$ can be decided in time $\mathcal{O}(n^2 m^2 e^{(m/e)} \cdot m')$.*

*Proof.* We generate at most $m'$ members of $\mathcal{F}^d$ by calling DUALIZE1 (if there are more then obviously $\mathcal{F}^d \neq \mathcal{G}$). Assuming that hyperedges are represented by bit vectors (defined by indicator functions), we can check whether $\mathcal{G}$ is identical to $\mathcal{F}^d$ by lexicographically ordering the hyperedges of both and simply comparing the two sorted lists. The time to sort and compare $m'$ hyperedges each one of size at most $\log n$ can be bounded by $\mathcal{O}(m' \log m' \log n)$. □

As a side remark we note an interesting implication of Lemma 2.3. For a hypergraph $\mathcal{F}$ with $|\mathcal{F}| \leq c \log n$ for a constant $c$, the algorithm DUALIZE1 finds all its minimal transversals with polynomial delay $\mathcal{O}(n^{c/e+2} \log^2 n)$ improving the previous best bound of $\mathcal{O}(n^{2c+6})$ by Makino [Mak03]. Similarly, if the number of minimal transversals is bounded by $\mathcal{O}(\log n)$, then DUALIZE1 can be used to find all these transversals in incremental polynomial time. Another implication which we will need in Section 5 is the following.

**Corollary 2.5.** *For a hypergraph $\mathcal{F} \subseteq 2^V$, we can generate the first $k$ minimal transversals in time $\mathcal{O}(n^2 k^3 e^{(k/e)} \cdot m)$, where $n = |V|$ and $m = |\mathcal{F}|$.*

*Proof.* We keep a partial list $\mathcal{G}$ of minimal transversals, initially empty. If $|\mathcal{G}| < k$, we call DUALIZE1 on $\mathcal{G}$ to generate at most $m+1$ elements of $\mathcal{G}^d$. If it terminates with $\mathcal{G}^d = \mathcal{F}$, then all elements of $\mathcal{F}^d$ have been generated. Otherwise, $X \in \mathcal{G}^d \setminus \mathcal{F}$ is a witness for the non-duality of $(\mathcal{G}, \mathcal{F})$, and so by symmetry $\bar{X}$ contains a new minimal transversal of $\mathcal{F}$ that extends $\mathcal{G}$. □

## 3  Maximum Degree as Parameter

Let $p$ be the maximum degree of a vertex in hypergraph $\mathcal{F} \subseteq 2^V$, i.e., $p = \max_{v \in V} |\{F \in \mathcal{F} : v \in F\}|$. We show that $\text{DUAL}(\mathcal{F}, \mathcal{G})$ is fixed-parameter tractable with parameter $p$ (a result which follows by similar techniques, but with weaker bounds, from [EGM03]).

For a labeling of vertices $V = \{v_1, v_2, \ldots, v_n\}$, let $\mathcal{F}_1, \mathcal{F}_2, \ldots, \mathcal{F}_n$ be a partition of hypergraph $\mathcal{F}$ defined as $\mathcal{F}_i = \{F \in \mathcal{F} : F \ni v_i, F \subseteq \{v_1, \ldots, v_i\}\}$. By definition the size of each set $\mathcal{F}_i$ in this partition is bounded by $p$. The algorithm is given in Figure 2 and essentially combines the technique of the previous section with the method of [LLK80] (see also [BEGK04]). We proceed inductively, for $i = 1, \ldots, n$, by finding $(\mathcal{F}_1 \cup \ldots \cup \mathcal{F}_{i-1})^d$. Then for each set $X$ in this transversal hypergraph we extend it to a minimal transversal to $(\mathcal{F}_1 \cup \ldots \cup \mathcal{F}_i)^d$ by finding $(\{F \in \mathcal{F}_i : F \cap X = \emptyset\})^d$, each set of which is combined with $X$, possibly also deleting some elements from $X$, to obtain a minimal transversal to $\mathcal{F}_1 \cup \ldots \cup \mathcal{F}_i$.

For a hypergraph $\mathcal{H}$ and its transversal $X$ (not necessarily minimal), let $\delta(X)$ denote a minimal transversal of $\mathcal{H}$ contained in $X$.

The following proposition states that with the partition $\mathcal{F}_1, \mathcal{F}_2, \ldots, \mathcal{F}_n$, the size of intermediate hypergraphs in this incremental algorithm never gets too large.

**Proposition 3.1 (cf. [EGM03, LLK80]).** *(i) $\forall S \subseteq V : |(\mathcal{F}_S)^d| \leq |\mathcal{F}^d|$, (ii) $|(\mathcal{F}_1 \cup \ldots \cup \mathcal{F}_i)^d| \leq |\mathcal{F}^d|$, (iii) For every $X \in (\mathcal{F}_1 \cup \ldots \cup \mathcal{F}_{i-1})^d$,*

$$|(\{F \in \mathcal{F}_i : F \cap X = \emptyset\})^d| \leq |(\mathcal{F}_1 \cup \ldots \cup \mathcal{F}_i)^d|.$$

*Proof.* All three follow from the fact that $(\mathcal{F}_S)^d$ is a truncation of $\mathcal{F}^d$ on $S$, where $S = \{v_1, \ldots, v_i\}$ in (ii) and $S = \{v_1, \ldots, v_i\} \setminus X$ in (iii). □

**Procedure DUALIZE2($\mathcal{F}, V$):**
    Input: A hypergraph $\mathcal{F} \subseteq 2^V$
    Output: The set $\mathcal{F}^d$

1. $\mathcal{X}_0 = \{\emptyset\}$
2. **for** $i = 1, \ldots, n$ **do**
3.     **for** each $X \in \mathcal{X}_{i-1}$ **do**
4.         Let $\mathcal{A} = \{F \in \mathcal{F}_i \; : \; F \cap X = \emptyset\}$
5.         Use DUALIZE1 to compute $\mathcal{A}^d$ if not already computed
6.         $\mathcal{X}_i \leftarrow \{\delta(X \cup Y) \; : \; Y \in \mathcal{A}^d\}$
7. **return** $\mathcal{X}_n$

**Fig. 2.** Sequential method for finding minimal transversals

Let $f(p, i)$ be the running time of algorithm DUALIZE1 when given a hypergraph with $p$ edges on $i$ vertices. Consider the $i$-th iteration. From Proposition 3.1 we have $|\mathcal{X}_{i-1}| \leq |\mathcal{F}^d|$ and since we only compute $\mathcal{A}^d$ in step 5 if not already computed, there are at most $\min\{2^p, |\mathcal{F}^d|\}$ calls to DUALIZE1. The size of $\mathcal{A}^d$ can also be bounded by Proposition 3.1, which gives us $|\mathcal{A}^d| \leq |\mathcal{F}^d|$. Furthermore it is easy to see that the minimal transversal in step 6 can be found in time $\mathcal{O}(n|\mathcal{F}|)$ by removing the extra vertices (at most $n$). Thus the time spent in the $i$-th iteration can be bounded by $\mathcal{O}(\min\{2^p, |\mathcal{F}^d|\} \cdot f(p, n) + n|\mathcal{F}| \cdot |\mathcal{F}^d|^2)$.

**Theorem 3.2.** *Let $\mathcal{F} \subseteq 2^V$ be a hypergraph on $|V| = n$ vertices in which the degree of each vertex $v \in V$ is bounded by $p$. Then all minimal transversals of $\mathcal{F}$ can be found in time $\mathcal{O}\left(n^2 mm' \cdot (\min\{2^p, m'\} \cdot np^2 e^{p/e} + m')\right)$, where $m = |\mathcal{F}|$ and $m' = |\mathcal{F}^d|$.*

## 4 Vertex Complementary Degree as Parameter

For a hypergraph $\mathcal{F} \subseteq 2^V$ and a vertex $v \in V$, consider the number of edges in $\mathcal{F}$ not containing $v$ for some vertex $v \in V$. Let $q$ be maximum such number, i.e., $q = \max_{v \in V} |\{F \in \mathcal{F} \; : \; v \notin F\}|$. We show that $\mathrm{DUAL}(\mathcal{F}, \mathcal{G})$ is fixed-parameter tractable with parameter $q$ and improve the running time of [Hag07].

The following proposition gives a decomposition rule originally due to [FK96] which for a vertex $v \in V$ divides the problem into two subproblems not containing $v$.

**Proposition 4.1 (cf. [FK96]).** *Let $\mathcal{F}, \mathcal{G} \subseteq 2^V$ be two hypergraphs satisfying (1), and $v \in V$ be a given vertex. Then $\mathcal{F}$ and $\mathcal{G}$ are dual if and only if the pairs $(\mathcal{F}_{V \setminus v}, \mathcal{G}^{V \setminus v})$ and $(\mathcal{F}^{V \setminus v}, \mathcal{G}_{V \setminus v})$ are dual.*

For a vertex $v \in V$, one of the subproblem $(\mathcal{F}_{V \setminus v}, \mathcal{G}^{V \setminus v})$ involves a hypergraph $\mathcal{F}_{V \setminus v}$ with at most $q$ edges. The algorithm solves it by calling DUALIZE1 resulting in time $\mathcal{O}(n^2 q^2 e^{(q/e)} \cdot |(\mathcal{F}_{V \setminus v})^d|)$. The other subproblem $(\mathcal{F}^{V \setminus v}, \mathcal{G}_{V \setminus v})$ is solved recursively. Since at least one vertex is reduced at each step of the algorithm, there are at most $n = |V|$ recursive steps.

**Theorem 4.2.** *Let $\mathcal{F}, \mathcal{G} \subseteq 2^V$ be two hypergraphs with $|\mathcal{F}| = m, |\mathcal{G}| = m'$ and $|V| = n$. Let $q = \max_{v \in V} |\{H \in \mathcal{F} : v \notin H\}|$. Then $\mathcal{F}^d = \mathcal{G}$ can be decided in time $\mathcal{O}(n^3 q^2 e^{(q/e)} \cdot m')$.*

## 5   Results Based on the Apriori Technique

Gunopulos et al. [GKM⁺03] showed (Theorem 23, page 156) that generating minimal transversals of hypergraphs $\mathcal{F}$ with edges of size at least $n - c$ can be done in time $\mathcal{O}(2^c poly(n, m, m'))$, where $n = |V|$, $m = |\mathcal{F}|$ and $m' = |\mathcal{F}^d|$. This is a fixed-parameter algorithm for $c$ as parameter. Furthermore, this result shows that the transversals can be generated in polynomial time for $c \in \mathcal{O}(\log n)$. The computation is done by an Apriori (level-wise) algorithm [AS94].

Using the same approach, we shall show below that we can compute all the minimal transversals in time $\mathcal{O}(\min\{2^c (m')^k poly(n, m), e^{k/e} n^{c+1} poly(m, m')\})$ if the union of any $k$ distinct minimal transversals has size at least $n - c$. Equivalently, if any $k$ distinct maximal independent sets of a hypergraph $\mathcal{F}$ intersect in at most $c$ vertices, then all maximal independent sets can be computed in the same time bound. As usual, an *independent set* of a hypergraph $\mathcal{F}$ is a subset of its vertices which does not contain any hyperedge of $\mathcal{F}$.

And again using the same idea, we show that the maximal frequent sets of an $m \times n$ database can be computed in $\mathcal{O}(2^c (nm')^{2^{k-1}+1} poly(n, m))$ time if any $k$ rows of it intersect in at most $c$ items, where $m'$ is the number of such sets.

Note that for $c \in \mathcal{O}(\log n)$ we have incremental polynomial-time algorithms for all four problems.

### 5.1   The Generalized Apriori Algorithm

Let $f : V \mapsto \{0, 1\}$ be a monotone Boolean function, that is, for which $f(X) \geq f(Y)$ whenever $X \supseteq Y$. We assume that $f$ is given by a polynomial-time evaluation oracle requiring maximum time $T_f$, given the input. The Apriori approach for finding all maximal subsets $X$ such that $f(X) = 0$ (maximal false sets of $f$), works by traversing all subsets $X$ of $V$, for which $f(X) = 0$, in increasing size, until all maximal such sets have been identified. The procedure is given in Figure 3.

**Lemma 5.1.** *If any maximal false set of $f$ contains at most $c$ vertices, then APRIORI finds all such sets in $\mathcal{O}(2^c m' n T_f)$ time, where $n = |V|$ and $m'$ is the number of maximal false sets.*

*Proof.* The correctness of this Apriori style method can be shown straightforwardly (cf. e. g. [AS94, GKM⁺03]). To see the time bound, note that for each maximal false set we check at most $2^c$ candidates (all the subsets) before adding it to $\mathcal{C}$. For each such candidate we check whether it is a false set and whether it cannot be extended by adding more vertices.                                    □

**Procedure APRIORI**$(f, V)$:

    Input: a monotone Boolean function $f : V \mapsto \{0, 1\}$

    Output: the maximal sets $X \subseteq V$ such that $f(X) = 0$

1. $\mathcal{C} \leftarrow \emptyset; \mathcal{C}_1 \leftarrow \{\{v\} : v \in V\}; i \leftarrow 1; \mathcal{C}_j \leftarrow \emptyset \; \forall j = 2, 3, \ldots$
2. **while** $\mathcal{C}_i \neq \emptyset$
3.     **for** $X, Y \in \mathcal{C}_i, |X \cap Y| = i - 1$
4.         $Z \leftarrow X \cup Y$
5.         **if** $f(Z) = 0$ **then**
6.             **if** $f(Z \cup \{v\}) = 1$, for all $v \in V \setminus Z$ **then**
7.                 $\mathcal{C} \leftarrow \mathcal{C} \cup \{Z\}$
8.         **else**
9.             $\mathcal{C}_{i+1} \leftarrow \mathcal{C}_{i+1} \cup \{Z\}$
10.    $i \leftarrow i + 1$
11. **return** $\mathcal{C}$

**Fig. 3.** The generalized Apriori algorithm

## 5.2 Maximal Independent Sets

Let $\mathcal{F} \subseteq 2^V$ be a hypergraph. An *independent set* of $\mathcal{F}$ is a subset of $V$ which does not contain any hyperedge of $\mathcal{F}$. It is easy to see that the hypergraph of maximal independent sets $\mathcal{F}^{dc}$ of $\mathcal{F}$ is the complementary hypergraph of the dual $\mathcal{F}^d$: $\mathcal{F}^{dc} = \{V \setminus T : T \in \mathcal{F}^d\}$.

Let $k$ and $c$ be two positive integers. We consider hypergraphs $\mathcal{F} \subseteq 2^V$ satisfying the following condition:

**(C1)** Any $k$ distinct maximal independent sets $I_1, \ldots, I_k$ of $\mathcal{F}$ intersect in at most $c$ vertices, i.e., $|I_1 \cap \cdots \cap I_k| \leq c$.

We shall derive below fixed-parameter algorithms with respect to either $c$ or $k$. We note that condition (C1) can be checked in polynomial time for $c = \mathcal{O}(1)$ and $k = \mathcal{O}(\log n)$. Indeed, (C1) holds if and only if every set $X \subseteq V$ of size $|X| = c + 1$ is contained in at most $k - 1$ maximal independent sets of $\mathcal{F}$. The latter condition can be checked in time $n^{c+1} \, \text{poly}(n, m, k) e^{k/e}$ as follows from the following lemma.

**Lemma 5.2.** *Given a hypergraph $\mathcal{F}$ with vertex set $V$ and a subset $S \subseteq V$ of vertices, we can check in polynomial time whether $S$ is contained in $k$ different maximal independent sets. Furthermore $k$ such sets can be generated in time $\mathcal{O}(\text{poly}(n, m, k) e^{k/e})$.*

*Proof.* Clearly, this check is equivalent to checking if $S$ does not contain an edge of $\mathcal{F}$ and if the truncated hypergraph $\mathcal{F}^{\bar{S}}$ has $k$ maximal independent sets, or equivalently $k$ minimal transversals. By Corollary 2.5, this can be done in $\mathcal{O}(\text{poly}(n, m, k) e^{k/e})$ time. □

For a set $S \subseteq V$, denote by $\mathcal{F}^{dc}[S]$ the set of maximal independent sets of $\mathcal{F}$ containing $S$.

**Theorem 5.3.** *If any $k$ distinct maximal independent sets of a hypergraph $\mathcal{F}$ intersect in at most $c$ vertices, then all maximal independent sets can be computed in time $\mathcal{O}(\min\{2^c(m')^k poly(n,m), e^{k/e}n^{c+1}poly(m,m')\})$, where $n = |V|$, $m = |\mathcal{F}|$ and $m' = |\mathcal{F}^{dc}|$.*

*Proof.* (i) $c$ as a parameter: we first use APRIORI to find the set $\mathcal{X}$ of all maximal subsets contained in at least $k$ distinct maximal independent sets of $\mathcal{F}$. By (C1) the size of each such subset is at most $c$. To do this we use APRIORI with the monotone Boolean function defined by $f(X) = 0$ if and only if $X \subseteq I_1 \cap \cdots \cap I_k$, for $k$ distinct maximal independent sets $I_1, \ldots, I_k$. The procedure is given in Figure 4. By Lemmas 5.1 and 5.2, all the intersections in $\mathcal{X}$ can be found in time $2^c poly(n,m,k)e^{k/e}|\mathcal{X}|$. Thus the total running time can be bounded by $2^c poly(n,m,k)e^{k/e}(m')^k$ since $|\mathcal{X}| \leq (m')^k$. It remains to argue that any maximal independent set $I \in \mathcal{F}^{dc}$ is generated by the procedure. To see this, let $Y$ be a maximal subset such that $Y = I \cap I_1 \cap \ldots \cap I_r$, where $I, I_1, \ldots, I_r$, are distinct maximal independent sets of $\mathcal{F}$ with $r \geq k - 1$, and let $v \in I \setminus (\cap_{j \in [r]} I_j)$. Note that such $v$ exists since $I \not\subseteq \cap_{j \in [r]} I_j$ since $I, I_1, \ldots, I_r$ are distinct maximal independent sets. Then by maximality of $Y$, $Y \cup \{v\}$ is contained in at most $k - 1$ maximal independent sets, one of which is $I$, and hence will be considered by the procedure in Step 7.

   (ii) $k$ as a parameter: Let $\mathcal{I}_1 = \{I \in \mathcal{F}^{dc} : |I| \leq c\}$ and $\mathcal{I}_2 = \mathcal{F}^{dc} \setminus \mathcal{I}_1$. Elements of $\mathcal{I}_1$ can be found using the APRIORI procedure with the monotone Boolean function, defined as $f(X) = 0$ if and only if $X \subseteq V$ is independent and has size at most $c$ (or by testing all subsets of size at most $c$ for maximal independence). Elements of $\mathcal{I}_2$ can be found by noting that each of them contains a set of size $c + 1$, and that each such set is contained in at most $k - 1$ elements of $\mathcal{I}_2$ by (C1). Thus for each set $X$ of size $c + 1$, we can use Lemma 5.2 to find all maximal independent sets containing $X$.                                                    □

**Corollary 5.4.** *Let $\mathcal{F} \subseteq 2^V$ be a hypergraph on $n = |V|$ vertices, and $k,c$ be positive integers.*

**Procedure MAX-INDP-GEN$(\mathcal{F},V)$:**
   Input: a hypergraph $\mathcal{F} \subseteq 2^V$
   Output: the set of maximal independent sets of $\mathcal{F}$

1. $\mathcal{C} \leftarrow \emptyset$
2. Use APRIORI to find the set of maximal $k$-independent set intersections $\mathcal{X}$
3. **for** each $X \in \mathcal{X}$ **do**
4.     **for** each $Y \subseteq X$ **do**
5.         **for** each $v \in V \setminus Y$ **do**
6.             **if** $|\mathcal{F}^{dc}[Y \cup \{v\}]| \leq k - 1$
7.                 $\mathcal{C} \leftarrow \mathcal{C} \cup \mathcal{F}^{dc}[Y \cup \{v\}]$ (obtained using Corollary 2.5)
8. **return** $\mathcal{C}$

**Fig. 4.** The fixed parameter algorithm for finding all maximal independent sets *(i)* If any $k$ distinct minimal transversals of $\mathcal{F}$ have a union of at least $n - c$ vertices, we can compute all minimal transversals in $\mathcal{O}(\min\{2^c (m')^k poly(n, m),$ $e^{k/e} n^{c+1} poly(m, m')\})$ time, where $m = \mathcal{F}$ and $m' = |\mathcal{F}^d|$.

*(ii)* If any $k$ distinct hyperedges of $\mathcal{F}$ have a union of at least $n - c$ vertices, we can compute all minimal transversals in time $\mathcal{O}(\min\{2^c m^k poly(n, m'),$ $e^{k/e} n^{c+1} poly(m, m')\})$, where $m = \mathcal{F}$ and $m' = |\mathcal{F}^d|$.

*Proof.* Both results are immediate from Theorem 5.3. (i) follows by noting that each minimal transversal is the complement of a maximal independent set, and hence any $k$ maximal independent sets are guaranteed to intersect in at most $c$ vertices. (ii) follows by maintaining a partial list $\mathcal{G} \subseteq \mathcal{F}^d$, and switching the roles of $\mathcal{F}$ and $\mathcal{G}$ in (i) to compute the minimal transversals of $\mathcal{G}$ using Theorem 5.4. Since condition (i) is satisfied with respect to $\mathcal{G}$, we can either verify duality of $\mathcal{F}$ and $\mathcal{G}$, or extend $\mathcal{G}$ by finding a witness for the non-duality (in a way similar to Corollary 2.5).    □

## 5.3   Maximal Frequent Sets

Consider the problem of finding the maximal frequent item sets in a collection of $m$ transactions on $n$ items, stated in the Introduction. Here, a transaction simply is a set of items. An item set is maximal frequent for a frequency $t$ if it occurs in at least $t$ of the transactions and none of its supersets does. As another application of the approach of the previous subsection we obtain the following.

**Theorem 5.5.** *If any $k$ distinct maximal frequent sets intersect in at most $c$ items, we can compute all maximal frequent sets in $\mathcal{O}(2^c (nm')^k poly(n, m))$ time, where $m'$ is the number of maximal frequent sets.*

*Proof.* The proof is analogous to that of Theorem 5.3. Just note that the set of transactions forms a hypergraph and replace "independent" by "frequent". To complete the proof, we need the following procedure to find $k$ maximal frequent sets containing a given set. For $1 \leq i \leq k$ and frequent set $X$, let $F_1, \dots, F_{i-1}$ be the maximal frequent sets containing $X$ and let $Y$ be the set with the property that $X \cup Y$ is frequent and $\forall j < i, \exists y \in Y : y \notin F_j$. Then any maximal frequent set containing $X \cup Y$ is different from $F_1 \dots F_{i-1}$ by construction and thus giving us a new maximal frequent set. The running time of the above procudure can be bounded by $\mathcal{O}(n^k poly(n, m))$. Combining it with Lemma 5.1 gives us the stated running time.    □

**Corollary 5.6.** *If any $k$ distinct transactions intersect in at most $c$ items, then all maximal frequent sets can be computed in time $\mathcal{O}(2^c (nm')^{2^{k-1}+1} poly(n, m))$, where $m'$ is the number of maximal frequent sets.*

*Proof.* Note that if $t \geq k$ then every maximal frequent set has size at most $c$ which in turn implies $\mathcal{O}(2^c poly(n, m) \cdot m')$ time algorithm using straightforward Apriori approach, so we may assume otherwise. Consider the intersection $X$ of $l$ distinct maximal frequent sets and let $|X| > c$, we bound the maximum such $l$. Since the intersection size is more then $c$, at most $k - 1$ transactions define these $l$ disticnt maximal frequent sets and so $l \leq \sum_{j=t}^{k-1} \binom{k-1}{j} \leq 2^{k-1}$.    □

# 6 Concluding Remarks

Giving an FPT algorithm for DUAL with respect to the parameter size $l$ of a largest edge remains open. Nevertheless, proving that DUAL is not FPT with respect to $l$ seems to be tough as this would imply that there is no polynomial time algorithm for DUAL assuming $W[1] \neq FPT$. Furthermore, this would be a strong argument for a separation of polynomial and quasi-polynomial time in "classical" computational complexity.

# References

[AIS93]      Agrawal, R., Imielinski, T., Swami, A.N.: Mining association rules between sets of items in large databases. In: Proc. SIGMOD 1993, pp. 207–216 (1993)

[AS94]       Agrawal, R., Srikant, R.: Fast algorithms for mining association rules in large databases. In: Proc. VLDB 1994, pp. 487–499 (1994)

[BEGK00]     Boros, E., Elbassioni, K.M., Gurvich, V., Khachiyan, L.: An efficient incremental algorithm for generating all maximal independent sets in hypergraphs of bounded dimension. Parallel Processing Letters 10(4), 253–266 (2000)

[BEGK03]     Boros, E., Elbassioni, K.M., Gurvich, V., Khachiyan, L.: An inequality for polymatroid functions and its applications. Discrete Applied Mathematics 131(2), 255–281 (2003)

[BEGK04]     Boros, E., Elbassioni, K.M., Gurvich, V., Khachiyan, L.: Generating maximal independent sets for hypergraphs with bounded edge-intersections. In: Farach-Colton, M. (ed.) LATIN 2004. LNCS, vol. 2976, pp. 488–498. Springer, Heidelberg (2004)

[BGH98]      Boros, E., Gurvich, V., Hammer, P.L.: Dual subimplicants of positive Boolean functions. Optimization Methods and Software 10(2), 147–156 (1998)

[BGKM02]     Boros, E., Gurvich, V., Khachiyan, L., Makino, K.: On the complexity of generating maximal frequent and minimal infrequent sets. In: Alt, H., Ferreira, A. (eds.) STACS 2002. LNCS, vol. 2285, pp. 133–141. Springer, Heidelberg (2002)

[BI95]       Bioch, J.C., Ibaraki, T.: Complexity of identification and dualization of positive Boolean functions. Inf. Comput. 123(1), 50–63 (1995)

[Dam06]      Damaschke, P.: Parameterized enumeration, transversals, and imperfect phylogeny reconstruction. Theoret. Comput. Sci. 351(3), 337–350 (2006)

[Dam07]      Damaschke, P.: The union of minimal hitting sets: Parameterized combinatorial bounds and counting. In: Thomas, W., Weil, P. (eds.) STACS 2007. LNCS, vol. 4393, pp. 332–343. Springer, Heidelberg (2007)

[DF99]       Downey, R.G., Fellows, M.R.: Parameterized Complexity. Springer, Heidelberg (1999)

[EG95]       Eiter, T., Gottlob, G.: Identifying the minimal transversals of a hypergraph and related problems. SIAM J. Comput. 24(6), 1278–1304 (1995)

[EG02]       Eiter, T., Gottlob, G.: Hypergraph transversal computation and related problems in logic and AI. In: Flesca, S., Greco, S., Leone, N., Ianni, G. (eds.) JELIA 2002. LNCS (LNAI), vol. 2424, pp. 549–564. Springer, Heidelberg (2002)

[EGM03]     Eiter, T., Gottlob, G., Makino, K.: New results on monotone dualization and generating hypergraph transversals. SIAM J. Comput. 32(2), 514–537 (2003)

[Eit94]     Eiter, T.: Exact transversal hypergraphs and application to Boolean $\mu$-functions. J. Symb. Comput. 17(3), 215–225 (1994)

[FK96]      Fredman, M.L., Khachiyan, L.: On the complexity of dualization of monotone disjunctive normal forms. J. Algorithms 21(3), 618–628 (1996)

[GB85]      Garcia-Molina, H., Barbará, D.: How to assign votes in a distributed system. J. ACM 32(4), 841–860 (1985)

[GKM+03]    Gunopulos, D., Khardon, R., Mannila, H., Saluja, S., Toivonen, H., Sharm, R.S.: Discovering all most specific sentences. ACM Trans. Database Syst. 28(2), 140–174 (2003)

[Hag07]     Hagen, M.: On the fixed-parameter tractability of the equivalence test of monotone normal forms. Inf. Process. Lett. 103(4), 163–167 (2007)

[JPY88]     Johnson, D.S., Papadimitriou, C.H., Yannakakis, M.: On generating all maximal independent sets. Inf. Process. Lett. 27(3), 119–123 (1988)

[KS03]      Kavvadias, D.J., Stavropoulos, E.C.: Monotone Boolean dualization is in coNP[$\log^2 n$]. Inf. Process. Lett. 85(1), 1–6 (2003)

[LLK80]     Lawler, E.L., Lenstra, J.K., Rinnooy Kan, A.H.G.: Generating all maximal independent sets: NP-hardness and polynomial-time algorithms. SIAM J. Comput. 9(3), 558–565 (1980)

[Mak03]     Makino, K.: Efficient dualization of $O(\log n)$-term monotone disjunctive normal forms. Discrete Applied Mathematics 126(2–3), 305–312 (2003)

[MR92]      Mannila, H., Räihä, K.-J.: On the complexity of inferring functional dependencies. Discrete Applied Mathematics 40(2), 237–243 (1992)

[Nie06]     Niedermeier, R.: Invitation to Fixed-Parameter Algorithms. Oxford University Press, Oxford (2006)

[RT75]      Read, R.C., Tarjan, R.E.: Bounds on backtrack algorithms for listing cycles, paths, and spanning trees. Networks 5, 237–252 (1975)

[SS98]      Sarkar, S., Sivarajan, K.N.: Hypergraph models for cellular mobile communication systems. IEEE Transactions on Vehicular Technology 47(2), 460–471 (1998)

# A Purely Democratic Characterization of W[1]

Michael Fellows[1,*], Danny Hermelin[2,**], Moritz Müller[3],
and Frances Rosamond[1,***]

[1] Parameterized Complexity Research Unit,
The University of Newcastle, Callaghan NSW 2308 - Australia
{michael.fellows,frances.rosamond}@newcastle.edu.au
[2] Department of Computer Science, University of Haifa,
Mount Carmel, Haifa 31905 - Israel
danny@cri.haifa.ac.il
[3] Mathematisches Institut, Albert Ludwigs Universität Freiburg,
Eckerstrasse 1, 79104 Freiburg - Germany
moritz.mueller@math.uni-freiburg.de

**Abstract.** We give a novel characterization of W[1], the most important fixed-parameter intractability class in the W-hierarchy, using Boolean circuits that consist solely of majority gates. Such gates have a Boolean value of 1 if and only if more than half of their inputs have value 1. Using majority circuits, we define an analog of the W-hierarchy which we call the $\widetilde{\text{W}}$-hierarchy, and show that $\text{W}[1] = \widetilde{\text{W}}[1]$ and $\text{W}[t] \subseteq \widetilde{\text{W}}[t]$ for all $t$. This gives the first characterization of W[1] based on the weighted satisfiability problem for monotone Boolean circuits rather than anti-monotone. Our results are part of a wider program aimed at exploring the robustness of the notion of weft, showing that it remains a key parameter governing the combinatorial nondeterministic computing strength of circuits, no matter what type of gates these circuits have.

## 1 Introduction

Arguably the most important class in the W-hierarchy is W[1]. From a theoretical point of view, it can be viewed as the parameterized analog of NP since it contains many parameterized variants of classical NP-complete problems such as $k$-INDEPENDENT SET, $k, \ell$-LONGEST COMMON SUBSEQUENCE, and, most importantly, because the $k$-STEP HALTING PROBLEM FOR TURING MACHINES OF UNLIMITED NONDETERMINISM is complete for W[1]. This is a parameterized analog of the generic NP-complete problem used in Cook's theorem [DFKHW94, DFS99]. From a practical standpoint, W[1] is the most important complexity class for showing fixed-parameter intractability results, providing an easy accessible platform for showing such results [DF95]. Indeed,

---

* Research supported by the Australian Research Council, and by an Alexander von Humboldt Foundation Research Award.
** Partially supported by the Caesarea Rothschild Institute.
*** Research supported by the Australian Research Council.

M. Grohe and R. Niedermeier (Eds.): IWPEC 2008, LNCS 5018, pp. 103–114, 2008.
© Springer-Verlag Berlin Heidelberg 2008

since the identification of the first complete problems for W[1], there has been a slew of fixed-parameter intractability results, reminiscent in some sense of the early days of NP-completeness.

The key combinatorial objects used to formulate W[1] and the W-hierarchy are constant depth logic circuits that model Boolean functions in the natural way. Their combinatorial nondeterministic computing strength is governed by their *weft*, defined to be the maximum number of unbounded in-degree gates in any path from the input gates to the output. The generic complete problem for the W-hierarchy is the $k$-WEIGHTED WEFT-$t$ CIRCUIT SATISFIABILITY problem, that takes as input a constant depth weft-$t$ circuit $C$ and a parameter $k$ and asks whether $C$ has a weight $k$ satisfying assignment (*i.e.* an assignment setting exactly $k$ input gates to 1). The class W[$t$] is then defined to be the class of parameterized problems which are parameterized reducible to $k$-WEIGHTED WEFT-$t$ CIRCUIT SATISFIABILITY.

In this paper, we explore an alternative, purely monotone characterization of W[1] and the W-hierarchy using a different type of Boolean circuit, namely, a majority circuit. In this type of circuit, we replace the role of logical gates by majority gates which have value 1 when more than half of their inputs have value 1. Using a majority circuit analog of $k$-WEIGHTED WEFT-$t$ CIRCUIT SATISFI-ABILITY, we obtain the $\widetilde{\text{W}}$-hierarchy. Our main results are:

**Theorem 1.** $\text{W}[1] = \widetilde{\text{W}}[1]$.

**Theorem 2.** $\text{W}[t] \subseteq \widetilde{\text{W}}[t]$ *for any positive integer* $t$.

Note that in proving Theorem 2, we use Theorem 1 for the case of $t = 1$. This is not uncommon, since most proofs involving the W-hierarchy require special treatment of W[1] (see [DF99, FG06] for many examples).

The importance of these results are twofold. First, Theorem 1 gives an alternative way of showing fixed-parameter intractability results. The complete problems of W[1] are usually antimonotone by nature, where a parameterized problem is antimonotone if when an instance with a given parameter is known to be a "yes"-instance, then it is also known to be a "yes"-instance for smaller parameter values (*e.g.* maximization problems). However, see [M06] for some exceptions. In circuit terms, it is known that one could use an antimonotone weft-1 circuit to show W[1]-hardness, but this is not known for monotone circuits. Majority circuits, however, are monotone circuits and so they might come in handy in reductions for monotone problems. This is what makes the proof of Theorem 1 so combinatorially challenging.

Second, our results suggest a robustness in the notion of weft. Indeed, there has long been quite a bit of informal criticism against the naturality of this notion. This work, along with another work by almost the same set of authors [FFHMR07], aims at showing that this is not necessarily so. We do so, by showing that if one replaces the role of logical gates in circuits by nontrivial combinatorial gates, then the notion of weft still generally remains the central property governing the nondeterministic combinatorial computing power of circuits. It seems that no matter what your favorite selection of combinatorial gates, the number of unbounded in-degree gates from the input layer to the output will still determine the parameterized complexity of finding weight $k$ satisfying assignments to your circuit.

The paper is organized as follows. In the next section we briefly review basic concepts of parameterized complexity, and formally introduce the notion of majority circuits. We then proceed in Sections 3 and 4 to prove Theorem 1, where in Section 3 we prove that $W[1] \subseteq \widetilde{W}[1]$, and in Section 4 we prove $\widetilde{W}[1] \subseteq W[1]$. Section 5 is devoted to proving Theorem 2. In Section 6 we give a brief summary of the paper, and discuss open problems.

## 2    Preliminaries

In the following we discuss notations and concepts that we use throughout the paper. In particular, we briefly review basic concepts from parameterized complexity, and define pure majority circuits, which play the leading role in this paper, and the corresponding $\widetilde{W}$-hierarchy. We will assume that the reader is familiar with basic concepts from classical complexity and graph theory.

### 2.1    Parameterized Complexity

A *parameterized problem* (or *parameterized language*) is a subset $L \subseteq \Sigma^* \times \mathbb{N}$, where $\Sigma$ is a fixed alphabet, $\Sigma^*$ is the set of all finite length strings over $\Sigma$, and $\mathbb{N}$ is the set of natural numbers. In this way, an input $(x, k)$ to a parameterized language consists of two parts, where the second part $k$ is the *parameter*. A parameterized problem $L$ is *fixed-parameter tractable* if there exists an algorithm which on a given input $(x, k) \in \Sigma^* \times \mathbb{N}$, decides whether $(x, k) \in L$ in $f(k)poly(n)$ time, where $f$ is an arbitrary computable function solely in $k$, and $poly(n)$ is a polynomial in the total input length $n = |(x, k)|$. Such an algorithm is said to run in FPT-*time*, and FPT is the class of all parameterized problems that can be solved by an FPT-time algorithm (*i.e.* all problems which are fixed-parameter tractable).

A framework for proving *fixed-parameter intractability* was developed over the years, using the notion of *parameterized reductions*. A parameterized reduction from a parameterized problem $L$ to another parameterized problem $L'$ is an FPT-time computable mapping that maps an instance $(x, k) \in \Sigma^* \times \mathbb{N}$ to an instance $(x', k') \in \Sigma^* \times \mathbb{N}$ such that $(x, k) \in L \iff (x', k') \in L$. Furthermore $k'$ is required to be bounded by some computable function in $k$.

### 2.2    Logical Circuits and the W-Hierarchy

A *(logical) circuit* $C$ is a connected directed acyclic graph with labels on its vertices, and a unique vertex with no outgoing edges. The vertices which are of in-degree 0 are called the *input gates* and they are labeled with Boolean variables $x_1, x_2, \ldots$. All other vertices are called the *logical gates* and are labeled with Boolean operators $\wedge$, $\vee$, and $\neg$, where vertices which are labeled $\neg$ have in-degree 1. The unique 0 out-degree vertex is the *output gate*. A *monotone logical circuit* is a logical circuit with no ¬-gates. An *antimonotone logical circuit* is a logical circuit where each input gate is connected to the the rest of the circuit via a ¬-gate, and there are no other occurrences of ¬-gates in the circuit.

An *assignment* $X$ for $C$ is an assignment of 0 or 1 to each of the variables in $C$. The *weight* of $X$ is the number of variables that it assigns a 1. The *value* of a logical gate in $C$ under $X$ is obtained straightforwardly according to the label of the gate and the value of its inputs. The *value* $C(X)$ of $C$ under $X$ is the value of the output gate of $C$. We say that $X$ *satisfies* a logical gate in $C$ if the value of this gate under $X$ is 1, and if it satisfies the output gate of $C$ (*i.e.* $C(X) = 1$), we say that $X$ satisfies $C$.

It is convenient to consider the vertices of $C$ as organized into *layers*. The input gates constitute the *input layer*, the logical gates which are directly connected to them are the *first layer*, and more generally, a vertex is in the $i$*'th layer* of $C$ if the length of the maximum-length path from it to the input layer equals $i$. The *depth* of a circuit is the length of the maximum-length path from the input layer to the output gate.

There is an important distinction between two types of logical gates in $C$. Small gates are gates which have in-degree bounded by a small constant (usually we can take this constant to be 3). Large gates are vertices with unbounded in-degree. The maximum number of large gates on a path between an input gate and the output of $C$ is the *weft* of $C$.

Constant depth logical circuits are used to define a hierarchy of fixed-parameter intractability known as the W-*hierarchy*. The generic problem for this hierarchy is $k$-WEIGHTED WEFT-$t$ CIRCUIT SATISFIABILITY, where $t$ is a problem-dependent positive integer constant. This problem takes as input a constant depth weft-$t$ circuit $C$ and a parameter $k$ and asks to determine whether $C$ has a weight $k$ satisfying assignment. The class W[$t$] is defined to be the class of parameterized problems which are parameterized reducible to $k$-WEIGHTED WEFT-$t$ CIRCUIT SATISFIABILITY. The W-hierarchy is then the hierarchy of classes FPT $\subseteq$ W[1] $\subseteq$ W[2] $\subseteq$ $\cdots$. It is well known that antimonotone logical circuits are sufficient to define the odd levels of the W-hierarchy, while monotone logical circuits suffice for defining the even levels of the hierarchy [DF99]. Hence, defining $k$-WEIGHTED WEFT-$t$ ANTIMONOTONE CIRCUIT SATISFIABILITY and $k$-WEIGHTED WEFT-$t$ MONOTONE CIRCUIT SATISFIABILITY in the natural manner, we get the former is complete for all odd $t$ and the latter for all even $t$.

## 2.3   Majority Circuits and the $\widetilde{W}$-Hierarchy

A majority gate is a gate which has value 1 if and only if more than half of its inputs are set to 1. They play the lead role in our novel characterization of W[1], via what we call (*purely*) *majority circuits*. These circuits have majority gates instead of logical gates, and thus have different expressive power in comparison to the ordinary logical circuits of the W-hierarchy. When speaking of majority circuits, we use the same terminology as for their logical counterparts, where all definitions and notions (in particular, the notion of weft) carry through to majority circuits. The only difference is that here we allow multiple edges, whereas in logical circuits these are always redundant.

We now define a hierarchy of complexity classes where majority circuits play an analogous role to the role played by logical circuits in the W-hierarchy. This is done via the analogous generic problem $k$-WEIGHTED WEFT-$t$ MAJORITY CIRCUIT SATISFIABILITY, which asks whether a given constant depth weft-$t$ majority circuit $C$ has a $k$ weight satisfying assignment, for parameter $k$. The class $\widetilde{W}[t]$ is defined to be the class of parameterized problems which are parameterized reducible to $k$-WEIGHTED WEFT-$t$ MAJORITY CIRCUIT SATISFIABILITY.

Finally, before proceeding, we show that we can always assume that when dealing with majority circuits, we have an additional input whose value is always set to 1. The will prove handy later on in proving both Theorem 1 and Theorem 2.

**Observation 1.** *Without loss of generality, we can assume a weft-$t$ majority circuit $C$, $t \geq 1$, is provided with an input gate which is always assigned the value 1.*

The reason is the following. To simulate a 1 in $C$, we replace the output gate of $C$ with an in-degree 2 majority gate which is connected to the old output gate and to a new input gate. Let $C'$ be $C$ after this modification. Then $C$ has a weight $k$ satisfying assignment if and only if $C'$ has a weight $k + 1$ satisfying assignment where the new input gate is assigned a value of 1.

## 3   W[1] $\subseteq$ $\widetilde{W}[1]$

In this section we prove the first part of the main result of this paper, namely that W[1] $\subseteq$ $\widetilde{W}[1]$. For this, we introduce an intermediate problem which we feel might also be of independent interest, the $k$-MAJORITY VERTEX COVER problem. After formally defining $k$-MAJORITY VERTEX COVER, we prove that it is in $\widetilde{W}[1]$, and also W[1]-hard. From this, it will immediately imply that W[1] $\subseteq$ $\widetilde{W}[1]$. We begin with a definition of $k$-MAJORITY VERTEX COVER.

**Definition 1.** Given a graph $G$ and a parameter $k$, the $k$-MAJORITY VERTEX COVER problem asks whether there exists a subset of $k$ vertices in $G$ which covers a majority of the edges of $G$. That is, whether there exists a $S \subseteq V(G)$ with $|S| = k$ and $|\{\{u, v\} \in E(G) \mid \{u, v\} \cap S \neq \emptyset\}| > |E(G)|/2$.

Recall that a problem is in $\widetilde{W}[1]$ if it can be parameterically reduced to $k$-WEIGHTED WEFT-1 MAJORITY CIRCUIT SATISFIABILITY, which is the problem of determining whether a purely majority circuit of weft 1 (and constant depth) has a weight $k$ satisfying assignment. The above definition, gives a clue to why $k$-MAJORITY VERTEX COVER is parametric reducible to $k$-WEIGHTED WEFT-1 MAJORITY CIRCUIT SATISFIABILITY. This is established in the following lemma.

**Lemma 1.** $k$-MAJORITY VERTEX COVER *is in* $\widetilde{W}[1]$.

*Proof.* Let $(G, k)$ be an instance of $k$-MAJORITY VERTEX COVER. We reduce $(G, k)$ to a weft-1 purely majority circuit $C$ which has a $k + 1$ weighted satisfying assignment if and only if $G$ has a subset of $k$ vertices that cover a majority of its edges.

The construction of $C$ is as follows. Let $n = |V(G)|$ and $m = |E(G)|$. The input layer of $C$ consists of $n$ input gates – one input gate $x_v$ for each vertex $v \in V(G)$. In addition, we use the construction of Observation 1 to ensure that the input layer has an additional input gate of value 1. The first layer of $C$ consists of $m$ small in-degree 3 majority gates, one for each edge in $G$. The gate associated with the edge $\{u, v\} \in E(G)$, is connected to $x_u$, $x_v$, and the constant 1. In this way, an assignment to the input layer corresponds to a subset of vertices in $G$, and a gate in the first layer is satisfied if and only if the edge associated with this gate is incident to a vertex selected by the assignment. The second layer consists of the output gate which is large majority gate which is connected to all the small majority gates of the first layer.

The above construction clearly runs in FPT-time, and the circuit constructed is of weft 1 and depth 2. Furthermore, its correctness can easily be verified. To see this, first assume that $G$ has a subset $S$ of $k$ vertices that cover a majority of its edges, and consider the assignment $X$ which assigns $x_v = 1$ to each gate $x_v$ with $v \in S$. Then $X$ is a weight $k$ assignment which satisfies, by construction, any gate in the first layer associated to an edge covered by $S$. Therefore, it satisfies a majority of the gates in the first layer, and also the output gate. In the other direction, if $X$ is a weight $k$ assignment which satisfies $C$, then $X$ satisfies more than half of the gates in the first layer, and so the subset of $k$ vertices $S := \{v \in V(G) \mid X \text{ assigns } x_v = 1\}$ covers more than half of the edges of $G$. $\qquad\square$

The next step is to show that $k$-MAJORITY VERTEX COVER is W[1]-hard. This might be a somewhat more surprising result than the previous lemma, since $k$-VERTEX COVER and other closely related variants are known to be in FPT [GNW05]. Our proof follows along a similar line of proof used in [GNW05] for showing that $k$-PARTIAL VERTEX COVER is W[1]-hard.

**Lemma 2.** *$k$-MAJORITY VERTEX COVER is* W[1]-*hard.*

*Proof.* We reduce the W[1]-complete parameterized independent set problem $k$-INDEPENDENT SET to $k$-MAJORITY VERTEX COVER.

Let $(G, k)$ be an instance of $k$-INDEPENDENT SET. We may assume that $k + 1 < n/4$, since otherwise the trivial brute-force algorithm runs in fpt-time. We construct an equivalent instance $(G', k')$ of $k$-MAJORITY VERTEX COVER as follows. Let $n = |V(G)|$ and for $v \in V(G)$ let $d(v)$ denote the number of vertices adjacent to $v$ in $G$.

The graph $G' = (V(G'), E(G'))$ is constructed from $G$ in two steps. First, for every $v \in V(G)$, we add $n - 1 - d(v)$ new vertices, which are connected only to $v$ and hence have degree one. This ensures that $v$ has degree exactly $n - 1$. Let $m_0$ be the number of edges of the graph obtained so far. Clearly

$$m_0 \geq \frac{n \cdot (n-1)}{2}. \tag{1}$$

We choose the smallest $s \in \mathbb{N}$ such that

$$k \cdot (n-1) + s > \frac{m_0 + s}{2}. \tag{2}$$

In the second step we add a vertex $v^*$ and $s$ further vertices to our graph and make $v^*$ adjacent to the $s$ many new vertices, which thus all have degree one. This finishes the construction of the graph $G'$. Note that $G'$ has $m_0 + s$ many edges. We set $k' := k + 1$. We show that $G$ contains an independent set of size $k$ if and only if $G'$ contains a set of size $k'$ covering a majority of its edges.

Assume first that $S$ is an independent set of $G$ of size $k$. Then $S \cup \{v^*\}$ covers $k \cdot (n-1) + s$ edges in $G'$, which by (2) is more than half of the edges of $G'$.

Conversely, let $S'$ be a subset of $k'$ vertices in $G'$, which cover more than $(m_0 + s)/2$ edges in $G'$. Then $v^* \in S'$, since all other vertices have degree at most $n-1$ and therefore at most $(k+1) \cdot (n-1)$ edges can be covered otherwise. However, as $k + 1 < n/4$ we get $(k+1) \cdot (n-1) < n \cdot (n-1)/4 \leq m_0/2$ (the last inequality holding by (1)); therefore, at most half of the edges would be covered. We set $S := S' \setminus \{v^*\}$. Thus $|S| = k$ and by the choice of $s$ the set $S$ must cover in $G'$ at least $k \cdot (n-1)$ edges. As vertices in $V(G)$ have degree $n-1$ in $G'$ and vertices in $V(G') \setminus (V \cup \{v^*\})$ have degree one, we see that $S \subseteq V(G)$. Moreover, in order to cover $k \cdot (n-1)$ edges, $S$ must be an independent set of $G$. □

**Corollary 1.** $W[1] \subseteq \widetilde{W}[1]$.

## 4   $\widetilde{W}[1] \subseteq W[1]$

We next prove that $\widetilde{W}[1] \subseteq W[1]$, completing the proof of the main result of this paper. In the interests of a clearer presentation, we prove this in two steps. In the first step, we introduce for positive integers $p, q \in \mathbb{N}$ a new problem called $k$-MAJORITY $(p, q)$-DNF SATISFACTION, and show that this problem reduces to the generic W[1]-complete $k$-STEP HALTING problem. Following this, we explain how this construction can be altered to show that $k$-WEIGHTED WEFT-1 MAJORITY CIRCUIT SATISFIABILITY also reduces to $k$-STEP HALTING, proving that $\widetilde{W}[1] \subseteq W[1]$.

We begin by introducing the $k$-MAJORITY $(p, q)$-DNF SATISFACTION problem. A *monotone $(p, q)$-DNF formula* is a Boolean formula of the form $\bigvee_{i=1}^{p} \bigwedge_{j=1}^{q} x_{i,j}$, where the $x_{i,j}$'s are Boolean variables which are not necessarily distinct. A *family of monotone $(p, q)$-DNF formulas* is a tuple $\mathcal{D} = (D_1, \ldots, D_m)$ of monotone $(p, q)$-DNF formulas, which are not necessarily distinct.

**Definition 2.** For a given family $\mathcal{D} = (D_1, \ldots, D_m)$ of monotone $(p, q)$-DNF formulas, and a parameter $k$, the $k$-MAJORITY $(p, q)$-DNF SATISFACTION problem asks to determine whether there exists a weight $k$ assignment to the variables of the family such that for more than $m/2$ indices $i$ the DNF $D_i$ is satisfied.

We now show that $k$-MAJORITY $(p, q)$-DNF SATISFACTION parameterically reduces to the $k$-STEP HALTING. Recall that the $k$-STEP HALTING problem, which is the parameterized analog of the classical HALTING PROBLEM, asks for a given non-deterministic single tape Turing machine (defined over an unbounded alphabet) $M$, and a parameter $k$, whether $M$ has a computation path on the empty string which halts after at most $k$ steps. The main idea of our proof for showing $\widetilde{W}[1] \subseteq W[1]$ is encapsulated in the following reduction.

**Lemma 3.** *There is a parameterized reduction from $k$-MAJORITY $(p, q)$-DNF* SATISFACTION *to $k$-STEP HALTING.*

*Proof.* Let $(\mathcal{D}, k)$ be a given instance for $k$-MAJORITY $(p, q)$-DNF SATISFACTION, with $\mathcal{D} = (D_1, \ldots, D_m)$ a family of monotone $(p, q)$-DNFs over variables $x_1, \ldots, x_n$, and $k$ the parameter. We construct a Turing machine $M$ such that for some appropriate computable function $f$ the machine $M$ halts in at most $k' := f(k)$ steps if and only if there exists a weight $k$ assignment to the variables $x_1, \ldots, x_n$ which satisfies more than half of the DNFs in $\mathcal{D}$. Then main idea is to encode in the state space of $M$ all relevant information needed for determining the number of DNFs satisfied by a given weight $k$ assignment. For ease of notation, we will assume that no DNF contains a conjunct with identical literals.

Consider some DNF $D_i \in \mathcal{D}$. There are $p$ possible choices of selecting $q$-conjuncts of variables to assign a 1 to, so as $D_i$ would be satisfied. We say that a subset of $q$ variables $X \subset \{x_1, \ldots, x_n\}$ *hits* the index $i$ if it is one of these possible choices. Moreover, we define the *neighborhood* $D(X)$ of $X$ to be the set of all indices $1 \leq i \leq m$ that it hits.

The information encoded in the state space of $M$ is as follows: For every $\ell$ subsets of variables $X_1, \ldots, X_\ell \subset \{x_1, \ldots, x_n\}$, $1 \leq \ell \leq p$ and $|X_1| = \cdots = |X_\ell| = q$, we encode the size of the intersection of their neighborhoods. That is, we encode $|D(X_1) \cap \cdots \cap D(X_\ell)|$, the number of indices hit by each of the $\ell$ subsets of variables. This requires $\mathcal{O}(n^{pq})$ state space, which is polynomial in $n$.

We next describe the computation of $M$ on the empty string in three different *phases*:

1. First, $M$ nondeterministically guesses a subset of $k$ variables.
2. Next, $M$ identifies all $q$-subsets of variables that are implicity selected by its $k$ variable guess in the previous step.
3. Finally, $M$ calculates the total number of indices $1 \leq i \leq m$ hit by all the $q$-subsets identified in the previous step. $M$ halts, if this number is more than $m/2$. Otherwise, $M$ enters an infinite loop.

Due to its construction, $M$ halts in at least one of its computation paths on the empty string if and only if there is a weight $k$ assignment that satisfies more than half of the DNF occurrences in $\mathcal{D}$. To complete the proof, we argue that $M$ halts after $k' = f(k)$ steps for some computable function $f$. The first phase requires $k$ nondeterministic steps since we can encode each variable by a single letter in the alphabet of $M$. In the second step, $M$ identifies $\mathcal{X} = X_1, \ldots, X_{|\mathcal{X}|}$, the set of all $q$-subsets of variables implicity selected by its $k$ variable guess.

Since $|\mathcal{X}| = \mathcal{O}(k^q)$, this phase requires $\mathcal{O}(k^q)$ steps. To compute the last phase, $M$ performs an exclusion\inclusion calculation using the information stored in its state space. The crucial observation is

$$\left| \bigcup_{X_i \in \mathcal{X}} D(X_i) \right| = \sum_{j=1}^{|\mathcal{X}|} (-1)^{j+1} \cdot \sum_{i_1 < \cdots < i_j} |D(X_{i_1}) \cap \cdots \cap D(X_{i_j})|$$

$$= \sum_{j=1}^{p} (-1)^{j+1} \cdot \sum_{i_1 < \cdots < i_j} |D(X_{i_1}) \cap \cdots \cap D(X_{i_j})|.$$

The second equation is due to the fact that any family of neighborhoods of more than $p$ many $q$-subsets of variables has empty intersection. Thus the information stored in the state space of $M$ suffices for this computation. The last phase therefore requires $\mathcal{O}(k^{pq})$ steps. □

We now turn to deal with weft-1 majority circuits, and our $\widetilde{W}[1]$-complete $k$-WEIGHTED WEFT-1 MAJORITY CIRCUIT SATISFIABILITY problem. We show that a similar construction used in the lemma above can be applied for reducing $k$-WEIGHTED WEFT-1 MAJORITY CIRCUIT SATISFIABILITY to $k$-STEP HALTING.

Let $C$ be a weft-1 majority circuit. First, we normalize $C$ so that its first layer consists only of in-degree $q$ ∨-gates, and its second layer consists only of in-degree $p$ ∧-gates for certain constants $p$ and $q$. This can be done as follows. Consider a large gate in $C$ and the portion of $C$ which is required to evaluate one of its inputs. Since this portion involves only small gates and has constant depth, it can be viewed as Boolean function over a constant number of variables. Also, this function is necessarily monotone, since majority gates can only compute monotone functions. We can therefore analyze its entire truth table, and convert it into a DNF using a disjunction of satisfying lines in the truth table in the straightforward manner. Since the function is monotone, any variable appearing in negation in some conjunct of the DNF, also appears positively in another conjunct with similar literals, and so both of these occurences can safely be removed. From this it follows that any portion of $C$ which is required to evaluate one of the inputs of one of its large gates can be modeled by a monotone DNF of constant size. Letting $p$ and $q$ be the largest disjunction and conjunct in all these DNFs, by an appropriate padding we model each such portion by a $(p, q)$-DNF.

Let us say that a weft-1 majority circuit is *simple* if it has only one big majority gate as its output gate. If $C$ is simple then the construction above suffices, and we can immediately apply the construction of $M$ used in Lemma 3. Otherwise, $C$ is logically equivalent to a constant size Boolean combination of simple sub-circuits. In this case, $M$ guesses $k$ variables, and computes the value of these constantly many simple sub-circuits as described in Lemma 3. It then computes in constant time the value of the Boolean combination under this assignment given by the computed values of the simple sub-circuits.

**Corollary 2.** $\widetilde{W}[1] \subseteq W[1]$.

Combining Corollaries 1 and 2, we complete our proof of Theorem 1.

## 5    Higher Levels of the Hierarchies

We now turn to consider higher levels of the W-and $\widetilde{W}$-hierarchies. We prove in this section that $W[t] \subseteq \widetilde{W}[t]$ for all positive integers $t$. For this, we will first show that this statement holds for even values of $t$. Following this, we will consider odd values greater or equal to 3, and thus by combining with Theorem 1 we will obtain our desired result.

Recall that a circuit is monotone if it does not have any $\neg$-gates, and that $k$-WEIGHTED WEFT-$t$ MONOTONE CIRCUIT SATISFIABILITY is complete for all even $t \geq 2$. To show that $W[t] \subseteq \widetilde{W}[t]$ for even values of $t$, we prove the following lemma.

**Lemma 4.** $k$-WEIGHTED WEFT-$t$ MONOTONE CIRCUIT SATISFIABILITY *parameterically reduces to* $k$-WEIGHTED WEFT-$t$ MAJORITY CIRCUIT SATISFIABILITY.

*Proof.* Let $(C, k)$ be an instance of $k$-WEIGHTED WEFT-$t$ MONOTONE CIRCUIT SATISFIABILITY, with $C$ a weft-$t$ monotone logical circuit and $k$ the parameter. Let $\ell$ be the maximum in-degree of any gate in $C$. We assume without loss of generality that any small gate in $C$ has in-degree at most 2.

We construct a majority circuit $C'$ of weft $t$ as follows. First, we add $\ell \cdot (k+1) - 1$ new input gates to $C'$ labeled with new pairwise distinct variables, and an additional new input gate labeled with the constant 1 (which we construct according to Observation 1). $C'$ simulates the gates of $C$ as follows. The simulation of a large $\vee$-gate, say of in-degree $\ell' \leq \ell$, is straightforward: relabel so it becomes a majority gate and add $\ell' - 1$ new edges coming from the input gate labeled 1. Small $\vee$-gates can be handled similarly. Small $\wedge$-gates are simply relabeled to become small majority gates. The interesting case is what to do when $g$ is a large $\wedge$-gate. Suppose $g$ has edges coming from $g_1, \ldots, g_{\ell'}$ (for some $\ell' \leq \ell$). We relabel $g$ to a majority gate. Then for all $1 \leq i \leq \ell'$, we replace each edge from $g_i$ to $g$ by $k + 1$ parallel edges. Additionally we wire $\ell'(k + 1) - 1$ many new inputs to $g$.

We show that for each gate $g$ in $C$ that an arbitrary weight $k$ assignment for the variables of $C'$ satisfies $g$ in $C'$ if and only if its restriction to the variables in $C$ satisfies $g$ in $C$. We proceed inductively on the layer $g$ lives in (in $C$ or $C'$). The rest being easy look at the case that $g$ is a large $\wedge$-gate. Then an assignment satisfying all $g_1, \ldots, g_{\ell'}$ clearly satisfies the majority gate $g$. Conversely if a weight $k$ assignment does not satisfy all $g_1, \ldots, g_{\ell'}$ then $g$ receives at most $(\ell' - 1)(k + 1)$ times a 1 from these gates plus possibly some from the new variables, but at most $k$. All together, $g$ receives no more than $(\ell' - 1)(k + 1) + k$ values 1 and this is less than half the in-degree of $g$ which is $2\ell'(k + 1) - 1$. In total we have that an assignment of weight $k$ to the variables of $C'$ satisfies $C'$ if and only if its restriction to the variables of $C$ satisfies $C$. It follows that $C$ has a satisfying assignment to its variables of weight $k$ if and only if $C'$ has a satisfying assignment to its variables of weight $k$. Why? To see necessity extend an assignment for the variables of $C$ by setting all new variables to 0 and use the above equivalence. For sufficiency, note that a satisfying weight $k$ assignment for $C'$ restricts to one of weight at most $k$ which satisfies $C$ by the above equivalence; but $C$ is monotone, so it has a satisfying assignment of weight $k$ if and only if it has a satisfying assignment of weight at most $k$.    □

**Corollary 3.** $W[t] \subseteq \widetilde{W}[t]$ *for any positive even integer* $t$.

We next show that $W[t] \subseteq \widetilde{W}[t]$ for odd values of $t$. For this, due to Theorem 1, it is enough to consider odd $t \geq 3$. In this case, we can consider a restricted type of antimonotone circuits which we call normalized antimonotone circuits. A *normalized antimonotone circuit* is an antimonotone circuit with its first layer containing only ¬-gates, its second layer containing only large ∧-gates, its third only large ∨-gates, and so on, alternating between layers of large ∧-gates and layers of large ∨-gates. Also, each gate is required to have incoming edges only from gates in the previous layer. It is known that for odd $t \geq 3$, $k$-WEIGHTED WEFT-$t$ CIRCUIT SATISFIABILITY restricted to normalized antimonotone circuits is complete for $W[t]$ [FG06].

**Lemma 5.** $k$-WEIGHTED WEFT-$t$ ANTIMONOTONE CIRCUIT SATISFIABILITY *restricted to normalized antimonotone circuits parameterically reduces to* $k$-WEIGHTED WEFT-$t$ MAJORITY CIRCUIT SATISFIABILITY.

*Proof.* Let $(C, k)$ be an instance of $k$-WEIGHTED WEFT-$t$ ANTIMONOTONE CIRCUIT SATISFIABILITY restricted to normalized antimonotone circuits. We transform $C$ to a majority circuit as follows: consider a large ∧-gate $g$ of the second layer, *i.e.* one wired only to negations of input gates, say $\neg x_1, \ldots, \neg x_\ell$. Relabel $g$ to become a majority gate and wire it to all inputs except those labeled $x_1, \ldots, x_\ell$, say this yields $\ell'$ edges.

We shall need that $\ell' \geq 2k$. We can assure this by the following preprocessing which is easily seen to run in FPT-time: that $\ell' < 2k$ means that $g$ conjuncts the negations of all but less than $2k$ many variables. We check for each weight $k$ assignment for these $< 2k$ variables whether it satisfies $C$. The number of these assignments is effectively bounded in terms of $k$. If we find a satisfying assignment, then $(C, k)$ is a "yes" instance. Otherwise we know that no weight $k$ assignment that satisfies $C$ also satisfies $g$. We then delete $g$. It is not hard to see that the resulting circuit is equivalent to $C$ with respect to weight $k$ assignments.

We add $\ell' - 2k + 1$ many parallel edges coming from a new input labeled with the constant 1 – this is the smallest number $s$ such that $k + s$ is bigger than half of $\ell' + s$. This means that this gate is satisfied by a weight $k$ assignment $X$ if and only if $X$ chooses $k$ variables whose negations are not wired into $g$ in $C$, *i.e.* $X$ satisfies $g$ in $C$. But if we replace all second layer gates in this manner, we end up with a *monotone* circuit containing logical as well as majority gates which is equivalent to the original circuit with respect to weight $k$ assignments. Hence, for the other gates we can proceed as in Lemma 4.    □

**Corollary 4.** $W[t] \subseteq \widetilde{W}[t]$ *for any positive odd integer* $t \geq 3$.

Combining Corollaries 1, 3, and 4, we complete our proof of Theorem 2.

## 6  Discussion

In this paper we presented an alternative characterization of W[1], using majority circuits instead of logical circuits. We also showed that this characterization holds in one direction for higher levels of the hierarchy. This gives the first monotone characterization of W[1], and is perhaps a first step in establishing a monotone characterization of the entire W-hierarchy. We believe our results may prove useful in showing fixed-parameter intractability results for monotone problems, as well as for other types of problems. Furthermore, our results exemplify the naturality of the notion of *weft*, showing that it remains the parameter governing the combinatorial nondeterministic computing strength of circuits, no matter what (nontrivial) type of gates they have.

The major open problem left by this paper is showing the other direction of Theorem 2, namely that $\widetilde{W}[t] \subseteq W[t]$ for all positive integers $t$. This, along with the results in this paper will prove that $W[t] = \widetilde{W}[t]$, giving a completely monotone characterization of the W-hierarchy. We conjecture that this is in fact the case.

**Acknowledgements.** Jörg Flum as well as some careful anonymous reviews pointed out a mistake in an earlier argument for Lemma 2.

## References

[DF99]     Downey, R.G., Fellows, M.R.: Parameterized Complexity. Springer, Heidelberg (1999)

[DF95]     Downey, R., Fellows, M.: Fixed parameter tractability and completeness II: Completeness for $W[1]$. Theoretical Computer Science A 141, 109–131 (1995)

[DFKHW94] Downey, R., Fellows, M., Kapron, B., Hallett, M., Wareham, H.T.: The parameterized complexity of some problems in logic and linguistics. In: Matiyasevich, Y.V., Nerode, A. (eds.) LFCS 1994. LNCS, vol. 813, pp. 89–100. Springer, Heidelberg (1994)

[DFS99]    Downey, R., Fellows, M., Stege, U.: Parameterized complexity: A framework for systematically confronting computational intractability. In: Graham, R., Kratochvil, J., Nesetril, J., Roberts, F. (eds.) Proceedings of the DIMACS-DIMATIA Workshop on the Future of Discrete Mathematics, Prague. Contemporary Trends in Discrete Mathematics 1997, AMS-DIMACS. Discrete Mathematics and Theoretical Computer Science, vol. 49, pp. 49–99 (1999)

[FG06]     Flum, J., Grohe, M.: Parameterized Complexity Theory. Springer, Heidelberg (2006)

[FFHMR07]  Fellows, M., Flum, J., Hermelin, D., Müller, M., Rosamond, F.: Parameterized complexity via combinatorial circuits (manuscript, 2007)

[GNW05]    Guo, J., Niedermeier, R., Wernicke, S.: Parameterized complexity of generalized vertex cover problems. In: Dehne, F., López-Ortiz, A., Sack, J.-R. (eds.) WADS 2005. LNCS, vol. 3608, pp. 36–48. Springer, Heidelberg (2005)

[M06]      Marx, D.: Parameterized complexity of independence and domination on geometric graphs. In: Bodlaender, H.L., Langston, M.A. (eds.) IWPEC 2006. LNCS, vol. 4169, pp. 154–166. Springer, Heidelberg (2006)

# Parameterized Complexity and Approximability of the SLCS Problem

S. Guillemot

LIFL/CNRS/INRIA, Bât. M3 Cité Scientifique, 59655,
Villeneuve d'Ascq cedex, France
Sylvain.Guillemot@lifl.fr

**Abstract.** We introduce the LONGEST COMPATIBLE SEQUENCE (SLCS) problem. This problem deals with *p-sequences*, which are strings on a given alphabet where each letter occurs at most once. The SLCS problem takes as input a collection of $k$ p-sequences on a common alphabet $L$ of size $n$, and seeks a p-sequence on $L$ which respects the precedence constraints induced by each input sequence, and is of maximal length with this property. We investigate the parameterized complexity and the approximability of the problem. As a by-product of our hardness results for SLCS, we derive new hardness results for the LONGEST COMMON SUBSEQUENCE problem and other problems hard for the W-hierarchy.

## 1 Introduction

The comparison of several sequences is an important task in several fields such as computational biology, pattern recognition, scheduling, data compression and data mining. Starting with [17], the computational complexity of several consensus problems on sequences has been investigated. As it was later realized, a natural framework to conduct these studies is the theory of parameterized complexity [7,19,13].

In this article, we initiate the study of a new consensus problem on collections of sequences, motivated by applications to the comparison of gene orders, and applications to the rank aggregation problem. The problem we introduce is called LONGEST COMPATIBLE SEQUENCE, and is abbreviated by SLCS. This problem deals with *p-sequences* [12]: we call *p-sequence* on an alphabet $L$ a string over $L$ where each letter occurs at most once. Given a collection $\mathcal{C} = \{s_1, ..., s_k\}$ of p-sequences on a common alphabet $L$ of size $n$, the LONGEST COMPATIBLE SEQUENCE problem seeks a longest *compatible sequence* for $\mathcal{C}$, which is a p-sequence $s$ on $L$ respecting the precedence constraints induced by each input sequence. We also consider the complementary minimization problem, denoted by CSLCS, where the goal is to minimize the number of labels missing from a compatible sequence. In addition of studying the approximability of these optimization problems, we also investigate the parameterized complexity of their natural parameterizations. The corresponding parameters are denoted by $q$ for SLCS, and by $p$ for CSLCS.

M. Grohe and R. Niedermeier (Eds.): IWPEC 2008, LNCS 5018, pp. 115–128, 2008.

We now discuss two potential applications of the problem. The first application is the *comparison of gene orders*. Identifying conservation among gene orders for several organisms is an important issue in bioinformatics, since it helps for gene prediction and can also be used in phylogenetic reconstruction as an alternative to DNA sequence analysis (see [18] for a survey). Given a set of $k$ organisms $S$ for which we have identified a set of $n$ homologous genes $\mathcal{G}$, each organism can be described by its gene order, which is a p-sequence on $\mathcal{G}$. Therefore, seeking a largest gene order which is conserved among all organisms under study amounts to seek a largest compatible sequence for the collection $\mathcal{C} = \{s_1, ..., s_k\}$ on the alphabet $\mathcal{G}$. The second application is the *aggregation of incomplete rankings*. This is a variant of the well-studied problem of rank aggregation, which consists in merging $k$ complete rankings of a same set of $n$ elements in a single ranking (see [16] for a survey). Rank aggregation is useful in several situations, such as combining answers from search engines [8,9] or searching similarities in databases [11]. This framework can be extended to handle incomplete rankings, which are rankings defined only on a subset of the elements. Here, the SLCS problem can be used to find a largest subset of the elements on which all the rankings agree.

Our results are as follows. On the positive side, we give polynomial, approximation, and FPT algorithms in Section 3. Among other results, we show that SLCS can be solved in $O(kn^k)$ time, and we give a $k$-approximation algorithm and a $O^*(k^p)$ FPT algorithm for the CSLCS problem. On the negative side, we describe parameterized intractability results in Section 4. We show W[1]-completeness of SLCS parameterized in $(q, k)$, and we show WNL-completeness of SLCS parameterized in $k$. The latter result relies on the definition of a new parameterized complexity class WNL, and turns out to have important consequences for the parameterized complexity of other problems. Namely, we show in Section 5 that this result implies WNL-hardness results for the LCS problem parameterized in the number of sequences, and for other problems previously classified as "hard for the W-hierarchy" [3,4,1,2]. Thus, the thesis of [3], which proposes to consider hardness for the W-hierarchy as a new intractability measure, can be recast with the more natural notion of WNL-hardness.

## 2   Definitions

We begin with definitions related to *sequences*. After [12], we call *p-sequence* (or simply *sequence*) on $L$ a string $s$ on $L$ where each letter occurs at most once. The letters are called *labels*, and the set of letters appearing in $s$ is called the *label set* of $s$, and is denoted by $L(s)$. The $i$th letter of $s$ is denoted by $s[i]$, and the length of $s$ is denoted by $|s|$. Given two elements $x, y \in L(s)$, $x <_s y$ means that $x$ precedes $y$ in $s$. Given $x \in L(s)$, we denote by $pred_s(x)$, resp. $succ_s(x)$, the predecessor, resp. successor, of $x$ in $s$, or $\perp$ if no such label exists. Given a set $L' \subseteq L$, the *restriction* of $s$ to $L'$, denoted by $s|L'$, is the subsequence obtained from $s$ by keeping labels in $L'$. Two sequences $s, s'$ *agree* iff $s|L(s') = s'|L(s)$. Equivalently, they must verify: for each $x, y \in L(s) \cap L(s')$, $x <_s y$ iff $x <_{s'} y$.

We now give definitions related to *collections* and *compatible sequences*. Given a set $L$ of $n$ elements, a *collection on* $L$ is a family $\mathcal{C} = \{s_1, ..., s_k\}$ of sequences on $L$; in the following, we will assume that each label of $L$ appears in at least one sequence, i.e. $L = \cup_{i \in [k]} L(s_i)$. Given a set $L' \subseteq L$, the *restriction* of $\mathcal{C}$ to $L'$ is the collection $\mathcal{C}|L'$ on $L'$ defined by: $\mathcal{C}|L' = \{s_1|L', ..., s_k|L'\}$. A *compatible sequence* for $\mathcal{C}$ is a sequence $s$ s.t. $L(s) \subseteq L$ and for each $i \in [k]$ the sequences $s, s_i$ agree. Equivalently, $s$ must verify: for each $x, y \in L(s)$, for each $i \in [k]$, if $x <_s y$ and $x, y \in L(s_i)$ then $x <_{s_i} y$. A *total compatible sequence* for $\mathcal{C}$ is a compatible sequence for $\mathcal{C}$ with label set $L$. $\mathcal{C}$ is said to be *compatible* iff it admits a total compatible sequence. A *conflict* in $\mathcal{C}$ is a set $L' \subseteq L$ s.t. $\mathcal{C}|L'$ is not compatible.

We now introduce definitions related to the SLCS problem. The SLCS problem asks: given a collection $\mathcal{C}$ on $L$, find a maximum cardinality subset $L' \subseteq L$ s.t. $\mathcal{C}|L'$ is compatible; observe that this is equivalent to seeking a longest compatible sequence for $\mathcal{C}$. We denote by $\text{SLCS}(\mathcal{C})$ the size of such an optimal solution. The complementary CSLCS problem asks: given a collection $\mathcal{C}$ on $L$, find a minimum cardinality subset $L' \subseteq L$ s.t. $\mathcal{C}|(L \backslash L')$ is compatible. To simplify notations, we also view these optimization problems as decision problems, i.e. we also denote by SLCS the decision problem which takes a collection $\mathcal{C}$ and an integer $q$, and asks if $\text{SLCS}(\mathcal{C}) \geq q$; similarly for CSLCS with the parameter $p$, asking if $\text{CSLCS}(\mathcal{C}) \leq p$. We use a bracket notation to denote parameterizations of the problem, e.g. $\text{SLCS}[q, k]$ stands for SLCS parameterized by the pair of parameters $(q, k)$.

The following definitions will be useful in Section 3. Consider a collection $\mathcal{C} = \{s_1, ..., s_k\}$ on $L$. For each $i$, we define $L^\top(s_i) = L(s_i) \cup \{\top\}$, and we extend $<_{s_i}$ to $L^\top(s_i)$ s.t. $x <_{s_i} \top$ for each $x \in L(s_i)$. A *position in* $\mathcal{C}$ is a tuple $\pi = (x_1, ..., x_k)$, where $x_i \in L^\top(s_i)$. The *initial position* is $\pi_\perp = (s_1[1], ..., s_k[1])$. The *final position* is $\pi_\top = (\top, ..., \top)$. The *index set* of a position $\pi$ is $I(\pi) = \{i \in [k] : \pi[i] \neq \top\}$. The *label set* of $\pi$ is $S(\pi) = \{\pi[i] : i \in I(\pi)\}$. Given a position $\pi$ in $\mathcal{C}$ and a label $x \in L$, we say that $x$ is *allowed* at $\pi$ iff for each $i \in [k]$ s.t. $x \in L(s_i)$, we have: $\pi[i] \leq_{s_i} x$. We denote by $L(\pi)$ the set of labels allowed at $\pi$. We define an order relation $\leq_{\mathcal{C}}$ on positions in $\mathcal{C}$ as follows: given $\pi, \pi'$ positions in $\mathcal{C}$, $\pi \leq_{\mathcal{C}} \pi'$ iff $\pi[i] \leq_{s_i} \pi'[i]$ for each $i \in [k]$. We also define two notions of *successor positions*. Let $\pi$ be a position in $\mathcal{C}$. Given $i \in I(\pi)$, we define $succ_i(\pi)$ as the position $\pi'$ s.t. (i) $\pi'[i] = succ_{s_i}(\pi[i])$, (ii) $\pi'[j] = \pi[j]$ for each $j \neq i$. Given $a \in S(\pi)$, we define $succ_a(\pi)$ as the position $\pi'$ s.t. (i) $\pi'[i] = succ_{s_i}(\pi[i])$ if $\pi[i] = a$, (ii) $\pi'[i] = \pi[i]$ otherwise.

# 3  Algorithmic Results for SLCS

We present polynomial-time and approximation algorithms for the SLCS problem (Section 3.1), then we describe FPT and approximation algorithms for the complementary CSLCS problem (Section 3.2).

## 3.1  Algorithms for SLCS

We first describe a polynomial-time solvable case. Given a collection $\mathcal{C} = \{s_1, ..., s_k\}$ on $L$, say that $\mathcal{C}$ is *complete* iff each label of $L$ occurs in every sequence;

say that $C$ is *precomplete* iff each label of $L$ occurs either in one sequence or in every sequence. We show that SLCS is efficiently solvable for such collections.

**Proposition 1.** SLCS *can be solved in* $O(kn^2)$ *time if the input is a complete (or precomplete) collection.*

*Proof.* If $C = \{s_1, ..., s_k\}$ is a complete collection on $L$, consider the acyclic digraph $G$ with vertex set $L$, and which contains an arc $(x, y)$ whenever $x <_{s_i} y$ for each $i \in [k]$. Then SLCS($C$) can be computed in $O(kn^2)$ time as the size of a longest directed path of $G$. If $C$ is a precomplete collection on $L$, let $L'$ be the set of labels which occur in every sequence, and let $L''$ be the other labels. We have SLCS($C$) = $|L''| +$ SLCS($C|L'$): indeed, if $s$ is a compatible sequence for $C|L'$, then for each $i \in [k]$ we can insert in $s$ the labels of $L'' \cap L(s_i)$, by respecting their relative order in $s_i$, as well as their order w.r.t. those elements of $L'$ which appear in $s$ and $s_i$; we then obtain a compatible sequence for $C$ of length $|s| + |L''|$. Since $C|L'$ is complete, SLCS($C|L'$) can be computed in $O(kn^2)$ time, hence SLCS($C$) can be computed in similar time bounds. □

The previous result yields an approximation algorithm for the general SLCS problem:

**Proposition 2.** SLCS *can be* $2^k$-*approximated in* $O(kn^2)$ *time.*

*Proof.* We use approximation via partitioning [14]. Let $C = \{s_1, ..., s_k\}$ be a collection on $L$. For each $X \subseteq [k]$ we define $L_X$ as the set of labels $x \in L$ which occurs exactly in the sequences $s_i$ for each $i \in X$. Clearly, the sets $L_X$ ($X \subseteq [k]$) form a partition of $L$. We define $n_X = |L_X|$ and $C_X = C|L_X$. Then $C_X$ is a complete collection, and by Proposition 1 we can solve SLCS on the instance $C_X$ in $O(|X|n_X^2)$ time. Consider the algorithm which computes a longest compatible sequence of $C_X$ for each $X \subseteq [k]$, and returns the longest of these sequences. This algorithm has running time $\sum_{X \subseteq [k]} O(|X|n_X^2) = O(kn^2)$, and is easily seen to be a $2^k$-approximation algorithm for SLCS. □

Finally, we show that SLCS can be solved in polynomial time for fixed $k$, using dynamic programming:

**Proposition 3.** SLCS *can be solved in* $O(kn^k)$ *time and* $O(n^k)$ *space.*

*Proof.* Given $\pi$ position in $C$, let SLCS($\pi$) denote the size of a longest compatible sequence of $C|L(\pi)$. We denote by $F(\pi)$ the set of *full* elements of $S(\pi)$, i.e. the elements $a \in S(\pi)$ s.t. for each $i \in [k]$, $a \in L(s_i) \Rightarrow a = \pi[i]$. Given two positions $\pi, \pi'$ and $a \in L(C)$, $\pi \rightarrow_a \pi'$ holds if $a \in F(\pi)$ and $\pi' \geq_C succ_a(\pi)$. A $\pi$-*chain* is a chain $\pi_1 \rightarrow_{a_1} \pi_2 \rightarrow_{a_2} \cdots \rightarrow_{a_m} \pi_{m+1}$, with $\pi_1 \geq_C \pi$ and $\pi_{m+1} = \pi_\top$. The *length* of the chain is $m$. It can be shown that: **Claim.** SLCS($\pi$) is the length of a longest $\pi$-chain.

The above claim yields a dynamic-programming algorithm for solving SLCS in $O(kn^k)$ time. The algorithm computes, for each position $\pi$, the size SLCS($\pi$) of a longest $\pi$-chain, using the following recurrence relations:

- if $\pi = \pi_\top$, $\mathrm{SLCS}(\pi) = 0$;
- if $\pi \neq \pi_\top$, $\mathrm{SLCS}(\pi) = \max(\mathrm{SLCS}_1(\pi), 1 + \mathrm{SLCS}_2(\pi))$, where:

$$\mathrm{SLCS}_1(\pi) = \max\{\mathrm{SLCS}(succ_i(\pi)) : i \in I(\pi)\}$$
$$\mathrm{SLCS}_2(\pi) = \max\{\mathrm{SLCS}(succ_x(\pi)) : x \in F(\pi)\}$$

At the end of the algorithm, $\mathrm{SLCS}(\mathcal{C})$ is obtained as $\mathrm{SLCS}(\pi_\perp)$. Since there are $O(n^k)$ positions $\pi$, and since each value $\mathrm{SLCS}(\pi)$ can be computed in $O(k)$ time from the values $\mathrm{SLCS}(\pi')$ ($\pi' >_{\mathcal{C}} \pi$), the algorithm runs in the claimed time and space bounds. □

## 3.2 Algorithms for CSLCS

We first describe a simple characterization of compatible collections. Given a collection $\mathcal{C} = \{s_1, ..., s_k\}$ on $L$, we define the digraph $G(\mathcal{C})$ as follows: (i) its vertex set is $L$, (ii) it contains an arc $(x, y)$ whenever there exists $i \in [k]$ s.t. $x, y \in L(s_i)$ and $x <_{s_i} y$. The following proposition characterizes compatible collections, and relates the SLCS and CSLCS problems to problems on digraphs.

**Proposition 4.** *(i) $\mathcal{C}$ is compatible iff $G(\mathcal{C})$ is acyclic; (ii) $\mathrm{SLCS}(\mathcal{C})$ is the size of a maximum acyclic subgraph of $G(\mathcal{C})$, and $\mathrm{CSLCS}(\mathcal{C})$ is the size of a minimum directed feedback vertex set of $G(\mathcal{C})$.*

We thus have a reduction from the CSLCS problem to the DIRECTED FEEDBACK VERTEX SET (DFVS) problem. Known FPT and approximation algorithms for DFVS [21,10,6] yield the following results for CSLCS:

**Proposition 5.** *(i) CSLCS can be solved in $O(2^{O(p \log p)} k n^4)$ time; (ii) CSLCS can be $O(\log n \log \log n)$-approximated in polynomial time.*

We now describe faster FPT and approximation algorithms for the problem when $k$ is bounded. Though the problem is solvable in $O(k n^k)$ time in this case, in practical applications it may be preferable to have algorithms with running time linear in $n$. This is the case for the algorithms we present (Proposition 7). They rely on the following result:

**Proposition 6.** *Given a collection $\mathcal{C}$, in $O(kn)$ time we can decide if $\mathcal{C}$ is compatible, return a total compatible sequence in case of positive answer, or return a conflict of size at most $k$ in case of negative answer.*

*Proof.* Before describing the algorithm, let us introduce the following notations. Suppose that $\mathcal{C} = \{s_1, ..., s_k\}$ is a collection on $L$. Given $\pi$ position in $\mathcal{C}$, and $x \in L$, let $n_\pi(x)$ denote the number of indices $i \in [k]$ s.t. ($x \in L(s_i)$ and $\pi[i] <_{s_i} x$).

The algorithm maintains a position $\pi$ in $\mathcal{C}$, and for each $x \in L$ a counter $n_x$ equal to $n_\pi(x)$. Additionally, it maintains a sequence $s$, which is the prefix of an hypothetical compatible sequence for $\mathcal{C}$. We start with $s = \epsilon$, $\pi = \pi_\perp$ and each $n_x$ initialized to the number of $i \in [k]$ s.t. $x$ is a non-initial label of $s_i$. While

$\pi \neq \pi_\top$, we seek $x \in S(\pi)$ s.t. $n_x = 0$. If no such $x$ exists, then the algorithm answers "no" and returns $S(\pi)$. Otherwise, we choose such an $x$, and: (i) for each $i \in [k]$ s.t. $\pi[i] = x$, let $y$ be the successor of $x$ in $s_i$ (if it exists), we decrement $n_y$, (ii) we set $\pi \leftarrow succ_x(\pi)$, (iii) we set $s \leftarrow sx$. When $\pi = \pi_\top$ is reached, the algorithm answers "yes" and returns $s$.

The correctness of the algorithm follows from the fact that: (i) if it answers negatively by returning $S(\pi)$, then $S(\pi)$ is a conflict of size $\leq k$ between $\mathcal{C}$, (ii) if it answers positively by returning $s$, then $s$ is a total compatible sequence for $\mathcal{C}$. The running time is easily seen to be $O(kn)$, since the initialization takes $O(kn)$ time, and since in each step finding an $x \in S(\pi)$ s.t. $n_x = 0$ and updating $s, \pi$ and the counters takes $O(k)$ time. $\qquad\square$

As a consequence of Proposition 6, we obtain:

**Proposition 7.** *(i)* CSLCS *can be solved in* $O(k^p \times kn)$ *time; (ii)* CSLCS *can be $k$-approximated in* $O(kn)$ *time.*

## 4   Hardness Results for SLCS

We present two parameterized intractability results for the SLCS problem. The problem is shown W[1]-complete w.r.t. the parameters $q, k$ (Section 4.2) and WNL-complete w.r.t. the parameter $k$ (Section 4.3).

### 4.1   The Clases W[1] and WNL

Let us consider the following problem:

**Name:** NONDETERMINISTIC TURING MACHINE COMPUTATION (NTMC)
**Instance:** a nondeterministic Turing machine $M$, integers $q, k$
**Question:** does $M$ accept the empty string in $q$ steps by examining at most $k$ cells?

Given two parameterized problems $\Pi, \Pi'$, we recall that a *parameterized reduction* (or *fpt-reduction*) from $\Pi$ to $\Pi'$ is an algorithm which maps each instance $I = (x, k)$ of $\Pi$ to an instance $I' = (x', k')$ of $\Pi'$, s.t. (i) the algorithm runs in $f(k)|x|^c$ time for some function $f$ and some constant $c$; (ii) there exists a function $g$ s.t. $k' \leq g(k)$; (iii) $I$ is a positive instance of $\Pi$ iff $I'$ is a positive instance of $\Pi'$. Given a parameterized problem $\Pi$, we denote by $[\Pi]^{fpt}$ the set of problems $\Pi'$ fpt-reducible to $\Pi$. We then define the classes W[1] and WNL as follows.

**Definition 1.** W[1] $= [\text{NTMC}[q]]^{fpt}$, WNL $= [\text{NTMC}[k]]^{fpt}$.

Our definition of W[1] is consistent with the results of [5], and shows the parallel between the classes W[1] and WNL, corresponding respectively to time-bounded and space-bounded computations of a nondeterministic Turing machine.

## 4.2   Complexity w.r.t. $q, k$

We need the following notations. If $s = a_1...a_m$ is a sequence, its mirror image is $\tilde{s} = a_m...a_1$. If $(V, <_V)$ is a total order and $\{s_x : x \in V\}$ is a family of sequences with disjoint label sets, we denote by $\prod_{x \in (V, <_V)}$ their concatenation in the order $<_V$. If $V$ is the interval of integers $[p, q]$ and $<_V$ is the natural order on $\mathbb{N}$, then $\prod_{x \in (V, <_V)} s_x$ is abbreviated as $\prod_{i=p}^q s_i$.

**Proposition 8.** SLCS$[q, k]$ *is* W[1]-*complete.*

*Proof.* Membership in W[1] can be shown by reduction to NTMC$[q]$. W[1]-hardness is proved by a parameterized reduction from PARTITIONED CLIQUE [20]. Let $I = (G, k)$ be an instance of PARTITIONED CLIQUE, where $G = (V, E)$ is a $k$-partite graph with partition $V_1, ..., V_k$. For $i, j \in [k]$ $(i < j)$, let $E_{i,j}$ denote the set of edges of $G$ having one endpoint in $V_i$ and one in $V_j$.

The corresponding instance $I' = (\mathcal{C}, k', q')$ of SLCS$[q, k]$ is defined as follows. We set $k' = 2k + 2, q' = 2k + k(k-1)/2$. We define the following collection $\mathcal{C}$:

- Label set: we introduce labels $a[v], b[v]$ for each $v \in V$, and a label $c[e]$ for each $e \in E$.
- Sequences: we create two sequences $s, s'$ and $2k$ sequences $t_1, t'_1, ..., t_k, t'_k$. We want to enforce that:

  1. a compatible sequence of length $q'$ has the form

$$s = \left(\prod_{i=1}^k a[v_i]\right) \left(\prod_{i=1}^k \prod_{j=i+1}^k c[e_{i,j}]\right) \left(\prod_{i=1}^k b[v'_i]\right)$$

     with $v_i, v'_i \in V_i$ (for each $i \in [k]$), $e_{i,j} \in E_{i,j}$ (for each $i, j \in [k], i < j$);
  2. in addition, we have $v_i = v'_i$ for each $i \in [k]$, and $e_{i,j} = \{v_i, v_j\}$ for each $i, j \in [k], i < j$.

The sequences $s, s'$ will be *control sequences*, whose role is to enforce point 1. The sequences $t_i, t'_i$ will be *selection sequences*, whose role is to enforce point 2.

These sequences are defined as follows. Let $<_V$ be an arbitrary total order on $V$, and let $<_E$ be an arbitrary total order on $E$. We first define the following sequences:

$$\forall i \in [k], \quad A[i] = \prod_{v \in (V_i, <_V)} a[v], \quad B[i] = \prod_{v \in (V_i, <_V)} b[v],$$

$$\forall i, j \in [k](i < j), \quad C[i, j] = \prod_{e \in (E_{i,j}, <_E)} c[e]$$

We then define $s, s'$ as follows:

$$s = \left(\prod_{i=1}^{k} A[i]\right) \left(\prod_{i=1}^{k} \prod_{j=i+1}^{k} C[i,j]\right) \left(\prod_{i=1}^{k} B[i]\right)$$

$$s' = \left(\prod_{i=1}^{k} \widetilde{A[i]}\right) \left(\prod_{i=1}^{k} \prod_{j=i+1}^{k} \widetilde{C[i,j]}\right) \left(\prod_{i=1}^{k} \widetilde{B[i]}\right)$$

Suppose that $i, j \in [k], i < j$. Given $v \in V_i \cup V_j$, let $E_{i,j}(v)$ denote the set of edges of $E_{i,j}$ incident to $v$. Now, for each $i \in [k]$, we define $t_i, t_i'$ as follows:

$$t_i = \prod_{v \in (V_i, <_V)} a[v] \left(\prod_{j=i+1}^{k} \prod_{e \in (E_{i,j}(v), <_E)} c[e]\right) b[v]$$

$$t_i' = \prod_{v \in (V_i, >_V)} a[v] \left(\prod_{j=1}^{i-1} \prod_{e \in (E_{i,j}(v), <_E)} c[e]\right) b[v]$$

The reduction is clearly computable in polynomial time, and its correctness is ensured by the fact that: $G$ has a partitioned clique iff $\mathcal{C}$ has a compatible sequence of length $q'$.     □

### 4.3   Complexity w.r.t. $k$

While NTMC[$k$] is the canonical WNL-complete problem, we introduce other complete problems for the purpose of this section. In the following, by a $q \times k$-*grid* we mean a directed grid with $q$ lines and $k$ columns, i.e. a digraph $G = (V, A)$ with vertex set $V = \{v_{i,j} : 1 \le i \le q, 1 \le j \le k\}$ and with arc set $A = \{(v_{i,j}, v_{i+1,j}) : 1 \le i < q, 1 \le j \le k\} \cup \{(v_{i,j}, v_{i,j+1}) : 1 \le i \le q, 1 \le j < k\}$. For $i \in \{1, 2\}$, we define the following problems.

**Name:** GRID LABELLING-$i$ (GL-$i$)
**Instance:** a $q \times k$-grid $G = (V, A)$, a set $S$, a partition $\{S_v : v \in V\}$ of $S$, for each $a = (u, v) \in A$ a function $f_a : S_u \cup S_v \to \mathbb{N}^*$.
**Question:** does there exist an $i$-admissible labelling of $G$?

A labelling of $G$ is an assignment of a value $l_v \in S_v$ to each $v \in V$. The labelling is 1-admissible iff for each $a = (u, v) \in A$, $f_a(l_u) = f_a(l_v)$. The labelling is 2-admissible iff for each $a = (u, v) \in A$, $f_a(l_u) \le f_a(l_v)$.

**Proposition 9.** *(i)* GL-1[$k$] *is* WNL-*complete; (ii)* GL-2[$k$] *is* WNL-*complete.*

*Proof.* Membership in WNL of these two problems can be shown by reduction to NTMC[$k$]. The WNL-hardness proof for GL-1[$k$] involves a sequence of reductions, and is deferred to the full version of the paper. WNL-hardness of GL-2[$k$]

(A)

(B)

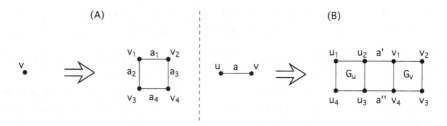

**Fig. 1.** Figure (A): the gadget $G_v$ associated to a vertex $v$ of $G$. Figure (B): interconnecting two gadgets $G_u, G_v$ corresponding to an horizontal arc $a = (u, v)$.

is shown by reduction from GL-1[$k$]. Given an instance $I$ of GL-1[$k$] involving a $q \times k$-grid $G$, we construct an instance $I'$ of GL $- 2[k]$ involving a $2q \times 2k$-grid $G'$.

For each vertex $v$ of $G$, we introduce the gadget $G_v$ depicted in Figure 1 (A). It consists of four vertices $v_1, v_2, v_3, v_4$ and four arcs $a_1 = (v_1, v_2), a_2 = (v_1, v_3), a_3 = (v_2, v_4), a_4 = (v_3, v_4)$. Given the set $S_v$ associated to $v$ in $I$, we associate to $v_1, v_2, v_3, v_4$ disjoint copies $S_1, ..., S_4$ of these sets, where $S_i = \{x_i : x \in S_v\}$. Let $s = |S_v|$, and let $\phi$ be a bijection of $S_v$ into $[s]$. For odd $i$, we set $f_{a_i}(x_j) = \phi(x)$; for even $i$ we set $f_{a_i}(x_j) = s + 1 - \phi(x)$. It can be checked that in a 2-admissible labelling of $G_v$, the four vertices must be labelled by the four copies of a same element.

We map each vertex $v$ to a gadget $G_v$, and we interconnect these gadgets by arcs to form a grid $G'$. For each horizontal arc $a = (u, v)$ of $G$, we create two horizontal arcs $a' = (u_2, v_1)$ and $a'' = (u_3, v_4)$, as depicted in Figure 1 (B). If the image of $f_a$ is $[n_a]$, we set $f_{a'}(x_j) = f_a(x)$ and $f_{a''}(x_j) = n_a + 1 - f_a(x)$. We proceed similarly for the vertical arcs. Then $G'$ is a $2q \times 2k$-grid, and it can be checked that $G$ has a 1-admissible labelling iff $G'$ has a 2-admissible labelling.

□

**Proposition 10.** SLCS[$k$] *is* WNL-*complete.*

*Proof.* Membership in WNL follows from the claim given in the proof of Proposition 3. This claim yields a reduction to NTMC[$k$]: given an instance $I = (\mathcal{C}, q, k)$ of SLCS[$k$], we construct a nondeterministic Turing machine $M$ which proceeds as follows. The first $k$ cells of the tape store a position $\pi$ in $\mathcal{C}$. The machine starts by nondeterministically choosing a position $\pi$. At each round, the machine seeks $a \in F(\pi)$, nondeterministically chooses a position $\pi'$, checks that $\pi \rightarrow_a \pi'$, and overwrites $\pi$ by $\pi'$. Then: SLCS($\mathcal{C}$) $\geq q$ iff $M$ accepts using time $q' = O(kq)$ and space $\leq 2k$.

WNL-hardness is shown by a parameterized reduction from GL-2[$k$]. Let $I$ be an instance of GL-2[$k$], consisting of a $q \times k$-grid $G = (V, A)$, of a set $S$, of a partition $\{S_v : v \in V\}$ of $S$, and for each $a = (u, v) \in A$ of a function $f_a : S_u \cup S_v \rightarrow \mathbb{N}^*$. For each $a \in A$, suppose that the image of $f_a$ is $[n_a]$. For each $i \in [q], j \in [k]$, let $v_{i,j}$ be the vertex of $G$ in line $i$, column $j$. For each $1 \leq i \leq q, 1 \leq j \leq k - 1$, let $a_{i,j}$ denote the horizontal arc $(v_{i,j}, v_{i,j+1})$. For each $1 \leq i \leq q - 1, 1 \leq j \leq k$, let $a'_{i,j}$ denote the vertical arc $(v_{i,j}, v_{i+1,j})$.

We construct an instance $I' = (\mathcal{C}, q', k')$ of SLCS[$k$] in the following way. We set $q' = 3kq - k - q$ and $k' = 2k + 4$. We define $\mathcal{C}$ as follows.

- Label set: we introduce labels $a[v, x]$ for each $v \in V, x \in S_v$, and $b[a, i]$ for each $a \in A, i \in [n_a]$.
- Sequences: we define two sequences $s, s'$, two sequences $t, t'$, and $2k$ sequences $u_1, u_1', ..., u_k, u_k'$. Let $<_S$ be a total order on $S$. We first define the following sequences.

$$\forall v \in V, A[v] = \prod_{x \in (S_v, <_S)} a[v, x], \forall a \in A, B[a] = \prod_{i=1}^{n_a} b[a, i]$$

$$\forall a = (u, v) \in A, i \in [n_a], C[a, i] = \prod_{x \in S_u : f_a(x) = i} a[u, x], C'[a, i] = \prod_{x \in S_v : f_a(x) = i} a[v, x]$$

The sequences of $\mathcal{C}$ are defined below.

- The sequences $s, s'$ are control sequences; they constrain the shape of a compatible sequence for $\mathcal{C}$.

$$s_i = \left( \prod_{j=1}^{k-1} A[v_{i,j}] B[a_{i,j}] \right) A[v_{i,k}], s_i' = \left( \prod_{j=1}^{k-1} \widetilde{A[v_{i,j}]} \widetilde{B[a_{i,j}]} \right) \widetilde{A[v_{i,k}]}$$

$$s = \left( \prod_{i=1}^{q-1} s_i \left( \prod_{j=1}^{k} B[a_{i,j}'] \right) \right) s_q, s' = \left( \prod_{i=1}^{q-1} s_i' \left( \prod_{j=1}^{k} \widetilde{B[a_{i,j}']} \right) \right) s_q'$$

- The sequences $t, t'$ are selection sequences; they ensure that the constraints corresponding to the horizontal arcs $a_{i,j}$ are satisfied.

$$t = \prod_{i=1}^{q} \prod_{j=1}^{k-1} \prod_{p=1}^{n_{a_{i,j}}} C[a_{i,j}, p] b[a_{i,j}, p], \quad t' = \prod_{i=1}^{q} \prod_{j=1}^{k-1} \prod_{p=1}^{n_{a_{i,j}}} b[a_{i,j}, p] C'[a_{i,j}, p]$$

- For each $j \in [k]$, the sequences $u_j, u_j'$ are selection sequences; they ensure that the constraints corresponding to the vertical arcs $a_{i,j}'$ are satisfied.

$$u_j = \prod_{i=1}^{q-1} \prod_{p=1}^{n_{a_{i,j}'}} C[a_{i,j}', p] b[a_{i,j}', p], \quad u_j' = \prod_{i=1}^{q-1} \prod_{p=1}^{n_{a_{i,j}'}} b[a_{i,j}', p] C'[a_{i,j}', p]$$

The reduction is computable in polynomial time, and its correctness follows by proving that $G$ has a 2-admissible labelling iff $\mathcal{C}$ has a compatible sequence of length $q'$.                                                      □

# 5    Consequences for Problems of Bounded Width

As a by-product of our hardness results for SLCS, we obtain new hardness results for problems of "bounded width": the LCS problem (Section 5.1), and other problems previously known to be "hard for the W-hierarchy" (Section 5.2).

## 5.1    Consequences for LCS

In this section, a sequence is a string *allowing* repetitions of letters, and we explicitly use the term p-sequence when considering strings with no repeated letter. We recall the definition of the LONGEST COMMON SUBSEQUENCE problem.

**Name:** LONGEST COMMON SUBSEQUENCE (LCS)
**Instance:** a collection of $k$ sequences $C = \{s_1, ..., s_k\}$ on an alphabet $\Sigma$ of size $m$, and an integer $q$
**Question:**, do the sequences admit a common subsequence of length $\geq q$?

There is a straightforward reduction from SLCS to LCS, which relies on a padding argument (inserting extra symbols at the beginning, at the end, and between each two consecutive labels of each input sequence).

**Lemma 1.** *There is a polynomial-time parameter-preserving reduction from* SLCS[q, k] *to* LCS[q, k].

Lemma 1, together with Propositions 8 and 10, implies the following for LCS:

**Proposition 11.** *(i)* LCS[q, k] *is* W[1]-*complete;* *(ii)* LCS[k] *is* WNL-*complete;* *(iii)* LCS[k, m] *is* WNL-*complete.*

*Proof.* For Point (i), the hardness result follows from Proposition 8 and Lemma 1, and the membership result is shown in [2]. For Point (ii), the hardness result follows from Proposition 10 and Lemma 1. The membership result is shown by a parameterized reduction to NTMC[k]. Let $I = (C, q, k)$ be an instance of LCS[k] with $C = \{s_1, ..., s_k\}$. We construct a nondeterministic Turing machine $M$ whose first $k$ cells of the tape store a tuple $(p_1, ..., p_k)$ with $p_i$ an integer between 0 and $|s_i|$. The machine starts with the tuple $(0, ..., 0)$ on its tape. At each round $i \leq q$, if the tuple on the tape is $t = (p_1, ..., p_k)$, the machine nondeterministically overwrites $t$ by a tuple $t' = (p'_1, ..., p'_k)$ s.t. $p'_j > p_j$ for each $j \in [k]$, and $s_1[p'_1] = ... = s_k[p'_k]$. If no such tuple $t'$ exists, the machine rejects. Then $M$ accepts the empty string in $q' = O(kq)$ steps using space $\leq 2k$ iff the sequences $s_i$ have a common subsequence of length $q$.

For Point (iii), hardness follows from a parameterized reduction from LCS[k] to LCS[k, m] described in [1], and membership follows from Point (ii).    □

Proposition 11 improves known hardness results in several ways. First, it gives an alternative proof for the W[1]-hardness of LCS[q, k], simpler than the original proof of [2]. Second, it classifies precisely the complexity of LCS[k]. The problem was only known to be W[t]-hard for each $t \geq 1$ [2], while we obtain a stronger WNL-completeness result.

## 5.2    Consequences for Other Problems

In [3,4], the W[t]-hardness result for LCS[k] shown in [2] was transferred to other problems by parameterized reductions. Hence, our stronger WNL-hardness result for LCS[k] also holds for these problems.

**Proposition 12.** *The problems* COLORED CUTWIDTH, FEASIBLE REGISTER ASSIGNMENT, DOMINO TREEWIDTH, TRIANGULATING COLORED GRAPHS *are* WNL-*hard.*

We now consider a problem on automata introduced in [22].

**Name:** BOUNDED DFA INTERSECTION (BDFA)
**Instance:** a family of $k$ dfa $\mathcal{A} = \{A_1, ..., A_k\}$ on an alphabet $\Sigma$, an integer $q$
**Question:** does there exist a word in $\Sigma^q$ that is accepted by every $A_i$?

The BDFA2 problem is the restriction of BDFA to instances with $|\Sigma| = 2$.

**Proposition 13.** *(i)* BDFA[k] *is* WNL-*complete; (ii)* BDFA2[k] *is* WNL-*complete.*

*Proof.* Let us show Point (i). To show WNL-hardness, we reduce from LCS[k]. Given an instance $I = (\mathcal{C}, q, k)$ of LCS[k], we create an instance $I' = (\mathcal{A}, q, k)$ of BDFA[k], where $\mathcal{A} = \{A_1, ..., A_k\}$ is such that for each $i$, $A_i$ is a dfa recognizing the set of subwords of $s_i$. To show membership in WNL, we reduce to NTMC[k]. Let $I = (\mathcal{A}, q, k)$ be an instance of BDFA[k] with $\mathcal{A} = \{A_1, ..., A_k\}$, $A_i = (Q_i, \Sigma, \delta_i, q_i^0, F_i)$. We construct a nondeterministic Turing machine $M$ whose first $k$ cells of the tape represents a tuple $t = (q_1, ..., q_k)$, with $q_i \in Q_i$. The machine starts with the tuple $t = (q_1^0, ..., q_k^0)$ on its tape. At each round $i \leq q$, if the tuple on the tape is $t = (q_1, ..., q_k)$, the machine nondeterministically chooses a letter $a \in \Sigma$, and overwrites $t$ by the new tuple $t' = (\delta_1(q_1, a), ..., \delta_k(q_k, a))$. At the end of round $q$, the machine accepts iff the tuple written on the tape has the form $t = (q_1, ..., q_k)$ with $q_i \in F_i$ for each $i$.

Point (ii) is easy to see: membership in WNL follows from Point (i), and WNL-hardness follows by a simple reduction from BDFA[k] described in [22].    □

# 6    Concluding Remarks

Let us mention two directions for further research. The first direction would consist in improving complexity results for SLCS and CSLCS.

1. What is the exact approximability threshold, as a function of $k$, for each problem? It is conjectured to be $2^{\Omega(k)}$ for SLCS, but may be $\Omega(\log k)$ for CSLCS.

2. Is the problem solvable in $p^{O(k)} n^c$ time? An algorithm with this running time would be preferable to the algorithm of Proposition 7, for small $k$.

Another direction for further work would be to settle the precise parameterized complexity of problems similar to LCS. Example of such problems are the SHORTEST COMMON SUPERSEQUENCE problem, and the LCS problem on bounded alphabets, parameterized by the number of sequences. These problems are currently known to be $W[t]$-hard for each $t \geq 1$ [15] and $W[1]$-hard [20], respectively.

## Acknowledgements

We are grateful to Vincent Berry and Eric Rivals for their careful proof-reading.

## References

1. Bodlaender, H.L., Downey, R.G., Fellows, M.R., Hallett, M.T., Wareham, H.T.: Parameterized complexity analysis in computational biology. Computer Applications in the Biosciences 11(1), 49–57 (1995)
2. Bodlaender, H.L., Downey, R.G., Fellows, M.R., Wareham, H.T.: The parameterized complexity of sequence alignment and consensus. Theoretical Computer Science 147(1–2), 31–54 (1994)
3. Bodlaender, H.L., Fellows, M.R., Hallett, M.T.: Beyond NP-completeness for problems of bounded width: Hardness for the W hierarchy (extended abstract). In: Proc. STOC 1994, pp. 449–458. ACM, New York (1994)
4. Bodlaender, H.L., Fellows, M.R., Hallett, M.T.: The hardness of perfect phylogeny, feasible register assignment and other problems on thin colored graphs. Theoretical Computer Science 244(1), 167–188 (2000)
5. Cai, L., Chen, J., Downey, R.G., Fellows, M.R.: On the parameterized complexity of short computation and factorization. Archive for Mathematical Logic 36(4–5), 321–337 (1997)
6. Chen, J., Liu, Y., Lu, S., O'Sullivan, B., Razgon, I.: A Fixed-Parameter Algorithm for the Directed Feedback Vertex Set Problem. In: Proc. STOC 2008 (to appear, 2008)
7. Downey, R.G., Fellows, M.R.: Parameterized Complexity. Springer, Heidelberg (1999)
8. Dwork, C., Kumar, R., Naor, M., Sivakumar, D.: Rank Aggregation Methods for the Web. In: WWW10 (2001)
9. Dwork, C., Kumar, R., Naor, M., Sivakumar, D.: Rank aggregation revisited (2001)
10. Even, G., Naor, J., Schieber, B., Sudan, M.: Approximating minimum feedback sets and multicuts in directed graphs. Algorithmica 20, 151–174 (1998)
11. Fagin, R., Kumar, R., Sivakumar, D.: Efficient similarity search and classification via rank aggregation. In: Proc. ACM SIGMOD 2003, pp. 301–312 (2003)
12. Fellows, M.R., Hallett, M.T., Stege, U.: Analogs & duals of the MAST problem for sequences & trees. Journal of Algorithms 49(1), 192–216 (2003)
13. Flum, J., Grohe, M.: Parameterized Complexity Theory. Springer, Heidelberg (2006)
14. Halldórsson, M.M.: Approximation via Partitioning. Technical Report IS-RR-95-0003F, School of Information Science, Japan Advanced Institute of Science and Technology, Hokuriku (1995)
15. Hallett, M.T.: An integrated complexity analysis of problems from computational biology. PhD thesis, Department of Computer Science, University of Victoria, Victoria, B.C., Canada (1996)
16. Hodge, J., Klima, R.E.: The Mathematics of Voting and Elections: A Hands-On Approach. In: Mathematical World, vol. 22. AMS (2000)

17. Maier, D.: The complexity of some problems on subsequences and supersequences. J. ACM 25(2), 322–336 (1978)
18. Moret, B.M.E., Tang, J., Warnow, T.: Reconstructing phylogenies from gene-content and gene-order data. In: Gascuel, O. (ed.) Mathematics of phylogeny and evolution. Oxford University Press, Oxford (2004)
19. Niedermeier, R.: Invitation to Fixed-Parameter Algorithms. Oxford University Press, Oxford (2006)
20. Pietrzak, K.: On the Parameterized Complexity of the fixed alphabet Shortest Common Supersequence and Longest Common Subsequence Problems. Journal of Computer and System Sciences 67(4), 757–771 (2003)
21. Seymour, P.D.: Packing directed circuits fractionally. Combinatorica 15, 281–288 (1995)
22. Wareham, H.T.: The parameterized complexity of intersection and composition operations on sets of finite-state automata. In: Yu, S., Păun, A. (eds.) CIAA 2000. LNCS, vol. 2088, pp. 302–310. Springer, Heidelberg (2001)

# FPT Algorithms for Path-Transversals and Cycle-Transversals Problems in Graphs

S. Guillemot

LIFL/CNRS/INRIA, Bât. M3 Cité Scientifique, 59655, Villeneuve d'Ascq cedex, France
Sylvain.Guillemot@lifl.fr

**Abstract.** In this article, we consider problems on graphs of the following form: given a graph, remove $p$ edges/vertices to achieve some property. The first kind of problems are *separation problems* on undirected graphs, where we aim at separating distinguished vertices in an graph. The second kind of problems are *feedback set problems* on group-labelled graphs, where we aim at breaking nonnull cycles in a group-labelled graph. We obtain new FPT algorithms for these different problems. A building stone for our algorithms is a general $O^*(4^p)$ algorithm for a class of problems aiming at breaking a set of paths in a graph, provided that the set of paths has a special property called *homogeneity*.

## 1 Introduction

It is well-known that the NODE MULTIWAY CUT problem has a half-integrality property [5]. The proof of this result relies on a property that we generalize under the name of *homogeneity* of a path system. We call *path system* a tuple $\sigma$ consisting of an undirected graph $G = (V, E)$, of a set $T \subseteq V$ of terminals, of a set $F \subseteq V$ of forbidden vertices, and of a set $\mathcal{P}_\sigma$ of paths between terminals. The generic PATH COVER problem aims at breaking each path $P \in \mathcal{P}_\sigma$ of a path system $\sigma$ by removing nonforbidden vertices.

We introduce the property of homogeneity of a path system, and show that it has consequences not only for the *approximability* of the path-cover problem, but also for its *parameterized complexity* [4,3]. Namely, we first show that for a homogeneous path system, the PATH COVER problem has a half-integrality property, which implies that the problem is 2-approximable and generalizes the result for NODE MULTIWAY CUT. We then devise a bounded-search algorithm which solves the problem in $O^*(4^p)$ time, by relying on half-integral solutions of the problem in order to guide the construction of a search tree.

As a first consequence of this general result, we obtain new algorithms for the MULTIWAY CUT and MULTICUT algorithms in their vertex-deletion and edge-deletion versions. For the MULTIWAY CUT problem, a $O^*(4^{p^3})$ algorithm problem was obtained by [8], then [1] presented an $O^*(4^p)$ algorithm, and we obtain a different algorithm with the same $O^*(4^p)$ running time. For the MULTICUT problem, where we aim at disconnecting $k$ pairs of terminals by removing $p$ edges/vertices, [8] presented a $O^*(2^{kp}p^p4^{p^3})$ algorithm, while we obtain simple FPT algorithms with running time $O^*((8k)^p)$.

M. Grohe and R. Niedermeier (Eds.): IWPEC 2008, LNCS 5018, pp. 129–140, 2008.
© Springer-Verlag Berlin Heidelberg 2008

As a second consequence of our general result, we obtain FPT algorithms for feedback set problems on group-labelled graphs. Path covering and packing problems on group-labelled graphs were for instance studied in [2], here we consider the following cycle-covering problem: given a digraph $G$ whose arcs are labelled by a group $\Gamma$, break each nonnull cycle of $G$. These problems are called GROUP FEEDBACK SET, and we consider them in their vertex and edge version. We show that both versions are solvable in $O^*((4|\Gamma|+1)^p)$ time. We also show that the edge-deletion version of the problem is solvable in $O^*((8p+1)^p)$ time, independent of $|\Gamma|$. These results generalize the FPT algorithms for the GRAPH BIPARTIZATION problem [9], which is a particular case of the GROUP FEEDBACK SET problem with $\Gamma = \mathbb{Z}_2$.

This article is organized as follows. Section 2 is devoted to our general result concerning homogeneous paths systems. Section 3 contains results for the MULTIWAY CUT and MULTICUT problems. Section 4 contains results for the GROUP FEEDBACK SET problems. Finally, in Section 5 we formulate some open questions and possible generalizations of the results.

## 2    Homogeneous Path Systems

### 2.1    Preliminaries

Let $G = (V, E)$ be an undirected graph. A *path* in $G$ is a sequence of vertices $P = x_1...x_m$ s.t. $x_i x_{i+1} \in E$ for each $1 \le i < m$; we say that $P$ is a path joining $x_1$ to $x_m$. A *cycle* in $G$ is a path $C = x_1 x_2 ... x_m$ with $x_1, x_m$ equal; we say that $C$ is a cycle at $x_1$. Consider a path $P = x_1 ... x_m$. The vertices $x_1, ..., x_{m-1}$ are the *initial vertices* of $P$. $P$ is *simple* iff the vertices $x_i$ are distinct. The *inverse* of $P$ is the path $\tilde{P} = x_m ... x_1$. Given a weight function $w$ on $V$, the *length* of $P$ is $w(x_1) + ... + w(x_m)$, and the *initial length* of $P$ is $w(x_1) + ... + w(x_{m-1})$.

A *path system* is a tuple $\sigma = (G, T, F)$ which consists of: (i) an undirected graph $G = (V, E)$, (ii) a set $T \subseteq V$ of *terminals*, (iii) a set $F \subseteq V$ of *forbidden vertices*, as well as a set $\mathcal{P}_\sigma$ of (not necessarily simple) paths in $G$ joining elements of $T$. A *transversal of $\sigma$* (or a *solution for $\sigma$*) is a set of vertices disjoint from $F$ and which meets each path of $\mathcal{P}_\sigma$.

The generic problem PATH COVER takes an instance $I = (\sigma, p)$ consisting of a path system $\sigma$, an integer $p$, and seeks a transversal of $\sigma$ of size at most $p$. In this section, we show that if $\sigma$ has a special property called *homogeneity*, then the PATH COVER problem is solvable in $O^*(4^p)$ time (Theorem 1). This result relies on a half-integrality property of an LP formulation of the problem.

We now define the property of homogeneity of a path system.

**Definition 1.** *The path system $\sigma = (G, T, F)$ is homogeneous iff the two following conditions hold:*

1. *for each path $P \in \mathcal{P}_\sigma$, there exists a simple path $P' \in \mathcal{P}_\sigma$ included in $P$;*
2. *for each path $P \in \mathcal{P}_\sigma$ joining $u, v \in T$, if $P = P_1 x P_2$ then: for each path $P'$ joining $x$ to $w \in T$, one of $P_1 P', \tilde{P}' P_2$ is in $\mathcal{P}_\sigma$.*

In the rest of the section, we consider an homogeneous path system $\sigma = (G, T, F)$.

## 2.2 LP Formulation and Half-Integrality

We first describe the LP formulation of the problem. Clearly, the optimization problem corresponding to PATH COVER can be formulated as an integer linear program. We consider the fractional relaxation of this program, denoted by $F_\sigma$, as well as its dual LP, denoted by $F'_\sigma$.

$$
(F_\sigma) \begin{cases} \text{minimize } \sum_{v \in V} d_v \\ \text{subject to} \\ \forall P \in \mathcal{P}_\sigma, \sum_{v \in P} d_v \geq 1 \\ \forall v \in V, d_v \geq 0, \forall v \in F, d_v = 0 \end{cases}
\qquad
(F'_\sigma) \begin{cases} \text{maximize } \sum_{P \in \mathcal{P}_\sigma} f_P \\ \text{subject to} \\ \forall v \in V \backslash F, \sum_{P \in \mathcal{P}_\sigma : v \in P} f_P \leq 1 \\ \forall P \in \mathcal{P}_\sigma, f_P \geq 0 \end{cases}
$$

By generalizing the results for MULTIWAY CUT established in [5], we demonstrate an half-integrality property of $F_\sigma$. Let $d$ be an optimal solution of $F_\sigma$, and let $f$ be the corresponding optimal solution of $F'_\sigma$. Consider the vertex-weighted graph $G'$ obtained from $G$ by weighting each $v \in V$ by $d_v$. Let $M$ be the set of $v \in V$ s.t. $d_v > 0$ and $v$ is reachable from $T$ by a path of initial length 0. Let $V_1 = \{v \in M : d_v = 1\}$, $V_{1/2} = \{v \in M : 0 < d_v < 1\}$, and $V_0 = V \backslash (V_1 \cup V_{1/2})$.

**Lemma 1.** *Let $P \in \mathcal{P}_\sigma$ s.t. $f_P > 0$. Then $P \cap M$ consists of: either an element of $V_1$, or two elements of $V_{1/2}$.*

*Proof.* Since $f_P > 0$, dual complementary slackness implies that $P$ has length 1. Let $u$ be the first vertex of $P$ s.t. $d_u > 0$, and let $v$ be the last vertex of $P$ s.t. $d_v > 0$. Then $u, v \in M$. If $u = v$, since $P$ has length one it follows that $d_u = 1$, and $P \cap M$ consists of an element of $V_1$. Suppose now that $u, v$ are distinct, then they belong to $V_{1/2}$. Suppose that there exists a third vertex $w \in P \cap M$. Then $P = P_1 w P_2$, where $P_1$ contains $u$ and $P_2$ contains $v$. Since $w \in M$, there exists a path $P'$ of initial length 0 joining $T$ to $w$. By Point 2 of Definition 1, one of $P_1 P'$, $\tilde{P}' P_2$ is in $\mathcal{P}_\sigma$. But these paths have length less than 1, contradicting the assumption that $d$ is a solution. □

Let $s$ be the solution of $F_\sigma$ which assigns the value $r$ to a vertex of $V_r$ for $r \in \{0, \frac{1}{2}, 1\}$.

**Lemma 2.** *$s$ is an optimal solution of $F_\sigma$.*

*Proof.* Let us first show that $s$ is a solution of $F_\sigma$. Consider $P \in \mathcal{P}_\sigma$, then since $P$ has length $\geq 1$ in $G'$, $P$ contains at least one vertex of $M$. Then either $P$ contains a vertex of $V_1$, or $P$ contains two vertices of $V_{1/2}$. In both cases, we obtain that $\sum_{v \in P} s_v \geq 1$.

We now show that $s$ is optimal by proving that $f$ has the same cost as $s$. Let $\mathcal{P}_1$ be the set of paths $P$ s.t. $f_P > 0$ and $P \cap M$ consists of an element of $V_1$, and let $\mathcal{P}_2$ be the set of paths $P$ s.t. $f_P > 0$ and $P \cap M$ consists of two elements of $V_{1/2}$. By primal complementary slackness and by the optimality of $d$, each vertex of $M$ is saturated. It follows that:

$$
\sum_{P \in \mathcal{P}_\sigma} f_P = \sum_{P \in \mathcal{P}_1} f_P + \sum_{P \in \mathcal{P}_2} f_P = |V_1| + \frac{1}{2}|V_{1/2}|
$$

We conclude by observing that this is exacly the cost of $s$. □

## 2.3  Some Technical Lemmas

The following lemma shows that an arbitrary optimal solution for $\sigma$ can be transformed into an optimal solution satisfying some additional properties. Let $U$ be the set of elements of $V_0$ reachable from $T$ by a path of length 0 in $G'$.

**Lemma 3.** *There is an optimal solution for $\sigma$ disjoint from $U$.*

*Proof.* Let $S$ be an optimal solution for $\sigma$. We define the set of bad vertices $B = S \cap U$. Our goal is to construct a solution $S'$ from $S$ by discarding the bad vertices and replacing them by some vertices outside of $U$.

Let $u \in V$. Say that $u$ is *accessible* iff it is reachable from $T$ by a path of initial length 0 going through an element of $B$. Say $u$ is *uniformly accessible* if each path of initial length 0 joining $T$ to $u$ goes through an element of $B$. We define $V_1'$ as the set of accessible elements of $V_1$. We define $V_{1/2}'$ as the set of uniformly accessible elements of $V_{1/2}$. We set $V' = V_1' \cup V_{1/2}'$.

Let $S' = S - B + V'$. We can show that $S'$ is an optimal solution for $\sigma$ disjoint from $U$. The proof of this result is technical, and is deferred to the full version of the paper, due to length limitations. $\qquad\square$

Let us introduce the following notations: given a path system $\sigma$, we denote the cost of an optimal solution of $F_\sigma$ by $opt_\sigma^*$, and we denote the cost of an optimal solution for $\sigma$ by $opt_\sigma$. By convention, these values are equal to $\infty$ when there is no solution.

Let $\sigma = (G, T, F)$ be a path system. A *frontier vertex* is a vertex $u$ s.t. (i) $u \notin F$, (ii) $u$ is reachable from $T$ by a path whose initial vertices are in $F$. Consider the new path system $\sigma' = (G, T, F \cup \{u\})$.

We first observe the following relations between solutions for $\sigma$ and for $\sigma'$.

**Lemma 4.**  *1. A solution for $\sigma'$ is also a solution for $\sigma$;*
 *2. A solution for $\sigma$ which does not contain $u$ is also a solution for $\sigma'$;*
 *3. A solution for $F_{\sigma'}$ is also a solution for $F_\sigma$.*

The following Lemma shows that the instances $\sigma$ and $\sigma'$ are equivalent whenever their optimal fractional solutions have the same cost. Its proof relies crucially on Lemma 3.

**Lemma 5.** *If $opt_\sigma^* = opt_{\sigma'}^*$, then $opt_\sigma = opt_{\sigma'}$.*

*Proof.* By Point 1 of Lemma 4, we have $opt_{\sigma'} \geq opt_\sigma$. We now show that $opt_{\sigma'} \leq opt_\sigma$. Let $s$ be an half-integral optimal solution of $F_{\sigma'}$, and let $V_0, V_{1/2}, V_1$ be defined accordingly. By Point 3 of Lemma 4, $s$ is a solution of $F_\sigma$. Moreover, it is an optimal solution of $F_\sigma$ since $opt_\sigma^* = opt_{\sigma'}^*$.

Let $G'$ be the weighted graph obtained from $G$ by assigning to each $v \in V$ the weight $s_v$. Let $U$ be the set of elements of $V_0$ reachable from $T$ by a path of length 0 in $G'$. By Lemma 3, there exists $S$ optimal solution for $\sigma$ disjoint from $U$. Since $u$ is a frontier vertex, it is reachable from $T$ by a path whose initial vertices are in $F$ (and thus have null weight in $G'$). Therefore, we have $u \in U$, hence $u \notin S$. It follows that $S$ is a solution for $\sigma'$ by Point 2 of Lemma 4. We conclude that $opt_{\sigma'} \leq |S| = opt_\sigma$. $\qquad\square$

## 2.4 The Main Result

We now describe the algorithm for PATH COVER, and justify its correctness and running time in Theorem 1.

---

SOLVEPATHCOVER$(\sigma, p)$

1: suppose that $\sigma = (G, T, F)$
2: $x \leftarrow opt_\sigma^*$
3: **if** $x \leq \frac{p}{2}$ **then**
4:     return true
5: **else if** $x > p$ **then**
6:     return false
7: **end if**
8: choose $u$ a frontier vertex
9: let $\sigma' = (G, T, F \cup \{u\})$
10: $x' \leftarrow opt_{\sigma'}^*$
11: **if** $x' = x$ **then**
12:     return SOLVEPATHCOVER$(\sigma', p)$
13: **else if** $x' > x$ **then**
14:     let $\sigma'' = (G \backslash u, T, F)$
15:     return (SOLVEPATHCOVER$(\sigma', p)$ or SOLVEPATHCOVER$(\sigma'', p - 1)$)
16: **end if**

---

**Theorem 1.** *The algorithm* SOLVEPATHCOVER *solves the* PATH COVER *problem in* $O^*(4^p)$ *time.*

*Proof.* We first justify the correctness. The case where $x \leq \frac{p}{2}$ or $x > p$ is handled correctly by the algorithm: indeed, we have $opt_\sigma^* \leq opt_\sigma \leq 2opt_\sigma^*$ by half-integrality; it follows that $opt_\sigma \leq p$ in the first case, and $opt_\sigma > p$ in the second case. Suppose now that $\frac{p}{2} < x \leq p$. Observe that there exists a frontier vertex: otherwise, each path in $\mathcal{P}_\sigma$ would only contain nodes in $F$, and we would have $x = \infty$. In Lines 11-12, the case $x' = x$ is handled correctly by the algorithm: because of Lemma 5, the instances $(\sigma, p)$ and $(\sigma', p)$ are equivalent. In Lines 13-16, the case $x' > x$ can be seen to be correct by Points 1 and 2 of Lemma 4. It follows that the algorithm is correct.

We now justify the running time. We view the execution of the algorithm as the construction of a search tree, where only recursive calls in Line 15 correspond to branches in the search tree. A node of the search tree is labelled by an instance $(\sigma, p)$. Moreover, let $P(n)$ denote the running time of the operations of Line 1-10, then the processing time of a node is bounded by $nP(n)$ and is thus polynomial.

Let $T_s$ denote the search tree. Let $u$ be a node of $T_s$. Suppose that $u$ is labelled by $(\sigma, p)$, let $p(u) = p$ and $k(u) = 2p + 1 - 2opt_\sigma^*$. Observe that if $u$ is an internal node of $T_s$, then $\frac{p}{2} < opt_\sigma^* \leq p$, and thus $0 < k(u) \leq p$. Let $S(u)$ denote the number of leaves of the subtree of $T_s$ rooted at $u$. Given $p, k$, let $T(p, k)$ denote the maximum value of $S(u)$ for $u$ node of $T_s$ s.t. $p(u) = p, k(u) \leq k$, or 0 if no such node exists. We claim that $T$ is such that:

$$\begin{cases} T(p,k) & \leq 1 \text{ if } p = 0 \text{ or } k = 0 \\ T(p,k) & \leq \max(1, T(p, k-1) + T(p-1, k)) \text{ otherwise} \end{cases} \quad (1)$$

Let $u$ be a node labelled by $(\sigma, p)$. We consider two cases. If $p = 0$ or $k = 0$, observe that $u$ must be a leaf: this is clear if $p = 0$, this results from the fact that $opt_\sigma^* > p$ if $k = 0$. In this case, we thus have $S(u) = 1$. Consider now the remaining cases. If $u$ is a leaf then $S(u) = 1$. Suppose now that $u$ is an internal node with two children $u', u''$, with $u'$ labelled by $(\sigma', p)$ and $u''$ labelled by $(\sigma'', p-1)$.

We first bound $S(u')$. Since $opt_{\sigma'}^* > opt_\sigma^*$ and since $opt_{\sigma'}^*$ is half-integral, we have $k(u') \leq k(u) - 1 \leq k - 1$, hence $S(u') \leq T(p, k-1)$.

We now bound $S(u'')$. Observe that $opt_{\sigma''}^* \geq opt_\sigma^* - 1$: indeed, given a solution $s$ for $F_{\sigma''}$ of cost $c$, we can extend this solution to $V$ by setting $s_u = 1$, obtaining a solution for $F_\sigma$ of cost $c + 1$. It follows that $k(u'') = 2(p-1) + 1 - 2opt_{\sigma''}^* \leq 2(p-1) + 1 - 2(opt_\sigma^* - 1) = 2p + 1 - 2opt_\sigma^* \leq k$. We obtain that $S(u'') \leq T(p-1, k)$.

We thus conclude that $S(u) = S(u') + S(u'') \leq T(p, k-1) + T(p-1, k)$, which completes the proof that $T$ satisfies the relations (1).

A straightforward induction then shows that $T(p,k) \leq 2^{p+k}$. Since $k(u) \leq p$ for an internal node $u$ of $T_s$, it follows that the number of leaves of $T_s$ is bounded by $T(p, p) \leq 2^{2p}$. We conclude that the algorithm has running time $O^*(4^p)$ as claimed. □

## 3   New Algorithms for the Multiway Cut and Multicut Problems

### 3.1   The Multiway Cut Problems

We first introduce some notations and definitions. Let $G = (V, E)$ be a graph and let $T \subseteq V$ be a set of terminals. Given two partitions $\mathcal{P}, \mathcal{P}'$ of $T$, $\mathcal{P} \sqsubseteq \mathcal{P}'$ means that $\mathcal{P}$ refines $\mathcal{P}'$. We denote by $C_T(G)$ the partition $\mathcal{P}$ of $T$ whose classes are the sets $C \cap T$ for $C$ connected component of $G$. If $F$ is a forest, we denote by $F|T$ the forest obtained from $F$ by removing the leaves not belonging to $T$, and contracting the nodes of degree 2 not belonging to $T$.

We consider the GENERALIZED VERTEX MULTIWAY CUT (GVMC) problem: given a graph $G = (V, E)$, a set $T \subseteq V$ of terminals, a set $F \subseteq V$ of forbidden vertices, a partition $\mathcal{P}$ of $T$, an integer $p$, can we find a set $S \subseteq V \backslash F$ of cardinal $\leq p$ s.t. $C_T(G \backslash S) \sqsubseteq \mathcal{P}$? We also consider the edge-version of the problem called GENERALIZED MULTIWAY CUT (GMC). While this was already shown in [1], we obtain another proof of the following result:

**Theorem 2.** GENERALIZED VERTEX MULTIWAY CUT *and* GENERALIZED MULTIWAY CUT *are solvable in* $O^*(4^p)$ *time.*

*Proof.* We formulate the GVMC problem as a path cover problem for a homogeneous path system and apply Theorem 1. We consider the path system

$\sigma = (G, T, F)$, where $\mathcal{P}_\sigma$ consists of the paths joining two vertices $u, v \in T$ belonging to different classes of $\mathcal{P}$. It is straightforward to verify that: (i) $\sigma$ is homogeneous, (ii) there is a polynomial-time separation oracle for $F_\sigma$.

An algorithm for GMC is obtained by a simple reduction to GVMC. Given $I = (G, T, F, \mathcal{P}, p)$ instance of GMC, we construct $I' = (G', T, F', \mathcal{P}, p)$ instance of GVMC as follows. For each edge $e$ of $G$ we introduce a new vertex $x_e$. We split each edge $e = uv$ in two edges $ux_e, vx_e$. We set $F' = V \cup \{x_e : e \in F\}$.  □

## 3.2   The Multicut Problem

We now consider the MULTICUT problem: given a graph $G = (V, E)$, a set $T \subseteq V$ of terminals, a set $P \subseteq [T]^2$ of $k$ pairs of terminals, a set $F \subseteq E$ of forbidden edges, an integer $p$, can we disconnect each pair of vertices in $P$ by removing at most $p$ edges of $E \backslash F$ ?

**Theorem 3.** MULTICUT *can be solved in* $O^*((8k)^p)$ *time.*

*Proof.* Let $I = (G, T, P, F, p)$ be an instance of MULTICUT. Say that a partition $\mathcal{P}$ of $T$ is *realizable* iff $I' = (G, T, F, \mathcal{P}, p)$ is a positive instance of GENERALIZED MULTIWAY CUT. Say that a partition $\mathcal{P}$ is *admissible* iff it separates each pair $uv$ in $P$. We will describe an algorithm that enumerates a set $\mathcal{S}$ of *good* partitions of $T$ s.t. (i) $\mathcal{S}$ is a superset of the set of realizable partitions; (ii) $\mathcal{S}$ has size $\leq (2k)^p$.

Given the set $\mathcal{S}$, we solve MULTICUT by seeking a partition $\mathcal{P}$ in $\mathcal{S}$ which is admissible and realizable. Deciding if the partition is admissible takes polynomial time, and deciding if the partition is realizable takes $O^*(4^p)$ time. Hence, we obtain a $O^*((8k)^p)$ time algorithm for MULTICUT. We now describe the construction of $\mathcal{S}$. The set $\mathcal{S}$ is the set of partitions returned by computation paths of the algorithm FINDGOODPARTITION described below.

---

FINDGOODPARTITION$(G, T, p)$

1: let $F$ be a spanning forest of $G$, let $F' = F|T$
2: choose a set $E$ of at most $p$ edges of $F'$
3: let $\mathcal{P} = C_T(F')$, let $\mathcal{P}' = C_T(F' \backslash E)$
4: choose a partition $\mathcal{P}''$ s.t. $\mathcal{P}' \sqsubseteq \mathcal{P}'' \sqsubseteq \mathcal{P}$
5: return $\mathcal{P}''$

---

The proof of Point (i) is deferred to the full paper. Consider Point (ii). Let $T_1(p)$ denote the maximum number of computation paths of the algorithm FINDGOODPARTITION$(G, T, p)$. We show that $T_1(p) \leq (2k)^p$. Since $F'$ is a forest whose leaves and nodes of degree 2 are elements of $T$, the number of edges of $F'$ is at most $2k$ (in fact $2k - 3$). It follows that the number of possible choices in Line 2 is at most $\frac{(2k)^p}{p!}$. Besides, it can be shown that the number of possible choices in Line 4 is at most $p!$. We conclude that the total number of choices made by the algorithm is at most $\frac{(2k)^p}{p!} \times p! = (2k)^p$.  □

### 3.3   The Vertex Multicut Problem

We finally consider the vertex version of the MULTICUT problem, called VERTEX MULTICUT: given a graph $G = (V, E)$, a set $T \subseteq V$ of terminals, a set $P \subseteq [T]^2$ of $k$ pairs of terminals, a set $F \subseteq V$ of forbidden vertices, can we disconnect each pair of vertices in $P$ by removing at most $p$ vertices in $V \backslash F$?

**Theorem 4.** VERTEX MULTICUT *can be solved in* $O^*((8k)^p)$ *time.*

*Proof.* Let $I = (G, T, P, F, p)$ be an instance of VERTEX MULTICUT. We now say that a partition $\mathcal{P}$ of $T$ is realizable iff $I' = (G, T, F, \mathcal{P}, p)$ is a positive instance of GENERALIZED VERTEX MULTIWAY CUT. As before, we describe an algorithm that enumerates a set $\mathcal{S}$ of *good* partitions of $T$ s.t. (i) $\mathcal{S}$ is a superset of the set of realizable partitions, (ii) $\mathcal{S}$ has size $\leq (p+1)(2k)^p$. Then, we use the set $\mathcal{S}$ to solve VERTEX MULTICUT in $O^*((8k)^p)$ time. The set $\mathcal{S}$ is obtained as the results of the computation paths of the algorithm FINDGOODPARTITION2 described below.

---

FINDGOODPARTITION2$(G, T, p)$

1: let $F$ be a spanning forest of $G$, let $F' = F|T$
2: let $N$ be the set of nodes of $F'$
3: choose $i \in \{1, 2\}$
4: **if** $i = 1$ or $p = 0$ **then**
5:     return FINDGOODPARTITION$(G, T, p)$
6: **else**
7:     choose $u \in N$
8:     return FINDGOODPARTITION2$(G \backslash u, T, p - 1)$
9: **end if**

---

The proof of Point (i) is deferred to the full paper. We show Point (ii). Let $T_2(p)$ be the maximum number of computation paths of the algorithm FINDGOODPARTITION2$(G, T, p)$. Since $F'$ has at most $2k$ vertices, there are at most $2k$ possible choices in Line 7; hence $T_2$ satisfies the following relation:

$$\begin{cases} T_2(0) & = 1 \\ T_2(p) & \leq T_1(p) + (2k)T_2(p-1) \end{cases}$$

Since $T_1(p) \leq (2k)^p$, a straightforward induction shows that $T_2(p) \leq (p+1)(2k)^p$.   $\square$

## 4   Feedback Set Problems on Group-Labelled Graphs

### 4.1   Preliminaries

Let $\Gamma$ be a group, with unit element $1_\Gamma$. In the following we consider problems involving $\Gamma$; though we fix $\Gamma$ for the rest of the section, we will in fact assume that the group is part of the input to the problem, being described by its multiplication table.

A *Γ-labelled graph* is a digraph with a labelling of its arcs by elements of $\Gamma$. Formally, this is a tuple $\mathcal{G} = (V, A, \Lambda)$, where $V$ is a set of vertices, $A \subseteq V^2$ is a set of arcs, and $\Lambda : A \to \Gamma$ is a labelling of the arcs; $\mathcal{G}$ does never contain both arcs $(x, y)$ and $(y, x)$. Given $x, y \in V$, we define $\Lambda(x, y)$ by setting:

$$\begin{cases} \Lambda(x,y) & = \Lambda(a) \text{ if } a = (x,y) \in A \\ \Lambda(x,y) & = \Lambda(a)^{-1} \text{ if } a = (y,x) \in A \\ \Lambda(x,y) & = \perp \text{ otherwise} \end{cases}$$

The *underlying graph* of $\mathcal{G}$ is the undirected graph $G = (V, E)$ obtained by forgetting the orientations of the arcs. By a path (or cycle) in $\mathcal{G}$, we will mean a path (or cycle) in $G$. Let $P = x_1...x_m$ be a path in $\mathcal{G}$, we set $\Lambda(P) = \Lambda(x_1, x_2)...\Lambda(x_{m-1}, x_m)$. A cycle $C$ in $\mathcal{G}$ is *null* if $\Lambda(C) = 1_\Gamma$, *nonnull* otherwise.

Let $\lambda : V \to \Gamma$. Let $F$ be a spanning forest of $\mathcal{G}$, we say that $\lambda$ is a *F-consistent labelling* of $\mathcal{G}$ iff for each path $P$ in $F$ joining $u$ to $v$, $\Lambda(P) = \lambda(u)^{-1}\lambda(v)$. We say that $\lambda$ is a *consistent labelling* of $\mathcal{G}$ iff for each path $P$ in $\mathcal{G}$ joining $u$ to $v$, $\Lambda(P) = \lambda(u)^{-1}\lambda(v)$.

**Lemma 6.** *Let $F$ be a spanning forest of $\mathcal{G}$. Then there exists a $F$-consistent labelling of $\mathcal{G}$.*

**Lemma 7.** *$\mathcal{G}$ has no nonnull cycle iff $\mathcal{G}$ has a consistent labelling.*

### 4.2   The Group Feedback Vertex Set Problem

Given a $\Gamma$-labelled graph $\mathcal{G}$, a *feedback vertex set* of $\mathcal{G}$ is a set of vertices which meets each nonnull cycle.

We consider the GROUP FEEDBACK VERTEX SET (GFVS) problem: given a $\Gamma$-labelled graph $\mathcal{G}$, a set $F \subseteq V$ of forbidden vertices, and an integer $p$, can we find a feedback vertex set of $\mathcal{G}$ disjoint from $F$ and of size at most $p$?

**Theorem 5.** GFVS *is solvable in $O^*((4|\Gamma| + 1)^p)$ time.*

The algorithm relies on iterative compression. We will consider the following auxiliary problem. The problem GFVS COMPRESSION takes

  - a $\Gamma$-labelled graph $\mathcal{G} = (V, A, \Lambda)$,
  - a feedback vertex set $S$ of $\mathcal{G}$,
  - a function $\phi : S \to \Gamma$,
  - a set $F \subseteq V$ of forbidden vertices,

and an integer $p$, and seeks a set $S'$ of $< p$ vertices disjoint from $F$ and which breaks each path $P$ joining two vertices $u, v \in S$ with $\Lambda(P) \neq \phi(u)^{-1}\phi(v)$.

We justify that the GFVS COMPRESSION problem is solvable in $O^*(4^p)$ time (Proposition 1), then we establish the relation with the GFVS problem (Proposition 2), which leads to the proof of Theorem 5.

**Proposition 1.** GFVS COMPRESSION *is solvable in* $O^*(4^p)$ *time.*

*Proof.* We formulate GFVS COMPRESSION as a path cover problem for a homogeneous path system, and apply Theorem 1.

Let $\sigma$ be the path system defined as follows: let $G$ be the underlying graph of $\mathcal{G}$, then $\sigma = (G, S, F)$, and $\mathcal{P}_\sigma$ consists of the paths $P$ joining two vertices $u, v \in S$ with $\Lambda(P) \neq \phi(u)^{-1}\phi(v)$. From the definitions, it is clear that GFVS-COMPRESSION is equivalent to solving the path cover problem for $\sigma$.

It can be shown that: (i) $\sigma$ is homogeneous, (ii) there is a polynomial-time separation oracle for $F_\sigma$. $\qquad\square$

The following proposition establishes the relation between the GFVS and GFVS COMPRESSION problems. Its proof relies on Lemma 7.

**Proposition 2.** *Let* $\mathcal{G} = (V, A, \Lambda)$, *and let* $F \subseteq V$. *Let* $S$ *be a feedback vertex set of* $\mathcal{G}$. *The following are equivalent:*

- *there exists a feedback vertex set* $S'$ *of* $\mathcal{G}$ *s.t.* $S'$ *is disjoint from* $F \cup S$ *and* $|S'| < |S|$;
- *there exists* $\phi : S \to \Gamma$ *s.t.* $I_\phi = (\mathcal{G}, S, \phi, F \cup S, |S|)$ *is a positive instance of* GFVS COMPRESSION.

We are now ready to prove Theorem 5.

*Proof of Theorem 5.* We solve GROUP FEEDBACK VERTEX SET using iterative compression. In the compression step, we are given a subset $V'$ of $V$, a feedback set $S$ of $\mathcal{G}[V']$ disjoint from $F$ and of size $p$, and we seek $S'$ feedback set of $\mathcal{G}[V']$ disjoint from $F$ and of size $< p$. We examine every possibility for $S \cap S'$: for each bipartition of $S = S_1 \cup S_2$, we seek $S' = S_1 \cup S_2'$ with $S_2 \cap S_2' = \emptyset$ and $|S_2'| < |S_2|$. Let $i = |S_2|$, then finding $S_2'$ is done in $O^*(|\Gamma|^i \times 4^i)$ time: by Proposition 2, we need to examine the $|\Gamma|^i$ functions $\phi : S_2 \to \Gamma$, and for each such function to solve GFVS COMPRESSION in $O^*(4^i)$ time. By summing on each possible value of $i$, we obtain that the total time required by the compression step is $\sum_{i=0}^p \binom{p}{i} O^*((4|\Gamma|)^i) = O^*((4|\Gamma| + 1)^p)$. Since there are at most $n$ compression steps, the running time of the algorithm is as claimed. $\qquad\square$

### 4.3   The Group Feedback Arc Set Problem

Given a $\Gamma$-labelled graph $\mathcal{G}$, a *feedback arc set* of $\mathcal{G}$ is a set of arcs which meets each nonnull cycle.

We consider the GROUP FEEDBACK ARC SET (GFAS) problem: given a $\Gamma$-labelled graph $\mathcal{G} = (V, A, \Lambda)$, a set $F \subseteq A$ of forbidden arcs, and an integer $p$, can we find a feedback arc set of $\mathcal{G}$ disjoint from $F$ and of size at most $p$?

**Theorem 6.** GFAS *is solvable in* $O^*((4|\Gamma| + 1)^p)$ *time, and in* $O^*((8p + 1)^p)$ *time.*

We rely on iterative compression in a similar fashion to the proof of Theorem 5. We consider the problem GFAS COMPRESSION which takes

- a $\Gamma$-labelled graph $\mathcal{G} = (V, A, \Lambda)$,
- a feedback vertex set $S$ of $\mathcal{G}$,
- a function $\phi : S \to \Gamma$,
- a set $F \subseteq A$ of forbidden arcs,

and an integer $p$, and seeks a set $S'$ of $< p$ arcs disjoint from $F$ and which breaks each path $P$ joining two vertices $u, v \in S$ with $\Lambda(P) \neq \phi(u)^{-1}\phi(v)$.

**Proposition 3.** GFAS COMPRESSION *is solvable in* $O^*(4^p)$ *time.*

*Proof.* We describe a simple reduction to GFVS COMPRESSION, and conclude using Proposition 1.

Let $I = (\mathcal{G}, S, \phi, F, p)$ be an instance of GFAS COMPRESSION. We create an instance $I' = (\mathcal{G}', S', \phi', F', p')$ of GFVS COMPRESSION as follows. For each arc $a$ of $\mathcal{G}$ we introduce two new vertices $x_a, y_a$. We split each arc $a = (u, v)$ of label $g$ in three arcs $(u, x_a), (x_a, y_a), (y_a, v)$ of respective labels $1_\Gamma, g, 1_\Gamma$. We set $F' = V \cup \{x_a : a \in A\} \cup \{y_a : a \in F\}$. We set $S' = S, \phi' = \phi$ and $p' = p$.

We verify that: $I$ if a positive instance of GFAS COMPRESSION iff $I'$ is a positive instance of GFVS COMPRESSION.    □

The following proposition is similar to Proposition 2, and establishes the relation between the GFAS and GFAS COMPRESSION problems.

**Proposition 4.** *Let* $\mathcal{G} = (V, A, \Lambda)$ *be a* $\Gamma$-labelled graph, and let $F \subseteq A$. Let $S \subseteq A$ be a feedback arc set of $\mathcal{G}$, and let $K \subseteq V$ be a vertex cover of the arcs of $S$. The following are equivalent:

- *there exists a feedback arc set* $S'$ *of* $\mathcal{G}$ *s.t.* $S'$ *is disjoint from* $F \cup S$ *and* $|S'| < |S|$;
- *there exists* $\phi : K \to \Gamma$ *s.t.* $(\mathcal{G}, K, \phi, F \cup S, |S|)$ *is a positive instance of* GFAS COMPRESSION.

Given $S$ feedback arc set of $\mathcal{G}$ of size $p$, we can choose a vertex cover $K$ of size $\leq p$, hence the above Lemma leads to consider $|\Gamma|^p$ functions $\phi$. Clearly, this leads to a $O^*((4|\Gamma| + 1)^p)$ algorithm for the problem. We now argue that the number of functions $\phi$ to examine can be reduced from $|\Gamma|^p$ to $(2p)^p$, which will yield a $O^*((8p + 1)^p)$ algorithm. We first introduce the following definitions.

For each connected component $C$ of $G$ s.t. $C \cap K \neq \emptyset$, choose an element $v_C \in C \cap K$. Let $K_1$ be the set of chosen elements of $K$. Say that a function $\phi : K \to \Gamma$ is *canonical* iff $\phi(x) = 1_\Gamma$ for each $x \in K_1$. Say that a function $\phi : K \to \Gamma$ is *realizable* iff it is canonical and there exists a spanning forest $F$ of $G$ and an $F$-consistent labelling $\lambda$ of $\mathcal{G}$ s.t. $\phi = \lambda | K$. Observe that Proposition 4 can be restated by requiring that $\phi$ is realizable.

The following Lemma gives an upper bound on the number of realizable functions.

**Lemma 8.** *The number of realizable functions is at most* $(2p)^p$.

The proof of Theorem 6 follows.

## 5    Concluding Remarks

The parameterized complexity of some problems considered in this article remains unsettled. We have seen that the VERTEX MULTICUT and MULTICUT problems were FPT w.r.t. $k, p$, and that the GROUP FEEDBACK VERTEX SET problem was FPT w.r.t. $|\Gamma|, p$. We conjecture that these problems are FPT for the single parameter $p$. A first step may be to study these problems e.g. on planar graphs.

We think that some variants of the problems on group-labelled graphs considered in Section 4 are also worth studying from the point of view of parameterized complexity. An interesting generalization of the GROUP FEEDBACK ARC SET problem is the UNIQUE LABEL COVER problem [7,6]. Other problems of interest are satisfiability problems for system of linear equations / inequations, parameterized by the maximum number of unsatisfied equations allowed. These problems may be FPT when restricted to instances with at most two variables per equation.

## References

1. Chen, J., Liu, Y., Lu, S.: An Improved Parameterized Algorithm for the Minimum Node Multiway Cut Problem. In: Dehne, F., Sack, J.-R., Zeh, N. (eds.) WADS 2007. LNCS, vol. 4619, pp. 495–506. Springer, Heidelberg (2007)
2. Chudnovsky, M., Geelen, J., Gerards, B., Goddyn, L.A., Lohman, M., Seymour, P.D.: Packing Non-Zero A-Paths In Group-Labelled Graphs. Combinatorica 5(26), 521–532 (2006)
3. Downey, R.G., Fellows, M.R.: Parameterized Complexity. Springer, Heidelberg (1999)
4. Flum, J., Grohe, M.: Parameterized Complexity Theory. Springer, Heidelberg (2006)
5. Garg, N., Vazirani, V., Yannakakis, M.: Multiway Cuts in Directed and Node Weighted Graphs. In: Shamir, E., Abiteboul, S. (eds.) ICALP 1994. LNCS, vol. 820, pp. 487–498. Springer, Heidelberg (1994)
6. Gupta, A., Talwar, K.: Approximating unique games. In: Proc. SODA 2006, pp. 99–106 (2006)
7. Khot, S.: On the power of unique 2-prover 1-round games. In: Proc. STOC 2002, pp. 767–775 (2002)
8. Marx, D.: Parameterized graph separation problems. Theoretical Computer Science (351), 394–406 (2006)
9. Reed, B., Smith, K., Vetta, A.: Finding odd cycle transversals. Operations Research Letters 32(4), 299–301 (2004)

# Wheel-Free Deletion Is $W[2]$-Hard

Daniel Lokshtanov

Department of Informatics, University of Bergen N-5020 Bergen, Norway
daniello@ii.uib.no

**Abstract.** We show that the two problems of deciding whether $k$ vertices or $k$ edges can be deleted from a graph to obtain a wheel-free graph is $W[2]$-hard. This immediately implies that deciding whether $k$ edges can be added to obtain a graph that contains no complement of a wheel as an induced subgraph is $W[2]$-hard, thereby resolving an open problem of Heggernes et al. [7] (STOC07) who ask whether there is a polynomial time recognizable hereditary graph class $\Pi$ with the property that computing the minimum $\Pi$-completion is $W[t]$-hard for some $t$.

## 1  Introduction

For a graph property $\Pi$ and an input graph $G$, a $\Pi$-*completion* of $G$ is a graph $H$ that has the property $\Pi$ and contains $G$ as a subgraph. We say that $H$ is a *minimum $\Pi$-completion* of $G$ if $H$ is a $\Pi$-completion of $G$ that minimizes the number of edges needed to add to $G$ in order to obtain $H$, and that the *minimum $\Pi$-completion problem* is the problem of obtaining such an $H$ when given $G$ as input. The first completion problem to be studied was the chordal-completion problem. This problem has been subjected to considerable scrutiny, due to a wide range of applications, such as sparse matrix computations [16], database management [17][1], knowledge based systems [10], and computer vision [3]. The computational complexity of finding minimum chordal-completions was settled when Yannakakis in [18] showed that the problem is NP-complete. Subsequently, it was shown that most interesting completion problems also are NP-complete [12][6][5].

Completion problems fall naturally within the class of *graph modification problems*. In a graph modification problem you are given a graph $G$ as input, and asked to convert $G$ into a graph with a property $\Pi$, modifying $G$ as little as possible. Specifically, you are given the graph $G$ together with three integers $i$, $j$, $k$ and asked whether $G$ can be made into a graph with the proberty $\Pi$ by deleting at most $i$ edges and $j$ vertices, and adding at most $k$ edges. When $i = j = 0$ it is easy to see that the problem reduces to the *minimum $\Pi$-completion problem*, whilst the cases where $j = k = 0$ and $i = k = 0$ are referred to as the *minimum $\Pi$-edge deletion* and *minimum $\Pi$-vertex deletion* problems respectively.

Graph modification problems have been studied extensively from the perspective of parameterized complexity. From the graph minor theory of Robertson and

M. Grohe and R. Niedermeier (Eds.): IWPEC 2008, LNCS 5018, pp. 141–147, 2008.

Seymour, it follows that the minimum $\Pi$-vertex deletion problem is fixed parameter tractable (FPT) if $\Pi$ is minor closed [15][14]. Kaplan, Shamir and Tarjan showed that the minimum chordal-completion, strongly chordal-completion, and proper interval-completion problems all are FPT using a bounded search tree approach [8]. The FPT algorithm for finding minimum chordal-completions was later improved by Cai, who also showed that the graph modification problem is fixed parameter tractable for all hereditary graph classes with a finite set of forbidden subgraphs [2]. More recent results include FPT algorithms for minimum interval-completion [7], bipartite-vertex deletion [13] and chordal-vertex and edge deletion [11]. One can also observe that two of the classical fixed parameter tractable problems in parameterized complexity, Vertex Cover and Feedback Vertex Set, can be seen as independent-set-vertex deletion and forest-vertex deletion respectively.

An interesting point about the above results, is that *they are all positive*. That is, to the author's best knowledge, all reasonable[1] graph modification problems that have been studied to this date have turned out to be fixed parameter tractable. This has given rise to speculation on whether it is possible that all graph modification problems of a certain kind could turn out to be FPT. Specifically, it was raised as an open problem by Heggernes et al. [7] whether it is possible that the $\Pi$-completion problem is FPT for every polynomial time recognizable hereditary graph class $\Pi$. We resolve this open problem by showing that this is *not* the case unless $FPT = W[2]$, by showing that the wheel-free-vertex deletion and wheel-free-edge deletion problems both are hard for $W[2]$.

Our proof of hardness is fairly simple, but contains an idea of how characterizations of the graph class $\Pi$ through "special" vertices can be employed to show that $\Pi$-modification problems are hard. The class of wheel-free graphs, while being constructed so as to make our hardness proof go through, is not so far fetched and therefore gives an indication that for other, more natural graph classes their corresponding graph modification problems might well turn out to be $W[2]$-hard. We hope that a refinement of our proof technique can yield a way to prove $W[2]$-hardness of vertex or edge-deletion into other, more "popular" graph classes, and potentially be a step towards a dichotomy of the parameterized complexity of graph modification problems.

## 2   Notation, Terminology and Preliminaries

A vertex $v$ in a graph $G$ is said to be *universal* if $v$ is adjacent to all other vertices of $G$. Given a graph $G = (V, E)$, the graph $G[S] = (S, E \cap \{uv : u \in S \wedge v \in S\})$ is the subgraph of $G$ *induced* by $S$. For a graph $H$ we say that $G = (V, E)$ contains $H$ as an induced subgraph if there is a set $S \subset V$ such that $G[S]$ is isomorphic to $H$. A graph family $\Pi$ is *hereditary* if for every $G \in \Pi$ all induced subgraphs of $G$ are in $\Pi$. For a graph family $\Pi$ and positive integer $k$, we define the two graph families $\Pi + kv$ and $\Pi + ke$ to be the families of all graphs that can be

---

[1] By *reasonable* we mean that the graph class considered is polynomial time recognizable and hereditary.

made into a graph in $\Pi$ by deleting at most $k$ vertices or edges respectively. Let $\overline{\Pi}$ be the class of all graphs whose complement belongs to $\Pi$.

A *wheel* is a graph $W$ that has a universal vertex $v$ such that $W \setminus v$ is a cycle. We say that $v$ is *apex* for this wheel, and that $W \setminus v$ is the cycle of the wheel. For a general graph $G$ we say that a vertex $v$ is apex if $v$ is apex for an induced wheel in $G$. The graph $W_k$ for $k \geq 3$ is the wheel such that the cycle of the wheel has $k$ vertices. We will refer to the family of graphs that do not contain any wheel as an induced subgraph as $\mathcal{W}$, the family of *wheel-free* graphs.

For a graph family $\Pi$ and positive integer $k$, we define the two graph families $\Pi + kv$ and $\Pi + ke$ to be the families of all graphs that can be made into a graph in $\Pi$ by deleting at most $k$ vertices or edges respectively. Let $\overline{\Pi}$ be the class of all graphs whose complement belongs to $\Pi$.

Before we turn to the main section of this paper, we observe that the class of wheel-free graphs is hereditary by definition. The graph class is also polynomial time recognizable, because of the following observation.

**Observation 1** *A graph $G$ is wheel-free if and only if no vertex of $G$ is apex.*

Using Observation 1 we can test whether a graph $G$ is wheel-free simply by iterating through every vertex $v$ and verifying that $N(v)$ induces a forest in $G$.

## 3    Wheel-Free Deletion Is $W[2]$ Hard

In this section we show that recognizing $\mathcal{W} + kv$ and $\mathcal{W} + ke$ graphs is hard for $W[2]$ when parameterized by $k$. We reduce from Hitting Set, and in fact, we reduce simultaneously to Wheel-free Vertex Deletion and to Wheel-free Edge Deletion. That is, given an instance of Hitting Set we will build a graph $G$ such that $G$ belongs to $\mathcal{W} + ke$ if and only if the instance to Hitting Set is a "yes" instance, and so that $G$ belongs to $\mathcal{W} + ke$ if and only if $G$ belongs to $\mathcal{W} + kv$. We proceed to formally define the problem we reduce from.

HITTING SET
INSTANCE: A tuple $(U, \mathcal{F}, k)$ where $\mathcal{F}$ is a collection of subsets of the finite universe $U$, and a positive integer $k$
PARAMETER: $k$
QUESTION: Is there a subset $X$ of $U$ of cardinality at most $k$ such that for every $Z \in \mathcal{F}$, $Z \cap X$ is nonempty?

**Lemma 2.** *[4] Hitting Set is $W[2]$-complete.*

If the answer to an instance of Hitting Set is yes, we say that $X$ is a $k$-Hitting Set, and that $(U, \mathcal{F}, k)$ has a $k$-Hitting Set. For an instance $(U, \mathcal{F}, k)$ of Hitting Set, let $n = |U|$ and $m = |\mathcal{F}|$. We build a graph $G' = (V', E')$ as follows. For every element $e$ in $U$ we make two vertices $e_1$ and $e_2$ and connect them by an edge. We say that the vertices $e_1$ and $e_2$ *correspond* to the element $e$ of $U$. Furthermore, for every set $S$ in $\mathcal{F}$ we make a $W_{3n}$ and distinguish an induced matching of

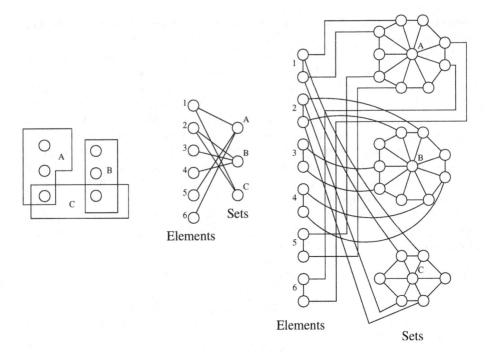

**Fig. 1.** On the left hand side we see an instance of Hitting Set. In the middle we have the element-set incidence graph of the instance, and on the right hand side the graph $G'$ as computed from the instance. On the left in $G'$ we see the vertices corresponding to the elements and on the right the wheels corresponding to sets. The special edges are the edges going from the element vertices to the wheels. We construct $G$ by contracting the special edges. In fact, the figure is not entirely accurate, as each wheel should have had 18 vertices in the cycle according to the construction. These omitted vertices are not drawn in order to to keep the size of the figure down, and they do not have any effect.

size $n$ in the cycle of new wheel. To each edge $uv$ of the distinguished induced matching we assign an element of $U$, say $e$. If $S$ contains $e$, we also add *special* edges between $u$ and $e_1$ and between $v$ and $e_2$. We say that the constructed wheel corresponds to the set $S$. This concludes the construction of $G'$. We are not done, however. To finalize the reduction we obtain the graph $G = (V, E)$ from $G'$ by contracting all the special edges. For a vertex $v$ in $G'$ we say that $\alpha(v)$ is the *image* of $v$ in $G$, that is the vertex of $G$ that $v$ gets contracted into. If a vertex $v$ is not incident to any special edges $\alpha(v) = v$ and $v$ is a vertex both in $G'$ and $G$. Finally, observe that if $x$ and $y$ are vertices of $G'$ that correspond to distinct elements of $U$, the images of $x$ and $y$ are nonadjacent in $G$.

**Lemma 3.** *The following are equivalent:* (1) $(U, \mathcal{F}, k)$ *has a $k$-Hitting Set,* (2) *$G$ is in $\mathcal{W} + ke$ and* (3) *$G$ is in $\mathcal{W} + kv$.*

*Proof.* We prove the equivalences by providing a circle of implications, namely that (1) implies (2), (2) implies (3) and that (3) implies (1).

*Claim.* If $(U, \mathcal{F}, k)$ has a $k$-Hitting Set then $G$ is in $\mathcal{W} + ke$.

*Proof.* Suppose $(U, \mathcal{F}, k)$ has a $k$-Hitting Set $X$. For every element $e \in X$ we remove the edge between the image of $e_1$ and the image of $e_2$ in $G$ to obtain a graph $H$. It remains to prove that $H$ is wheelfree. We do this by proving that no vertex is apex. Let $E_D$ be $E(G) \setminus E(H)$. Consider a vertex $v$ in $H$ that was the apex vertex of a $W_{3n}$ in $G'$. In $G$ the neighbourhood of $v$ induces a cycle, and since $X$ is a Hitting Set, $E_D$ contains at least one of the edges of this cycle. Hence $v$ is not apex in $H$. Consider now a vertex $v$ in $H$ that was in the cycle of some $W_{3n}$ in $G'$ and that had no special edges incident to it. In $G'$ the neighbourhood of $v$ induces a $P_3$ and since the image of vertices that correspond to different elements of $U$ is nonadjacent in $G$, the neighbourhood of $v$ induces a $P_3$ also in $G$. As $H \subseteq G$, it follows that $v$ is not apex in $H$. Finally, consider a vertex $v$ in $H$ that is the image of a vertex of $G'$ that was adjacent to a special edge. In this case $v$ must be the image of a vertex of $G'$ that corresponded to an element $e$ of $U$. Without loss of generality we can assume that $v = \alpha(e_1)$. The neighbourhood of $v$ is the union of the neighbourhoods of $v$ in all the $W_{3n}$'s. If $\alpha(e_1)\alpha(e_2) \notin E_D$ the neighbourhood of $v$ in each $W_{3n}$ induces a $P_3$ with $\alpha(e_2)$ being one of the endpoints. Thus, if $\alpha(e_1)\alpha(e_2) \notin E_D$ the neighbourhood of $v$ in $H$ induces a tree, and if $\alpha(e_1)\alpha(e_2) \in E_D$ the neighbourhood of $v$ in $H$ induces a matching. In both cases $v$ is not apex, and we are done.

*Claim.* If $G$ is in $\mathcal{W} + ke$ then $G$ is in $\mathcal{W} + kv$.

*Proof.* Observe that if $|E_D| = k$ and $H = G \setminus E_D$ is a wheel-free graph, there is a set $V_D$ of cardinality at most $k$ such that every edge in $E_D$ is incident to some edge in $V_D$. Thus, as $G \setminus V_D = G \setminus E_D[V \setminus V_D]$ and the class of wheel-free graphs is hereditary, $G \setminus V_D$ is wheel-free.

*Claim.* If $G$ is in $\mathcal{W} + kv$ then $(U, \mathcal{F}, k)$ has a $k$-Hitting Set.

*Proof.* For a given set $S$ in $\mathcal{F}$, let $V'_S$ be the vertex set in $G'$ of the wheel corresponding to $S$. Let $V_S$ be the image of $V'_S$. Clearly, $V_S$ induces a wheel in $G$. Without loss of generality, we can assume that every element of $U$ is contained in some set of $\mathcal{F}$ and that every set in $\mathcal{F}$ is nonempty. From this it follows that $\bigcup_{S \in \mathcal{F}} V_S = V(G)$. Furthermore, from the construction of $G$, it follows that any vertex $v$ that is contained in $V_S \cap V_{S'}$ for a pair of distinct sets $S$ and $S'$ in $\mathcal{F}$ must correspond to an element $e \in U$. Having this in mind, we construct a mapping $f : V(G) \to U$ as follows: if $v$ corresponds to an element $e$ of $U$, then $f(v) = e$. Otherwise we let $f$ map $v$ to an *arbitrary* element of the *unique* set $S \in \mathcal{F}$ such that $v \in V_S$.

Now, suppose there is a set of vertices $V_D$ of cardinality at most $k$ such that $G \setminus V_D$ is wheel-free. We prove that $X = \{f(v) : v \in V_D\}$ is a $k$-hitting set. First, observe that by construction $|X| \leq k$. Finally, for any set $S \in \mathcal{F}$ we have that $V_S \cap V_D \neq \emptyset$. Let $v$ be a vertex in $V_S \cap V_D$. From the construction of the mapping $f$ it follows that $f(v) \in S$, and that $f(v) \in X$. Thus $X \cap S$ is nonempty for every set $S \in \mathcal{F}$ so $X$ must be a $k$-hitting set.

Together, the three claims complete the proof of Lemma 3.

**Theorem 1.** *Recognizing $\mathcal{W} + ke$ and $\mathcal{W} + kv$ graphs is $W[2]$ hard when parameterized by $k$.*

*Proof.* The proof follows directly from the construction of $G$ and Lemma 3.

From the above theorem it immediately follows that completing into the class of graphs that do not contain the complement of a wheel as an induced subgraph is $W[2]$ hard. Thus we get a corollary that answers the question posed by Heggernes et al. by providing the first polynomial time recognizable hereditary graph class $\Pi$ such that completing into $\Pi$, that is recognizing $\Pi - ke$ is $W[t]$-hard for some $t$.

**Corollary 1.** *Recognizing $\overline{\mathcal{W}} - ke$ graphs is $W[2]$ hard when parameterized by $k$.*

## 4   Conclusions

In this paper we have shown that graph modification problems indeed can be hard from a parameterized point of view. Hopefully, this result is a step towards understanding the parameterized complexity of completion and deletion problems for polynomial time recognizable, hereditary graph classes. While obtaining a dichotomy for these problems might turn out to be a daunting task, it might also be that general results are achievable through clever use of combinatorics or algorithmical tricks. For instance, Khot and Raman gave a dichotomy for the parameterized complexity of $(n - k)$-vertex-deletion problems [9], the *parametric duals* of minimum vertex-deletion problems by using Ramsey numbers in a smart way. If general results turn out to be too difficult to obtain, it would be interesting to see whether all of the "popular" graph classes, such as permutation graphs, AT-free graphs and perfect graphs turn out to have fixed parameter tractable graph modification problems, or if some of these graph modificatoion problems turn out to be hard for $W[t]$ for some $t$.

## References

1. Beeri, C., Fagin, R., Maier, D., Yannakakis, M.: On the desirability of acyclic database systems. J. Assoc. Comput. Mach. 30, 479–513 (1983)
2. Cai, L.: Fixed-parameter tractability of graph modification problems for hereditary properties. Inf. Process. Lett. 58(4), 171–176 (1996)
3. Chung, F.R.K., Mumford, D.: Chordal completions of planar graphs. J. Comb. Theory 31, 96–106 (1994)
4. Downey, R.G., Fellows, M.R.: Parameterized Complexity (Monographs in Computer Science). Springer, Heidelberg (1998)
5. Garey, M.R., Johnson, D.S.: Computers and Intractability: A Guide to the Theory of NP-Completeness. W.H. Freeman, New York (1979)
6. Hakimi, S.L., Schmeichel, E.F., Young, N.E.: Orienting graphs to optimize reachability. Information Processing Letters 63(5), 229–235 (1997)

7. Heggernes, P., Paul, C., Telle, J.A., Villanger, Y.: Interval completion with few edges. In: STOC 2007: Proceedings of the thirty-ninth annual ACM symposium on Theory of computing, pp. 374–381. ACM Press, New York (2007)

8. Kaplan, H., Shamir, R., Tarjan, R.E.: Tractability of parameterized completion problems on chordal, strongly chordal, and proper interval graphs. SIAM J. Comput. 28(5), 1906–1922 (1999)

9. Khot, S., Raman, V.: Parameterized complexity of finding subgraphs with hereditary properties. Theor. Comput. Sci. 289(2), 997–1008 (2002)

10. Lauritzen, S.L., Spiegelhalter, D.J.: Local computations with probabilities on graphical structures and their applications to expert systems. J. Royal Statist. Soc., ser. B 50, 157–224 (1988)

11. Marx, D.: Chordal deletion is fixed-parameter tractable (manuscript, 2007)

12. Natanzon, A., Shamir, R., Sharan, R.: Complexity classification of some edge modification problems. Discrete Applied Mathematics 113(1), 109–128 (1999)

13. Reed, B.A., Smith, K., Vetta, A.: Finding odd cycle transversals. Oper. Res. Lett. 32(4), 299–301 (2004)

14. Robertson, N., Seymour, P.D.: Graph minors. xiii. the disjoint paths problem. J. Combin. Theory Ser. B. 63, 65–110 (1995)

15. Robertson, N., Seymour, P.D.: Graph minors. xx. wagners conjecture. J. Combin. Theory Ser. B. 92, 325–357 (2004)

16. Rose, D.J.: A graph-theoretic study of the numerical solution of sparse positive definite systems of linear equations. Graph Theory and Computing, 183–217 (1972)

17. Tarjan, R.E., Yannakakis, M.: Simple linear-time algorithms to test chordality of graphs, test acyclicity of hypergraphs, and selectively reduce acyclic hypergraphs. SIAM J. Comput. 13, 566–579 (1984)

18. Yannakakis, M.: Computing the minimum fill-in is np-complete. SIAM Journal on Algebraic and Discrete Methods 2, 77–79 (1981)

# Parameterized Derandomization

Moritz Müller

Mathematisches Institut, Albert-Ludwigs-Universität Freiburg,
Eckerstrasse 1, 79104 Freiburg - Germany
moritz.mueller@math.uni-freiburg.de

**Abstract.** The class W[P] is a parameterized analogue of NP. Chen et al. [4] have given a machine characterization of W[P]. The corresponding machine model gives rise to a parameterized analogue of BPP. What is the connection between parameterized and classical derandomization?

## 1   Introduction

*Parameterized complexity:* In the parameterized setting instances of problems come along with a parameter, a natural number $k$. The parameter is expected to be small compared to the instance in typical situations. It is intended to encode some knowledge we have about typical instances – e.g. that a certain part of it is small compared to the rest or that it has a simple structure. To allow full exploitation of this knowledge the notion of tractability is adjusted accordingly. A parameterized problem is *fixed-parameter tractable* if and only if it is solvable in time $f(k) \cdot n^{O(1)}$ for some computable function $f$. Here $n$ denotes the length of the input and $k$ its associated parameter.

While this tractability notion led to a host of algorithmic techniques (see [13] for a survey), also a big variety of seemingly intractable classes have been found. On the positive turnside these classes allow for a fine-grained intractability analysis and only a handful of them play a dominant role. Important are the classes W[1], W[P] and paraNP as they can be seen as parameterized analogues of NP.

*Randomization:* In classical complexity theory a randomized polynomial time algorithm can be viewed as a binary "NP-machine", where a run on some input is determined by a sequence in $\{0, 1\}$. This sequence can be interpreted as the outcome of independent "coin tosses". The amount of randomness, classically that is, the number of coins, is a computational resource we want to spare.

The machine characterizations of parameterized intractable classes [4], specifically those of the mentioned NP analogues, motivate different modes of parameterized randomization [10]. E.g. we get a notion of W[P]-randomized computations by replacing "NP-machine" in the classical definition by "W[P]-machine". This amounts to Turing machines using at most $f(k) \cdot \log n$ coins. Implemented on a nondeterministic Random Access Machine we may say that the algorithm uses few, $f(k)$ many, but large, $n^{O(1)}$-sided, dice. We denote the corresponding analogue of BPP by W[P]-BPFPT.

M. Grohe and R. Niedermeier (Eds.): IWPEC 2008, LNCS 5018, pp. 148–159, 2008.
© Springer-Verlag Berlin Heidelberg 2008

This way, parameterized complexity theory provides us with a genuine view on randomization – a direction shown up by Downey et al. in [5].

*Randomness in parameterized complexity:* Downey et.al. [5] have introduced a parameterized analogue of the classical BP operator. Within this frame Montoya originated the theoretical investigation of W[P]-randomized computations. In particular he proved some strong form of probability amplification [9] for the class W[P]-BPFPT.

Examples of parameterized randomized algorithms as the approximate counting algorithm of Arvind and Raman [2] or the reduction giving an analogue of the Valiant-Vazirani Lemma in [5] use paraNP-randomized computations. Assumptions that certain hard parameterized problems are also intractable in the one or the other randomized sense have been employed by Alekhnovich and Razborov [1] to show that short resolution refutations are hard to find, or by Chen and Flum [3] to disprove #W[1]-hardness of certain counting problems.

Concerning these examples the quest for derandomization has led to some positive results. W[P]-randomized approximate counting is considered in [10]. The Valiant-Vazirani Lemma [5] has been derandomized to W[1]-randomized reductions [11] and applied in [3]. And recently Eickmeyer et al. [6] derandomized Alekhnovich's and Razborov's result.

However, up to now little work has been done on a theoretical frame for randomness in the parameterized world.

*This paper:* It is widely believed that P = BPP. In this case all parameterized modes of randomization are derandomizable. In this situation one should know what derandomization of the more restricted modes of parameterized randomized computation means in terms of classical complexity theory. This is the question considered in this paper. Specifically we ask how difficult it is to derandomize W[P]-BPFPT. Our main result reads

**Theorem 1.** *The following statements are equivalent.*

1. *W[P]-BPFPT has a weakly uniform derandomization.*
2. *There is a polynomial time computable, nondecreasing, unbounded function*
   $c : \mathbb{N} \to \mathbb{N}$ *such that* $\mathrm{BPP}[c] = \mathrm{P}$.

Here $\mathrm{BPP}[c]$ is BPP restricted to at most $c(n) \cdot \log n$ many random bits. Observe that for bounded $c$ we trivially have $\mathrm{BPP}[c] = \mathrm{P}$. Thus nontrivial classical derandomization would mean to show $\mathrm{BPP}[c] = \mathrm{P}$ for some unbounded $c$. Weak uniformity is some extra condition on W[P]-BPFPT = FPT which is mathematically strong, but, as we will argue, philosophically weak.

Informally Theorem 1 states that (weakly uniform) derandomization of W[P]-randomized computations means the same as any nontrivial classical derandomization. Note that $c$ can grow extremly slow.

The mentioned statements form the content of Section 4. In fact we show more than Theorem 1 (see Theorem 15 in Section 4). First we define parameterized randomized classes in Section 2. Section 3 contains a technical lemma stating how to transform a given W[P]-randomized program to one using larger dice. The lemma turns out to be very useful. We examplify this by giving two applications, first how to get programs using small dice and second how to characterize W[P]-randomized computations by Turing machines.

## 2    Parameterized Randomization

*Parameterized problems:* Fix a finite alphabet $\Sigma$ containing at least two letters. *Parameterized problems* are pairs $(Q, \kappa)$ of classical problems $Q \subseteq \Sigma^*$ and polynomial time computable *parameterizations* $\kappa : \Sigma^* \to \mathbb{N}$. A parameterized problem $(Q, \kappa)$ is *fixed-parameter tractable* if and only if it can be solved by an algorithm with *fpt running time* (with respect to $\kappa$), i.e. time $f(\kappa(x)) \cdot |x|^{O(1)}$ for some computable $f : \mathbb{N} \to \mathbb{N}$. The class of parameterized problems $(Q, \kappa)$ solvable in time $f(\kappa(x)) \cdot |x|^{O(1)}$ is denoted by $O^*(f)$.

*Machines for intractable classes:* We define W[1] and W[P] by their machine characterizations [4]. The machine model is that of a *nondeterministic RAM*. These are usual RAMs [12] using registers $0, 1, \ldots$, whose contents are natural numbers $r_0, r_1, \ldots$. Additionally to the usual instructions they can GUESS:

"guess a natural number $< r_0$ and store it in register 0."

A *program* is a finite sequence of instructions. Runs and acceptance are defined as usual. An execution of GUESS is a *nondeterministic step*. We use the uniform cost measure.

Let $\kappa$ be a parameterization. A program $\mathbb{P}$ is $\kappa$-*restricted* if and only if there are computable functions $f, g$ and a polynomial $p$ such that for all $x \in \Sigma^*$ and each run of $\mathbb{P}$ on $x$ the program $\mathbb{P}$ performs at most $g(\kappa(x))$ many nondeterministic steps and the number $f(\kappa(x)) \cdot p(|x|)$ upper bounds the number of steps, the registers used and the numbers stored in any register at any time. If additionally for some computable $h : \mathbb{N} \to \mathbb{N}$ all nondeterministic steps occur within the last $h(\kappa(x))$ many steps, we call the program *tail-nondeterministic*.

**Definition 2**   – W[1] is the class of all parameterized problems $(Q, \kappa)$ decidable by some $\kappa$-restricted tail-nondeterministic program.
   – W[P] is the class of all parameterized problems $(Q, \kappa)$ decidable by some $\kappa$-restricted program.
   – paraNP is the class of all parameterized problems $(Q, \kappa)$ decidable by some nondeterministic Turing machine with fpt running time (with respect to $\kappa$).

*Modes of parameterized randomization:* Instead of flipping coins our programs "roll dice" (execute GUESS). In order to get these rolls induce the uniform measure on runs, the program should besides being exact "always use the same die".

**Definition 3**   – A program $\mathbb{P}$ has *uniform guess bounds* if and only if for all $x \in \Sigma^*$ the content $r_0$ of register 0 is the same for any two nondeterministic steps in (possibly distinct) runs of $\mathbb{P}$ on $x$.

– A program or a Turing machine is *exact* if and only if for every $x$ it performs the same number of nondeterministic steps in every run on $x$.

This parellels in the classical frame the restriction to exact binary machines.

**Definition 4.** Let $\kappa$ be a parameterization.

– An exact binary nondeterministic Turing machine with fpt running time with respect to $\kappa$ is *paraNP-randomized* (with respect to $\kappa$).
– An exact $\kappa$-restricted program with uniform guess bounds is *W[P]-randomized* (with respect to $\kappa$).
– An exact $\kappa$-restricted tail-nondeterministic program with uniform guess bounds is *W[1]-randomized* (with respect to $\kappa$).

Given a (with respect to $\kappa$) W[P]- or W[1]-randomized program $\mathbb{P}$, assume that for some computable $g$ it performs on input $x \in \Sigma^*$ exactly $g(\kappa(x))$ many non-deterministic steps[1], say, each with $r_0 = n_x$. Then we say, that $\mathbb{P}$ *on $x$ uses* $g(\kappa(x))$ *many $n_x$-sided dice.*

**Notation:** By $\mathbb{P}(x)$ we denote the output of $\mathbb{P}$ on $x$, that is the function mapping a random seed (outcome of dice rolls) for $\mathbb{P}$ on $x$ to the output of $\mathbb{P}$ on $x$ of the run determined by the random seed. $\mathbb{P}(x)$ is a random variable on the Laplace space of random seeds $\{0, \ldots, n_x - 1\}^{g(\kappa(x))}$. As usual we sloppily denote various probability measures always by Pr and let the context determine what is meant. E.g. when talking about $\mathbb{P}(x)$, by Pr we refer to the uniform measure on $\{0, \ldots, n_x - 1\}^{g(\kappa(x))}$.

We get a parameterized randomized class by selecting a mode of randomization, choosing one- or two-sided error and choosing a bound on the error. For notational simplicity we always choose the error bound $1/|x|$. In this work, our arguments are not sensible to this choice, e.g. $1/4$ would be equally good.

**Definition 5.** A parameterized problem $(Q, \kappa)$ is in W[P]-BPFPT if and only if there is a W[P]-randomized program $\mathbb{P}$ with respect to $\kappa$ such that for all $x \in \Sigma^*$

$$\Pr\left[\mathbb{P}(x) \neq \chi_Q(x)\right] < 1/|x|.$$

Here $\chi_Q$ is the characteristic function of $Q$. We then say that $\mathbb{P}$ *decides $Q$ (with two-sided error $1/|x|$)*. We apply a similar mode of speech for Turing machines. W[1]-BPFPT and paraNP-BPFPT are similarly defined.

**Remark 6.** It is not hard to see that W[P]-BPFPT is contained in XP, because the number of random seeds of a W[P]-randomized program on an input $x$ can be bounded by $|x|^{f(\kappa(x))}$ for some computable $f : \mathbb{N} \to \mathbb{N}$.

---

[1] It is not hard to see that for each $\mathbb{P}$ there is a $\mathbb{P}'$ with this property and identically distributed outputs.

# 3   The Dice Lemma

*Large dice for* W[P]-BPFPT: Say we run a W[1]-randomized algorithm with larger dice having a desired number of sides by interpreting in some way each outcome of a roll with a large die as an outcome of a roll with a small die. The new program can be seen as the old one using a defective random source, i.e. a biased die. We need to estimate the loss of the success probability. First, intuitively, the larger the new die (compared with the old one) the better. Secondly the more dice the worse. This is made precise by the "Dice Lemma".

Recall that the *distance in variation* of random variables $X, Y$ both with range, say, $E$ is

$$d_V(X,Y) := \sup_{A \subseteq E} \big| \Pr[X \in A] - \Pr[Y \in A] \big|.^2$$

As usual we write $X \sim Y$ if $d_V(X,Y) = 0$, i.e. if $X$ and $Y$ are identically distributed.

**Lemma 7 (Dice Lemma).** *Let $\kappa$ be a parameterization and let $g : \mathbb{N} \to \mathbb{N}$ be computable. Let $\mathbb{P}$ be a W[P]-randomized program which on $x \in \Sigma^*$ uses $g(\kappa(x))$ many $n_x$-sided dice. Let $(q_x)_{x \in \Sigma^*} : \Sigma^* \to \mathbb{N}$ be computable in fpt time with respect to $\kappa$ (output coded in unary) such that $n_x \leq q_x$ for all $x \in \Sigma^*$.*

*Then there is a W[P]-randomized program $\mathbb{P}'$ which on $x \in \Sigma^*$ uses $g(\kappa(x))$ many $q_x$-sided dice such that for all $x \in \Sigma^*$*

$$d_V(\mathbb{P}'(x), \mathbb{P}(x)) \leq \begin{cases} 0 & \text{if } n_x \text{ divides } q_x \\ g(\kappa(x)) \cdot n_x/q_x & \text{else} \end{cases}.$$

We omit the technical proof (see the Appendix) and give two applications.

*Small dice for* W[P]-BPFPT:

**Corollary 8.** *Let $(Q, \kappa)$ be a parameterized problem that can be decided by a W[P]-randomized program which on $x \in \Sigma^*$ uses $g(\kappa(x))$ many dice. Then $(Q, \kappa)$ can be decided by a W[P]-randomized program which on $x \in \Sigma^*$ uses $O(g(\kappa(x)))$ many $|x|$-sided dice.*

*Proof.* Choose a program $\mathbb{P}$ deciding $(Q, \kappa)$ according to the assumption. Say, $\mathbb{P}$ on $x$ uses $g(\kappa(x))$ many $n_x$-sided dice. By $\kappa$-restrictedness we have $n_x \leq h(\kappa(x)) \cdot |x|^c$ for some computable $h$ and some $c \in \mathbb{N}$.

We want a program $\mathbb{P}'$ which on $x$ uses $|x|$-sided dice and decides $Q$ with two-sided error at most $2/|x|$. We get an error below $1/|x|$ by running our program some constant number of times and taking a majority vote.

Furthermore it is enough if $\mathbb{P}'$ works as desired on inputs $x$ with $|x| \geq h(\kappa(x)) \cdot g(\kappa(x))$, because on shorter inputs a deterministic XP algorithm for $(Q, \kappa)$ needs time which can be effectively bounded in terms of the parameter.

---

[2] By the usual sloppy convention the two occurences of Pr refer respectively to the possibly different measures underlying the spaces of $X$ respectively $Y$.

Apply the Dice Lemma to get a program $\mathbb{P}''$ which on $x$ uses $g(\kappa(x))$ many $|x|^{c+1}$-sided dice. Then on inputs $x$ with $|x| \geq h(\kappa(x)) \cdot g(\kappa(x))$ the program $\mathbb{P}''$ answers incorrectly with probability at most

$$1/|x| + g(\kappa(x)) \cdot h(\kappa(x)) \cdot |x|^c/|x|^{c+1} \leq 2/|x|.$$

Let $B : \{0, \ldots, |x| - 1\}^{c+1} \to \{0, \ldots, |x|^{c+1} - 1\}$ be some polynomial time computable bijection. Whenever $\mathbb{P}''$ rolls one of its $|x|^{c+1}$-sided die $\mathbb{P}'$ rolls $(c + 1)$ many $|x|$-sided dice, say with outcome $(a_1, \ldots, a_{c+1})$, and continues the simulation with $B(a_1, \ldots, a_{c+1})$. It is clear that $\mathbb{P}'(x) \sim \mathbb{P}''(x)$ and that $\mathbb{P}'$ is a W[P]-randomized program using $(c + 1) \cdot g(\kappa(x))$ many $|x|$-sided dice. $\qquad\square$

In the following sense improving the above lower bound on the size of dice amounts to derandomization. This may serve as a derandomization lemma.

Recall that $f \in o^{\mathrm{eff}}(g)$ if and only if there is a computable, nondecreasing, unbounded $\iota : \mathbb{N} \to \mathbb{N}$ such that for all sufficiently large $n \in \mathbb{N}$

$$f(n) \leq g(n)/\iota(n).$$

**Lemma 9.** *Let $(Q, \kappa)$ be a parameterized problem. If $(Q, \kappa)$ can be decided by a W[P]-randomized program which on $x \in \Sigma^*$ uses $|x|^{o^{\mathrm{eff}}(1)}$-sided dice, then $(Q, \kappa) \in \mathrm{FPT}$.*

*Proof.* Choose an algorithm $\mathbb{A}$ according to the assumption. Choose a computable $g : \mathbb{N} \to \mathbb{N}$ and a computable, nondecreasing and unbounded $\iota : \mathbb{N} \to \mathbb{N}$ such that $\mathbb{A}$ on input $x \in \Sigma^*$ uses $g(\kappa(x))$ many dice with at most $|x|^{1/\iota(|x|)}$ sides. Let $h : \mathbb{N} \to \mathbb{N}$ be some computable, nondecreasing function such that $h(\iota(n)) \geq n$ for all $n \in \mathbb{N}$ (see e.g. Lemma 16 for how to find such an $h$).

Let $\mathbb{A}'$ on $x \in \Sigma^*$ simulate $\mathbb{A}$ on $x$ exhaustively on all its random seeds. We show that $\mathbb{A}'$ runs in fpt time. This is the case if the number of random seeds of $\mathbb{A}$ on $x$ obeys an fpt bound. Write $n := |x|$ and $k := \kappa(x)$. We distinguish two cases:

- if $g(k) < \iota(n)$, then there are at most $(n^{1/\iota(n)})^{g(k)} < n$ many random seeds.
- if $g(k) \geq \iota(n)$, then $h(g(k)) \geq h(\iota(n)) \geq n$, and hence there are at most $h(g(k))^{g(k)}$ many random seeds. $\qquad\square$

*Turing machines for W[P]-BPFPT:* As a second application of the Dice Lemma we characterize W[P]-BPFPT by Turing machines. This shows a certain robustness of the class and provides a link to classical classes which are usually defined in terms of Turing machines.

**Corollary 10.** *Let $(Q, \kappa)$ be a parameterized problem. The following statements are equivalent.*

1. *$(Q, \kappa) \in$ W[P]-BPFPT.*
2. *There are $c \in \mathbb{N}$, computable $f, h : \mathbb{N} \to \mathbb{N}$ and an exact randomized Turing machine $\mathbb{A}$ such that for all $x \in \Sigma^*$ and every run of $\mathbb{A}$ on $x$ the machine $\mathbb{A}$ performs at most $f(\kappa(x)) \cdot |x|^c$ many steps, tosses at most $h(\kappa(x)) \cdot \lceil \log |x| \rceil$ many coins and decides $Q$ with two-sided error at most $1/|x|$.*

*Proof.* To show that (1) implies (2) let $(Q, \kappa) \in$ W[P]-BPFPT and choose a program $\mathbb{P}$ witnessing this. By Corollary 8 we can assume that $\mathbb{P}$ on $x$ uses, say, $g(\kappa(x))$ many $|x|$-sided dice for some computable $g : \mathbb{N} \to \mathbb{N}$. Apply the Dice Lemma to get a program $\mathbb{P}'$ which on $x$ uses $q_x$-sided dice for $q_x := \left(2^{\lceil \log |x| \rceil}\right)^3 \geq |x|^3$. Then $\mathbb{P}'$ on $x$ errs with probability at most

$$1/|x| + g(\kappa(x)) \cdot |x|/q_x.$$

If $|x| \geq g(\kappa(x))$ this is at most $2/|x|$. As in the proof of Corollary 8 we get a program $\mathbb{P}''$ with error at most $1/|x|$ and $O(g(\kappa(x)))$ many $q_x$-sided dice.

The Turing machine $\mathbb{A}$ on $x$ simply simulates $\mathbb{P}''$ on $x$, thereby tossing $\log q_x = 3 \cdot \lceil \log |x| \rceil$ many coins whenever $\mathbb{P}''$ rolls one of its dice. Since this happens at most $O(g(\kappa(x)))$ many times, it follows that $\mathbb{A}$'s number of coins obeys the claimed bound. Clearly $\mathbb{A}$ is exact and $\mathbb{A}(x) \sim \mathbb{P}''(x)$.

The converse is seen by standard simulation: Given a Turing machine $\mathbb{A}$ as in (2) a program $\mathbb{P}$ on $x$ simply simulates $\mathbb{A}$ using as random bits the binary expansions of $h(\kappa(x))$ many rolls with $2^{\lceil \log x \rceil}$-sided dice. $\mathbb{P}$ may produce more random bits than actually needed by $\mathbb{A}$. But because $\mathbb{A}$ is exact, this surplus is the same in any run on $x$. Thus $\mathbb{P}(x) \sim \mathbb{A}(x)$. $\qquad \square$

An interesting subclass of W[P]-BPFPT is formed by the parameterized versions of problems in BPP. Using Corollary 10 it is not hard to give the following characterization of this part:

**Corollary 11.** *Let $(Q, \kappa)$ be a parameterized problem. The following statements are equivalent.*

1. *$(Q, \kappa) \in$ W[P]-BPFPT and $Q \in$ BPP.*
2. *There is a computable $h : \mathbb{N} \to \mathbb{N}$ and an exact, randomized, polynomially time bounded Turing machine $\mathbb{A}$ such that for all $x \in \Sigma^*$ and every run of $\mathbb{A}$ on $x$ the machine $\mathbb{A}$ tosses at most $h(\kappa(x)) \cdot \lceil \log |x| \rceil$ many coins and decides $Q$ with two-sided error at most $1/|x|$.*

*Proof.* That (2) implies (1) follows by Corollary 10. Conversely let $\mathbb{P}$ witness that $(Q, \kappa) \in$ W[P]-BPFPT and let $\mathbb{A}_Q$ witness that $Q \in$ BPP. By classical probability amplification we can assume that $\mathbb{A}_Q$ has two-sided error at most $1/|x|$. Choose for $\mathbb{P}$ according to Corollary 10 (2) a Turing machine $\mathbb{A}$, a constant $c \in \mathbb{N}$ and functions $f, h$. We can assume $f$ to be time constructible.

Let $\mathbb{A}'$ be the following Turing machine: on $x$ it checks if $f(\kappa(x)) \leq |x|$. This can be done in polynomial time because $f$ is time constructible. If this is the case $\mathbb{A}'$ simulates $\mathbb{A}$ and otherwise $\mathbb{A}_Q$. Then $\mathbb{A}'$ is an exact randomized Turing machine deciding $Q$ in polynomial time with two-sided error $1/|x|$. $\mathbb{A}$ on $x$ uses at most $h(\kappa(x)) \cdot \lceil \log |x| \rceil$ many coins if $|x| \geq f(\kappa(x))$ and at most $f(\kappa(x))^{O(1)}$ many coins otherwise. $\qquad \square$

## 4   Derandomization

*Derandomization of* paraNP-BPFPT*:* The following has first been observed by Grohe [8]:

**Proposition 12.** *The following statements are equivalent.*

*1.* paraNP-BPFPT = FPT.
*2.* BPP = P.
*3.* paraNP-BPFPT = W[P]-BPFPT

*Scetch of proof:* Clearly (1) implies (3). That (2) implies (1) follows by standard arguments (as e.g. in [7, Proposition 3.7]). We show that (3) implies (2).

Let $Q \in$ BPP. Then $(Q, 1)$, i.e. $Q$ with the parameterization which is constantly one, is in paraNP-BPFPT. By assumption $(Q, 1) \in$ W[P]-BPFPT. Choose a Turing machine $\mathbb{A}$ according to Corollary 10 (2). Then $\mathbb{A}$ on an input $x \in \Sigma^*$ runs in polynomial time and tosses $O(\log |x|)$ many coins. Thus simulating $\mathbb{A}$ on all $|x|^{O(1)}$ many random seeds requires only polynomial time. So $Q \in$ P.

$\square$

*Uniform derandomization of* W[P]-BPFPT: Derandomization of W[P]-BPFPT means that each parameterized problem with a W[P]-randomized algorithm can be solved in time $f(\kappa(x)) \cdot |x|^{O(1)}$ for some computable function $f$. In general this function $f$ may depend on the problem.

In the classical setting, derandomizing a BPP algorithm is done by running it on pseudorandom strings which it is not able to distinguish from truly random seeds. So, intuitively, if we prove derandomization by providing a method for simulating dice, then we may expect $f$ to be determined by the number and the size of the dice we simulate, i.e. that $f$ depends only on the size of the sample space used by the algorithm. But the size of dice can be assumed to be $|x|$ by Corollary 8. This may make one expect that successful derandomization is "strongly uniform": $f$ depends only on the number of dice we simulate.

However, while in the classical setting the length of the random seed can be assumed to be polynomially related to the running time, in the parameterized setting we are asked to produce short pseudorandom sequences of length, say, $g(k) \cdot \log n$ which fool algorithms running in time, say, $g(k) \cdot n^d$. It may be conceivable that the running time of a suitable pseudorandom generator increases with this running time, say, it depends on $d$ - e.g. we find determinizations running in time $O^*(f)$ for $f(k) = 2^{2^{2^{\cdots^{2^k}}}}$ a tower of height $d$.

So instead of a single $f$ we may only expect a successful derandomization to produce a family $(f_d)_d$ of functions such that randomized algorithms running in time $g(k) \cdot n^d$ have determinizations running in time $O^*(f_d)$. This relaxes the assumption of "strong uniformity" to "weak uniformity".

As usual we call a family $(f_d)_d = (f_d)_{d \in \mathbb{N}}$ of functions $f_d : \mathbb{N} \to \mathbb{N}$ *computable* if and only if $(d, k) \mapsto f_d(k)$ is computable.

**Definition 13.** Let $t, c : \Sigma^* \to \mathbb{N}$. We say that a classical or parameterized problem *has a* $(t, c)$-*machine* if and only if it can be decided with two-sided error $1/|x|$ by an exact randomized Turing machine which on $x$ runs for at most $t(x)$ many steps and tosses at most $c(x) \cdot \log |x|$ many coins. For $t, c : \mathbb{N} \to \mathbb{N}$ by a $(t, c)$-*machine* we mean a $(t \circ |\cdot|, c \circ |\cdot|)$-machine.

For $c : \mathbb{N} \to \mathbb{N}$ we let BPP[c] denote the class of all classical problems with a $(t, c)$-machine for some polynomial $t$.

**Definition 14** – W[P]-BPFPT *has a weakly uniform derandomization if and only if for all computable $g : \mathbb{N} \to \mathbb{N}$ there is a computable family of functions $(f_d)_d$ such that for all $d \in \mathbb{N}$ and all parameterized problems $(Q, \kappa)$:*
*if $(Q, \kappa)$ has a $\big(g(\kappa(x)) \cdot |x|^d, g(\kappa(x))\big)$-machine, then $(Q, \kappa) \in O^*(f_d)$.*

– *If we weaken the last condition to:*
*if $(Q, \kappa)$ has a $\big(|x|^d, g(\kappa(x))\big)$-machine, then $(Q, \kappa) \in O^*(f_d)$.*
*we say that the BPP part of W[P]-BPFPT has a weakly uniform derandomization.*[3]

– *If the family is constant (i.e. there is a $f$ such that $f_d = f$ for all $d$) we say that W[P]-BPFPT has a strongly uniform derandomization.*

**Theorem 15.** *The following statements are equivalent.*

1. *W[P]-BPFPT has a strongly uniform derandomization.*
2. *W[P]-BPFPT has a weakly uniform derandomization.*
3. *The BPP part of W[P]-BPFPT has a weakly uniform derandomization.*
4. *There is a polynomial time computable, increasing, time-constructible function $g : \mathbb{N} \to \mathbb{N}$ and there is a computable family of functions $(f_d)_d$ such that for all $d \in \mathbb{N}$ and all parameterized problems $(Q, \kappa)$:*
*if $(Q, \kappa)$ has a $\big(|x|^d, g(\kappa(x))\big)$-machine, then $(Q, \kappa) \in O^*(f_d)$.*
5. *There is a polynomial time computable, nondecreasing, unbounded function $c : \mathbb{N} \to \mathbb{N}$ such that $\mathrm{BPP}[c] = \mathrm{P}$.*

For nondecreasing, unbounded $f : \mathbb{N} \to \mathbb{N}$ we define the *inverse of $f$* to be the function $\iota_f : \mathbb{N} \to \mathbb{N}$ given by

$$\iota_f(n) := \max\big(\{i \in \mathbb{N} \mid f(i) \leq n\} \cup \{1\}\big).$$

Further we set $\iota_f^+ := \iota_f + 1$. We need the following simple lemma, more or less the same as [7, Lemma 3.23].

**Lemma 16.** *$\iota_f$ and $\iota_f^+$ are nondecreasing, unbounded and $\iota_f \circ f = \mathrm{id}$, $f \circ \iota_f \leq \mathrm{id}$ and $f \circ \iota_f^+ \geq \mathrm{id}$ for any nondecreasing, unbounded $f$. Furthermore $\iota_f$ and $\iota_f^+$ are computable in polynomial time for any increasing and time constructible $f$.*

*Proof of Theorem 15:* The rest being trivial it suffices to show that (4) implies (5) and that (5) implies (1). First we show that (4) implies (5). So assume (4) and choose a polynomial time computable, increasing, time-constructible $g : \mathbb{N} \to \mathbb{N}$ and a computable family of functions $(f_d)_d$ accordingly. We can assume that for all $d \in \mathbb{N}$ the function $f_d : \mathbb{N} \to \mathbb{N}$ is increasing and time constructible.

Because the family $(f_d)_d$ is computable there is a time constructible increasing $\widetilde{f} : \mathbb{N} \to \mathbb{N}$ such that for all $n \in \mathbb{N}$

$$\widetilde{f}(n) \geq \max\{f_1(n), \dots, f_n(n)\}.$$

By lemma 16 the inverse $\iota_{\widetilde{f}}$ is polynomial time computable, nondecreasing and unbounded. Because for all $d \in \mathbb{N}$ and for all $n \geq d$ we have $\widetilde{f}(n) \geq f_d(n)$, it follows that

---

[3] These modes of speech rely on Corollaries 10 and 11.

$$\text{for all } d \in \mathbb{N} \text{ for all } n \geq \tilde{f}(d) : \iota_{\tilde{f}}(n) \leq \iota_{f_d}(n). \tag{1}$$

We denote subtraction of 1 by $s$, that is $s(n) := \max\{n - 1, 0\}$. Set

$$c := g \circ s \circ \iota_{\tilde{f}}.$$

$c$ is polynomial time computable, nondecreasing and unbounded, because it is a composition of such functions.

Let $Q \in \mathrm{BPP}[c]$. Then there is a constant $d \in \mathbb{N}$ such that $Q$ has a $(|x|^d, c(|x|))$-machine $\mathbb{A}$. We aim to show $Q \in \mathrm{P}$. For $\kappa_c(x) := c(|x|)$ define

$$\kappa := \iota_g^+ \circ \kappa_c.$$

Then $\kappa$ is polynomial time computable by Lemma 16. Thus $(Q, \kappa)$ is a parameterized problem. Because

$$c(|x|) \leq g \circ \iota_g^+ \circ \kappa_c(x) \leq g(\kappa(x)),$$

$\mathbb{A}$ is also a $(|x|^d, g(\kappa(x)))$-machine. By assumption (3) we get $(Q, \kappa) \in O^*(f_d)$. Thus to show $Q \in \mathrm{P}$ it suffices to show that $f_d(\kappa(x)) \leq |x|$ for all sufficiently long $x$. To see this, first note that for all $n > 0$

$$\iota_g^+ \circ g \circ s(n) = \iota_g(g(s(n)) + 1 = s(n) + 1 = n, \tag{2}$$

so $\iota_g^+ \circ g \circ s$ is the identity on positive numbers. Then for all $x$ with $|x| \geq \tilde{f}(d)$

$$f_d(\kappa(x)) = f_d \circ \iota_g^+ \circ g \circ s \circ \iota_{\tilde{f}}(|x|) = f_d \circ \iota_{\tilde{f}}(|x|) \leq f_d \circ \iota_{f_d}(|x|) \leq |x|.$$

The first equality holds by definition of $\kappa$, the second equality follows with (2) from $\iota_{\tilde{f}} > 0$ and the first inequality follows with (1) from $f_d$ being increasing.

We now show that (5) implies (1). Assume (5) and choose a polynomial time computable, nondecreasing, unbounded $c : \mathbb{N} \to \mathbb{N}$, such that $\mathrm{BPP}[c] = \mathrm{P}$. For (1) we have to show for all computable $g : \mathbb{N} \to \mathbb{N}$ how to decide deterministically problems with a $(g(\kappa(x)) \cdot |x|^{O(1)}, g(\kappa(x)))$-machine. Clearly it suffices to do so for all computable nondecreasing $g$. So let such a $g : \mathbb{N} \to \mathbb{N}$ be given.

First note:

*Claim I:* There is a computable $r : \mathbb{N} \to \mathbb{N}$ such that any problem $(Q, \kappa)$ with a $(g(\kappa(x)) \cdot |x|^{O(1)}, g(\kappa(x)))$-machine can be solved in (deterministic) time $r(|x|) \cdot |x|^{O(1)}$.

*Proof of Claim I:* Simulating a $(g(\kappa(x)) \cdot |x|^{O(1)}, g(\kappa(x)))$-machine on all possible random seeds and taking a majority vote, decides $Q$ in time

$$|x|^{g(\kappa(x))} \cdot g(\kappa(x)) \cdot q(|x|).$$

for some polynomial $q$. Because $\kappa$ is polynomial time computable, there is a constant $d_\kappa \in \mathbb{N}$ (depending on $\kappa$) such that $\kappa(x) \leq 2^{|x|^{d_\kappa}}$ and hence $\kappa(x) \leq 2^{2^{|x|}}$

for all but finitely (depending on $\kappa$) many $x$. Since $g$ is nondecreasing, we can hence decide $Q$ in time

$$|x|^{g(2^{2^{|x|}})} \cdot g(2^{2^{|x|}}) \cdot p(|x|)$$

for some polynomial $p$ (depending on $(Q, \kappa)$). ⊣

Clearly $\iota_c^+$ is computable. Choose some time constructible $f : \mathbb{N} \to \mathbb{N}$ with

$$f \geq \max\left\{\iota_c^+ \circ g, g\right\}. \tag{3}$$

Choose $r$ according to Claim I. Without loss of generality we assume $r$ to be nondecreasing. Let $(Q, \kappa)$ have a $(g(\kappa(x)) \cdot |x|^{O(1)}, g(\kappa(x)))$-machine $\mathbb{A}$. We aim to show

$$(Q, \kappa) \in O^*(r \circ f).$$

*Claim II:* $Q_{\geq f} := \{x \in Q \mid |x| \geq f(\kappa(x))\} \in \mathrm{BPP}[c]$.

*Proof of Claim II:* Define the machine $\mathbb{A}'$ as follows. On $x$ it first checks if $f(\kappa(x)) > |x|$. This can be done in polynomial time since $f$ is time constructible. If this is the case, it rejects. If $f(\kappa(x)) \leq |x|$ it simulates $\mathbb{A}$. But then $\mathbb{A}$ needs time at most

$$g(\kappa(x)) \cdot |x|^{O(1)} \leq f(\kappa(x)) \cdot |x|^{O(1)} \leq |x| \cdot |x|^{O(1)},$$

where the first inequality follows from $g \leq f$ by (3). Thus $\mathbb{A}'$ runs in polynomial time. Clearly $\mathbb{A}'$ decides $Q_{\geq f}$ with two-sided error at most $1/|x|$. If $f(\kappa(x)) > |x|$, then $\mathbb{A}'$ uses no coins at all. If $f(\kappa(x)) \leq |x|$, then $\mathbb{A}'$ uses at most $g(\kappa(x)) \cdot \lceil \log |x| \rceil$ many coins. But then

$$g(\kappa(x)) \leq c \circ \iota_c^+ \circ g \circ \kappa(x) \leq c \circ f \circ \kappa(x) \leq c(|x|).$$

Here the second inequality holds because $c$ is nondecreasing and $f \geq \iota_c^+ \circ g$ by (3). The third inequality holds because $c$ is nondecreasing and $f(\kappa(x)) \leq |x|$. ⊣

By Claim II and our assumption we get $Q_{\geq f} \in \mathrm{P}$. We get the following algorithm solving $Q$: on $x$ it first checks in polynomial time if $f(\kappa(x)) > |x|$. If this is the case it simulates a decision procedure for $Q$ running in time (recall that $r$ is nondecreasing)

$$r(|x|) \cdot |x|^{O(1)} \leq r(f(\kappa(x))) \cdot |x|^{O(1)}.$$

Otherwise it runs a polynomial time procedure deciding $Q_{\geq f}$. This shows that $(Q, \kappa) \in O^*(r \circ f)$. □

## 5   Questions

We characterized "uniform" derandomization of W[P]-BPFPT in terms of classical complexity theory. An obvious question is: Can you get rid of the uniformity assumption? Another: Can you similarly characterize derandomization of W[1]-BPFPT in terms of classical complexity theory?

For W[1]-BPFPT no strong probability amplification is known, that is, it is not known if it is possible to amplify a success probability of $1/2 + 1/|x|$ to $3/4$ or of $3/4$ to $1 - 1/|x|$. Classical arguments using Chernoff bounds allow to amplify a success probability of $3/4$ to, say, $1 - 2^{2^{-k}}$. This question is related to the struggle for parameterized analogues of Todas Theorem [14] like $Wt] \subseteq \mathrm{FPT}^{\#W[1]}$ or $A[t] \subseteq \mathrm{FPT}^{\#W[P]}$, as has been asked for in [5,7].

...to mention some of the yet unresolved theoretical questions. On the practical side we hope to have provided definitions sufficiently handy for designing randomized solutions for parameterized problems.

**Acknowledgements.** I thank Martin Grohe for proposing the concept of weakness for uniformity and I thank Jörg Flum for his comments on an earlier draft of this paper.

# References

1. Alekhnovich, M., Razborov, A.A.: Resolution is Not Automatizable Unless W[P] is Tractable. In: Proceedings of the 41th IEEE Symposium on Foundations of Computer Science, pp. 210–219 (2001)
2. Arvind, V., Raman, V.: Approximation Algorithms for Some Parameterized Counting Problems. In: Bose, P., Morin, P. (eds.) ISAAC 2002. LNCS, vol. 2518, pp. 453–464. Springer, Heidelberg (2002)
3. Chen, Y., Flum, J.: The Parameterized Complexity of Maximality and Minimality Problems. Annals of Pure and Applied Logic 151(1), 22–61 (2008); In: Bodlaender, H.L., Langston, M.A. (eds.) IWPEC 2006. LNCS, vol. 4169, pp. 22–61. Springer, Heidelberg (2006)
4. Chen, Y., Flum, J., Grohe, M.: Machine-based methods in parameterized complexity theory. Theoretical Computer Science 339, 167–199 (2005)
5. Downey, R.G., Fellows, M.R., Regan, K.W.: Parameterized Circuit Complexity and the W Hierarchy. Theoretical Computer Science 191(1–2), 97–115 (1998)
6. Eickmeyer, K., Grohe, M., Grübner, M.: Approximisation of W[P]-complete minimisation problems is hard. In: 23rd CCC (2008)
7. Flum, J., Grohe, M.: Parameterized Complexity Theory. Springer, Heidelberg (2006)
8. Grohe, M.: Communication
9. Montoya, J.A.: Communication
10. Müller, M.: Randomized Approximations of Parameterized Counting Problems. In: Bodlaender, H.L., Langston, M.A. (eds.) IWPEC 2006. LNCS, vol. 4169, pp. 50–59. Springer, Heidelberg (2006)
11. Müller, M.: Valiant-Vazirani Lemmata for Various Logics (manuscript)
12. Papadimitriou, C.H.: Computational Complexity. Addison Wesley, Reading (1994)
13. Sloper, C., Telle, J.A.: Towards a Taxonomy of Techniques for Designing Parameterized Algorithms. In: Bodlaender, H.L., Langston, M.A. (eds.) IWPEC 2006. LNCS, vol. 4169, pp. 251–263. Springer, Heidelberg (2006)
14. Toda, S.: PP is as Hard as the Polynomial Hierarchy. SIAM Journal on Computing 20(5), 865–877 (1991)

# A Linear Kernel for Planar Feedback Vertex Set

Hans L. Bodlaender and Eelko Penninkx

Department of Information and Computing Sciences, Utrecht University
P.O. Box 80.089, 3508 TB Utrecht, The Netherlands
hansb@cs.uu.nl, penninkx@cs.uu.nl

**Abstract.** In this paper we show that FEEDBACK VERTEX SET on planar graphs has a kernel of size at most $112k^*$. We give a polynomial time algorithm, that given a planar graph $G$ finds a equivalent planar graph $G'$ with at most $112k^*$ vertices, where $k^*$ is the size of the minimum Feedback Vertex Set of $G$. The kernelization algorithm is based on a number of reduction rules. The correctness of most of these rules is shown using a new notion: bases of induced subgraphs. We also show how to use this new notion to automatically prove safeness of reduction rules and obtain tighter bounds for the size of the kernel.

## 1 Introduction

The FEEDBACK VERTEX SET problem is one of the classic graph optimization problems, with several applications [18]. In this paper, we focus on kernelization algorithms for this problem, when restricted to planar undirected graphs. We assume the reader to be familiar with standard terminology of fixed parameter complexity, see e.g., [17, 30, 19].

Much research has been done on FPT algorithms for FEEDBACK VERTEX SET. The problem was first shown to be in FPT by Downey and Fellows [16]. Recently Chen et al. [10] showed that the problem on directed graphs is also in FPT, solving a long standing open problem.

Several papers gave different and increasingly faster FPT algorithms for the FEEDBACK VERTEX SET problem on undirected graphs [6, 17, 4, 31, 27, 32, 24, 12]. The currently fastest algorithms are a probabilistic algorithm using $O(4^k kn)$ time [4] and a deterministic algorithm using $O(5^k kn^2)$ time [9].

Work has also done on exact algorithms for FEEDBACK VERTEX SET [33, 20]; the current fastest algorithm by Fomin et al. [20] uses $O(1.7548^n)$ time.

Several papers have been written on approximating the FEEDBACK VERTEX SET problem. Bar-Yehuda et al. [3] gave an algorithm with a ratio $O(\log n)$. Independently, Becker and Geiger [5] and Bafna et al. [2] gave polynomial time approximation algorithms with ratio two. Chudak et al. [11] gave an interesting interpretation of these two algorithms with the primal-dual method, and obtained a simpler version of the algorithm from [2].

The FEEDBACK VERTEX SET problem is still NP-complete when restricted to planar graphs [22]. Demaine and Hajiaghayi [13] give, amongst others, a PTAS

M. Grohe and R. Niedermeier (Eds.): IWPEC 2008, LNCS 5018, pp. 160–171, 2008.

for planar FEEDBACK VERTEX SET. Some work has been done on the directed variant of on planar graphs, see [26, 23, 34].

In this paper, we derive a kernelization algorithm for FEEDBACK VERTEX SET on planar graphs that gives a linear kernel. We present an algorithm, that given a planar graph $G$ and an integer $k$, finds a planar graph $G'$ with at most $112k^*$ vertices and an integer $k' \leq k$, such that $G$ has a feedback vertex set of size $k$, if and only if $G'$ has a feedback vertex set of size $k'$. Here $k^*$ denotes the size of some minimum FVS for $G$. This is an improvement of the cubic kernel on planar graphs given by [28]. For general graphs, it is known that there exists a polynomial kernel [8, 7]. A cubic kernel for FEEDBACK VERTEX SET on general graphs was derived in [7]. For a good overview paper on kernelization, see [25].

Recently, Van Dijk [35] has experimentally evaluated the kernelization algorithm from [7]. It appears that in a practical setting, this algorithm runs very fast and often gives substantial size reductions of input graphs.

Once a kernel has been found, some exact algorithm for FEEDBACK VERTEX SET on planar graphs can be used to solve the problem on the kernel. See [15, 14, 21].

In many cases, one would expect that kernels for problems on planar graphs are small. In some cases this is trivial (e.g., the four colour theorem directly implies that INDEPENDENT SET on planar graphs has a kernel of size at most $4k$); in other cases, involved analysis appears to be necessary. An interesting result is for instance the kernel for DOMINATING SET on planar graphs of size at most $335k$, see [1].

This paper is organized as follows. Section 2 gives some preliminary definitions. In Section 3, we present some notions that are useful for proving correctness of reduction rules for the FEEDBACK VERTEX SET problem. In Section 4, we present a number of reduction rules. In Section 5 we show how to use an automated procedure to prove reduction rules. In Section 6 we prove a first linear size bound and in Section 7 we prove the $112k^*$ size bound. Some final comments are made in Section 8.

## 2    Preliminaries

All graphs in this paper are undirected, planar, and can have parallel edges and self-loops. Consider a graph $G = (V, E)$. A *path* is a sequence of vertices and edges $v_0 e_1 v_1 e_2 \ldots e_n v_n$ such that $e_i = \{v_{i-1}, v_i\}$ and $v_i \neq v_j$ for all $1 \leq i < j \leq n$. A *cycle* is a path where $v_0 = v_n$ and all other vertices are different. The length of a path or cycle is the number of edges used. By $G[W]$ we denote the graph induced by $W \subseteq V$, defined as $G[W] = (W, E \cap W \times W)$. The *neighbourhood* $N(v) = \{u \in V \setminus v : \{u, v\} \in E\}$ of a vertex $v \in V$ is the set of vertices adjacent to $v$. The neighbourhood of a set $U \subseteq V$ is defined as $N(U) = (\cup_{u \in U} N(u)) \setminus U$. The *border* $\beta(U)$ of a set $U \subseteq V$ is defined as $\beta(U) = \{u \in U : N(u) \setminus U \neq \emptyset\}$.

The degree $d(v, W) = |\{e \in E : e = \{v, u\}, u \in W\}|$ of a vertex $v$ with respect to $W \subseteq V$ is defined as the number of edges between $v$ and vertices from $W$. We use $d(v)$ as a shorthand for $d(v, V)$. We sometimes use single vertices $v$ instead

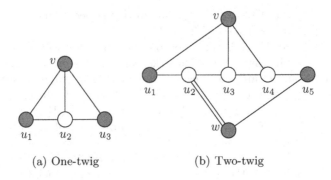

**Fig. 1.** Twigs, grey vertices are connected to the rest of the graph

of the singleton set $\{v\}$ for readability. E.g., $d(v,w) = 2$ implies that there are exactly two parallel edges $\{v,w\}$.

A *forest* is a graph without cycles, and a *tree* is a connected forest. A *feedback vertex set (FVS)* of $G = (V, E)$ is a subset of vertices $F \subseteq V$ such that $G[V \setminus F]$ is a forest. The FEEDBACK VERTEX SET problem asks for a FVS of minimum size.

Let $A$ be some FVS of $G$. The graph $G[V \setminus A]$ is a forest consisting of $t$ trees. By $T_i \subseteq V, 1 \le i \le t$ we denote all trees in this forest. A tree $T_i$ is called a *singleton tree* if $|T_i| = 1$.

A *twig* of size $n$ consists of vertices $U = \{u_1, \ldots, u_n\}$ such that $G[U]$ is a path, and a minimal non-empty set of *observers* $W \subset V \setminus U$ of size at most two such that the following three rules hold: (i) $d(u_i, u_{i+1}) = 1$ if $1 \le i \le n-1$ and $d(u_i, u_j) = 0$ if $|i - j| \ne 1$, (ii) $d(u_i, W) > 0$, (iii) $N(u_i) \subseteq W \cup \{u_{i-1}, u_{i+1}\}$ for $i = 2 \ldots n-1$. We use the terms *one-twig* and *two-twig* to denote twigs where $|W| = 1$ and $|W| = 2$ respectively. See Figure 1 for examples. Grey colored vertices can be connected to other vertices in the graph.

## 3   Domination

Consider a graph $G = (V, E)$, vertex set $U \subseteq V$ and the induced graph $G[U]$. Now consider a cycle $C$ such that $C \cap U \ne \emptyset$ and $C \setminus U \ne \emptyset$. Each component of the cycle in $G[U]$ either *touches* $G[U]$ in one vertex on the border, or it *passes through* $G[U]$ by following a path between two different vertices in $\beta(U)$. Given some subset $X \subseteq U$ we define the sets $A_1(X)$ and $A_2(X)$ that contain all possible ways a cycle $C$ can interact with $G[U \setminus X]$:

- $A_1(X) = \{u \in U : d(u, V \setminus U) \ge 2\} \setminus X$ contains all vertices in $G[U]$ that are on the border, have at least two edges to the outside of $G[U]$, and are not in $X$.
- $A_2(X) = \{\{u,v\} \in 2^{\beta(U) \setminus X} : u \ne v$ and there is a path in $G[U \setminus X]$ between $u$ and $v\}$ contains all unordered pairs $\{u,v\}$ of vertices on the border that are not in $X$ having a path between $u$ and $v$ in $G[U \setminus X]$.

Note that if $A_1(X) = A_2(X) = \emptyset$ and $X$ is a FVS in $G[U]$ then all cycles in $G$ that use a vertex in $U$ are blocked by some vertex in $X \subseteq U$. We define that $X$ *dominates* $Y$ in $G[U]$, denoted as $X \models Y$, if $X, Y \subseteq U$, $X$ is a FVS in $G[U]$ and:

- $Y$ is not a FVS in $G[U]$, or
- $|X| \leq |Y|$, $A_1(X) \subseteq A_1(Y)$ and $A_2(X) \subseteq A_2(Y)$.

It is easy to see that our definition of domination is intuitive: if $X \models Y$ in some $G[U]$ then every cycle that is broken by $Y$ is also broken by $X$, or $Y$ is not even a FVS in $G[U]$.

A *basis* $\mathcal{B} \subseteq 2^U$ of $G[U]$ is a subset of the powerset of $U$ such that every $Y \subseteq U$ is dominated by some $X \in \mathcal{B}$. A trivial basis can be constructed by taking every FVS $F$ of $G[U]$. A basis $\mathcal{B}$ is *minimal* if there are no $X, Y \in \mathcal{B}$ with $X \neq Y$ such that $X \models Y$. Note that $\{F^*\}$ is a minimal basis for $G = G[V]$ (in which case $A_1 \equiv A_2 \equiv \emptyset$) if and only if $F^*$ is a minimum FVS of $G$. We have the following lemma:

**Lemma 1.** *Every minimal basis $\mathcal{B}$ for $G[U]$ contains exactly one $X \in \mathcal{B}$ such that $|X| > |Y|$ for all $Y \in \mathcal{B}$ where $Y \neq X$, and $A_1(X) = A_2(X) = \emptyset$.*

*Proof.* For every basis $\mathcal{B}$ for $G[U]$ at least one element $X \in \mathcal{B}$ has the property $A_1(X) = A_2(X) = \emptyset$. If $\mathcal{B}$ is minimal then no $Y \in \mathcal{B}, Y \neq X$ exists such that $X \models Y$, so $|Y| < |X|$ for all $Y \in \mathcal{B}, Y \neq X$. Also note that there is no $Y \subseteq U$ such that $Y \models X$ because the basis is minimal. Thus, $X$ is a minimum subset of $U$ such that $A_1(X) = A_2(X) = \emptyset$. □

**Lemma 2.** *Consider a basis $\mathcal{B}$ of $G[U]$. Let $C = \bigcap_{X \in \mathcal{B}} X$. Then there exists an optimal FVS $F^*$ of $G$ such that $C \subseteq F^*$.*

*Proof.* Take an optimal FVS $F$. Consider the set $Y = F \cap U$ of vertices that are inside $G[U]$. By definition of a basis there must be some $X \in \mathcal{B}$ such that $X \models Y$. Replace $Y$ by $X$ to obtain an optimal FVS $F^*$ such that $C \subseteq F^*$. □

## 4    Rules

In this section we present reduction rules. A *safe* reduction rule transforms an instance of FVS $(G, k)$ to another instance $(G', k')$ such that $G$ has a FVS of size at most $k$ if and only if $G'$ has a FVS of size $k'$. Safeness of most rules is trivial.

Rules transform $G$, mostly by *selecting* vertices that are included in some optimal FVS. Selection of a vertex $v \in V$ results in a new instance $(G[V \setminus v], k-1)$. Multiple selected vertices may be processed in any order.

For vertices $U$ observed in a reduction rule we assume without loss of generality that for all vertices $u \in \beta(U)$ on the border of $G[U]$ we have $d(u, V \setminus U) \geq 2$, meaning that cycles can *touch* $G[U]$ in all vertices on the border. This generalization is allowed because every basis $\mathcal{B}$ also forms a basis if we lower the out-degree of the vertices $u \in \beta(U)$. This can be seen by definition of dominance. When testing for rule $i$ we assume that rules $j < i$ do not apply.

## 4.1 Simple Rules

We present the first six rules. Correctness of Rules 1–5 is trivial.

**Rule 1 (Self Loop Rule).** *If there is an edge $\{v, v\} \in E$ then select $v$.*

**Rule 2 (Triple Edge Rule).** *If there are more than two parallel edges $\{v, w\}$ in $G$ then remove all but two.*

**Rule 3 (Degree Zero Rule).** *If there is a vertex $v \in V$ such that $d(v) = 0$ then remove $v$.*

**Rule 4 (Degree One Rule).** *If there is a vertex $v \in V$ such that $d(v) = 1$ then remove $v$.*

**Rule 5 (Degree Two Rule).** *Suppose there is a vertex $v$ such that $d(v) = 2$ with incident edges $\{v, w\}$ and $\{v, x\}$. Then remove $v$ and add an edge $\{w, x\}$. Note that if $w = x$ a self-loop is added.*

**Rule 6 (Degree Three Rule).** *Suppose there is a vertex $v$ such that $d(v) = 3$ having only two neighbours $w$ and $x$ where $d(v, w) = 2$. Then select $w$.*

This rule is safe because $\mathcal{B} = \{\{w\}\}$ is a basis for $\{v, w, x\}$. We can select $w$ because of Lemma 2.

The following two lemmas are necessary to prove safeness of the other rules. We use the short term *tree* for a set of vertices that induces a subtree of $G$.

**Lemma 3.** *Suppose Rules 1–6 do not apply in a planar graph $G$. Consider a tree $T \subseteq V$ with leafs $L$. Then $|N(l) \setminus T| \geq 2$ for all $l \in L$, meaning that leaf has at least two neighbours outside of $T$.*

*Proof.* We have $d(l, T) = 1$ and $d(l, N(T)) \geq 2$ because the Degree Two Rule does not apply. If $|N(l)| = 2$ then the Degree Three Rule would apply, so $|N(l)| \geq 3$ proving the lemma. □

**Lemma 4.** *Suppose Rules 1–6 do not apply in graph $G$. Consider a tree $T \subseteq V$ with leafs $L$. If $|N(T)| = 2$ then $|L| \leq 2$.*

*Proof.* Suppose $|L| = 3$ and consider the set $M = N(T) \cup \{v\}$ where $v \in T$ is the unique node in $T$ with $d(v, T) = 3$. We have a disjoint path between every node in $L$ and $M$, which implies a $K_{3,3}$ minor in $G$ contradicting the planarity of $G$. If $|L| > 3$, the case $|L| = 3$ is contained as a minor. □

**Lemma 5.** *Consider a singleton tree $T$ (i.e., a single vertex) in graph $G$ where Rules 1–6 do not apply. If $|N(T)| = 2$ then $d(T, N(T)) = 4$ and $T$ is connected with double edges to both its neighbours.*

*Proof.* Because the Degree Three Rule does not apply we have $d(T) > 3$. As $|N(T)| = 2$ we directly get $d(T) = 4$. □

## 4.2  Less Simple Rules

**Rule 7 (Small Tree Rule I).** *Consider two nodes $v, w \in V$ and a tree $T$ such that $N(T) = \{v, w\}$ and $|T| \geq 3$. Then select $v$ and $w$.*

Because of Lemma 4 we know that $T$ has exactly two leafs $l_1, l_2$. Because of Lemma 3 we know that $l_1, l_2$ are connected to both $v$ and $w$. The Degree Two Rule does not apply, so every $t \in T \setminus \{l_1, l_2\}$ is connected to at least one of $v, w$. Observe that there is no FVS of size 1 in $G[T \cup N(T)]$, and that $A_1(\{v, w\}) = A_2(\{v, w\}) = \emptyset$. So $\mathcal{B} = \{\{v, w\}\}$ is a basis for $G[T \cup N(T)]$ by Lemma 1, and we can select both because of Lemma 2.

**Rule 8 (Small Tree Rule II).** *Consider two nodes $v, w \in V$ and a tree $T = \{t_1, t_2\}$ such that $N(T) = \{v, w\}$ and $d(t_1) \geq d(t_2)$. If $d(t_2) = 3$ and $d(v, w) = 0$ then contract edge $\{t_1, t_2\}$ and add an edge $\{v, w\}$, otherwise select $v$ and $w$.*

If $d(t_2) = 3$ and $d(v, w) = 0$ then $\mathcal{B} = \{\{t_1\}, \{v, w\}\}$ is a basis for $G[T \cup N(T)]$. For the new subgraph we have a basis $\mathcal{B}' = \{\{t\}, \{v, w\}\}$ where $t$ is the vertex created by contracting $\{t_1, t_2\}$. As $A_1(t_1) = A_1(t) = \{v, w\}$ and $A_2(t_1) = A_2(t) = \{\{v, w\}\}$ we see that this transformation is safe. Note that this leads to the creation of a singleton tree $T$ with $|N(T)| = 2$. In the other case $\mathcal{B} = \{\{v, w\}\}$ is a basis so we can select $v$ and $w$.

**Rule 9 (Parallel Singleton Trees Rule).** *If there are four different vertices $v_1, v_2, w_1, w_2$ such that $d(v_i, w_j) = 2$ and $d(w_i) = 4$ then select $v_1$ and $v_2$.*

This rule is safe because $\mathcal{B} = \{\{v_1, v_2\}\}$ is a basis for $G[\{v_1, v_2, w_1, w_2\}]$. This rule prevents an arbitrary number of singleton trees with $|N(T)| = 2$ to emerge.

**Rule 10 (One-Twig Rule).** *If there is a one-twig of size 6 with path $U = \{u_1, \ldots, u_6\}$ and observer $v$ then select $v$.*

If $v$ is not selected we have to select at least 3 vertices from $U$ because $d(v, U) \geq 6$, but this choice is dominated by $\{v, u_1, u_6\}$. This implies that there is a basis where $v$ is always included, so we select $v$ by Lemma 2.

**Rule 11 (Two-Twig Rule).** *If there is a two-twig of size 15 with path $U = \{u_1, \ldots, u_{15}\}$ and observers $v, w$ with $d(v, U) \geq d(w, U)$ then select $v$.*

If $v$ is not selected we have to select at least 4 vertices from $U$ because $d(v, U) \geq 8$, but this choice is dominated by $\{v, w, u_1, u_{15}\}$. This implies that there is a basis where $v$ is always included, so we select $v$ by Lemma 2.

With Rules 7–11 we can prove another result.

**Lemma 6.** *Consider a graph $G$ where Rules 1-11 do not apply. Consider two vertices $v, w \in V$. Then at most one maximal tree $T$ exists in $G$ such that $N(T) = \{v, w\}$ and $|T| = 1$.*

*Proof.* If $|N(T)| = 2$ and $|T| \geq 2$ then one of the Small Tree Rules applies, so $|T| = 1$. Because of the Degree Three Rule we have $d(T, v) = d(T, w) = 2$, so $T$ is a singleton tree with $|N(T)| = 2$. Because of the Parallel Singleton Tree Rule we know that there is only one such $T$, proving the lemma.     $\square$

### 4.3  The Algorithm

The algorithm to kernelize a given instance of PLANAR FVS $(G, k)$ is to apply all rules in order until no rule applies. It is clear that all rules use time polynomial in $|V|$. Applying a rule results in either a decrease in the number of vertices, edges or $k$, so the complete kernelization algorithm also uses time and memory polynomial in $|V|$. We are only interested in the size of the remaining kernel, so we omit a more detailed analysis of the complexity of the algorithm.

## 5  Automated Proofs

In this section we will explain how to use the concept of a basis to automatically find or test reduction rules that *select* some vertex for a given subgraph. Consider some graph $G = (V, E)$ and vertex set $U \subseteq V$. Note that $\mathcal{B} = \{F \subseteq V : F$ is a FVS for $G[U]\}$ is a basis for $G[U]$. We first shrink $\mathcal{B}$ by removing every $Y \in \mathcal{B}$ for which there is some $X \in \mathcal{B}$ such that $X \models Y$ and $Y \not\models X$. Next we partition the basis $\mathcal{B}$ in different groups $\mathcal{B}_1, \ldots, \mathcal{B}_n$ such that if $X \models Y$ and $Y \models X$ then $X, Y \in \mathcal{B}_i$. Note that every minimal basis can be constructed by taking exactly one item from every $\mathcal{B}_i$, which implies that every minimal basis has equal cardinality.

Once the basis is partitioned in groups it is straightforward to check if we can construct a minimal basis such that some $v \in V$ is included in every item of this basis such that Lemma 2 applies.

Correctness of this algorithm trivial. Time and memory complexity of this algorithm are $O(c^{|U|})$, with $c > 1$. In our implementation $c = 2$ because we just generate every possible subset of $U$ for step 1. In Section 7 we use this method to find improved versions of the Twig-Rules.

## 6  A Linear Bound on the Size of the Kernel

In this section we will prove the first result of this paper. Note that there is a simple polynomial time algorithm that executes all rules until none is possible.

**Theorem 1.** *Consider a planar graph $G = (V, E)$ where none of the reduction rules can be applied. Let $A \subseteq V$ be a minimum FVS of size $k^*$. Then $|V| = O(k^*)$.*

Consider some embedding of $G$ on the plane, and a maximal tree $T$ in the graph $G[V \setminus A]$. We order the edges between $T$ and $A$ as we would encounter them in a clockwise walk around the tree at an infinitesimally distance, ignoring parallel edges. We start right before the last edge of $T$ outward from some leaf. From this sequence of edges we construct a sequence $S = s_1 s_2 \ldots s_m$ of vertices that are the endpoint of the edges in $A$. See Figure 2(a) for an example, where $s_1 = s_2 = s_3 = v_1$, $s_2 = v_2$ and $s_m = v_n$. We define that $s_0 = s_m$ and $s_{m+1} = s_1$. A *switch* occurs when $s_i \neq s_{i+1}$. We claim the following:

**Lemma 7.** *The number of switches $s(T)$ of a tree $T$ is at most $2|N(T)| - 2$.*

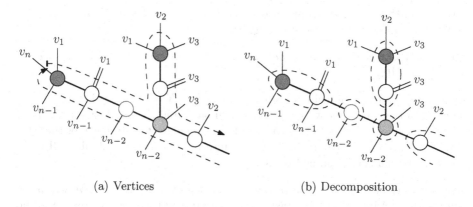

(a) Vertices                         (b) Decomposition

**Fig. 2.** An example tree (leafs are dark grey, internal vertices light gray)

*Proof.* Because the graph is planar there are no $1 \leq i < j < k < l \leq m$ such that $s_i = s_k$ and $s_j = s_l$. This implies that if $s_i \neq s_{i+1}$ then (i) there is no $j > i$ such that $s_j = s_i$, or (ii) there is no $j < i$ such that $s_j = s_{i+1}$, or (iii) both. This implies that we have at most $2|N(T)| - c$ values of $i$ such that $s_i \neq s_{i+1}$ for $1 \leq i \leq m$. Because $s(T) = 0$ for $|N(T)| = 1$ we have $c = 2$ proving the lemma.                                                                                    □

**Lemma 8.** *The number of leafs $|L|$ of a tree $T$ is at most $s(T)$.*

*Proof.* Because of Lemma 3, every leaf contains at least one switch.        □

We will now show how to decompose a tree $T$ in *internal vertices* and *twigs*. A vertex $v \in T$ is an internal vertex if $d(v, T) \geq 3$. When these vertices are removed from $T$ we obtain a collection of paths, which we will break apart in maximal subpaths that form the path of some twig. Consider some path $P = p_1 p_2 \ldots p_n$. Note that for every vertex $p_i \in P$ we have $d(p_i, T) \leq 2$ and $d(p_i) \geq 3$ because of the Degree Two Rule, so $p_i$ is connected to at least one vertex in $A$. We now find the maximum $j$ such that $p_1 \ldots p_j$ is the path of some twig. Note that $j \geq 2$ if $n \geq 2$. If $j < n$ then a switch occurs between $p_j$ and $p_{j+1}$ by definition of a twig. In this case we remove $p_1 \ldots p_j$ from $P$ to obtain $P'$, and the procedure is repeated. After all paths are processed we have decomposed $T$ in twigs and internal vertices. See Figure 2(b) for an example.

**Lemma 9.** *Consider a maximal tree $T$ in the graph $G[V \setminus A]$ with $s(T)$ switches. The number of twigs in this tree is at most $2s(T) - 3$, and this bound is obtained by placing exactly one switch at every leaf.*

*Proof.* First note that it is suboptimal to place more than one switch at a leaf. To maximize the number of twigs we either place a switch at a leaf, or at some internal path. We call the number of switches at a leaf $x$, and at an internal path $y$. Hence we have $x$ leafs, at most $x - 2$ internal vertices $U$, and the graph $G[T \setminus U]$ contains at most $2x - 3$ paths. We have $y$ switches left to create extra twigs from these paths, yielding a total of $2x + y - 3$ paths, with $x + y = s(T)$. This is of course maximized by taking $x = s(T)$.     □

**Lemma 10.** *Consider a maximal tree $T$ in the graph $G[V \setminus A]$. Then $|T| \leq 58|N(T)| - 102$.*

*Proof.* The number of switches $s(T)$ is at most $2|N(T)| - 2$ by Lemma 7, the number of twigs is at most $2s(T) - 3$ by Lemma 9, and the number of leafs $|L|$ of $T$ is at most $s(T)$ by Lemma 8. A twig has size at most 14 because of the Two Twig Rule, and we have at most $|L| - 2$ internal vertices. Combining all these ingredients yields the claimed result.     □

We now only need a bound on the sum of $|N(T)|$ over all trees $T$. We will use a simple geometric argument. Consider the planar graph $H$ with vertex set $A$, no double edges, and a face for every tree $T_i$ in $G[V \setminus A]$ having incident vertices $N(T_i)$. We claim the following:

**Lemma 11.** $\sum_{i=1}^{n} |N(T_i)| \leq 2|E(H)|$.

*Proof.* Let $f_i$ be the face in $H$ that corresponds to $T_i$, and let $|f_i|$ be the number of vertices incident to $f_i$. Then $|N(T_i)| = |f_i|$. But $f_i$ is also equal to the number of incident edges to $f_i$. Hence $\sum_{i=1}^{n} |N(T_i)|$ counts every edge of the faces corresponding to $T_i$ twice, which proves the lemma.     □

**Lemma 12.** $\sum_{i=1}^{n} |N(T_i)| \leq 6k - 12$

*Proof.* Because $H$ is planar we can use Euler's formula to obtain $|E(H)| \leq 3|V(H)| - 6$. Using $|V(H)| = |A| = k$ and Lemma 11 we prove the result.     □

Combining Lemmas 10 and 12 almost proves Theorem 1, we only have to add the possible singleton trees with $|N(T)| = 2$. Because of Lemma 6 there can be at most one between every pair of nodes in $A$. Because $G$ is planar we know that the number of these nodes is at most the number of edges in a planar graph containing $|A| = k$ nodes. This adds at most $3k - 6$, bringing the exact constant in Theorem 1 to 352. In the following section we will show how to prove a lower constant.

## 7   Improving the Bound on the Size of the Kernel

We can improve the result from the previous section by using improved rules for the one- and two-twigs. Due to space limitations we only state the new rules and the final result. Both rules were found using the algorithm from Section 5. We ran the algorithm on a computer with a 2GHz AMD Athlon processor and it took 406 seconds for all 262144 cases necessary for $Z(2+, 1)$. Implementation was done in C++ using the LEDA graph library [29].

**Rule 12 (Improved One-Twig Rule).** *If there is a one-twig of size 3 with path-nodes $u_1, u_2, u_3$ and observer $v$ then remove $u_2$ and add edges $\{v, u_1\}, \{v, u_3\}$ and $\{u_1, u_3\}$.*

**Rule 13 (Improved Two-Twig Rule).** *Consider a twig with path $U = u_1 \ldots u_n$ and observers $W = \{v, w\}$. Let $q_i = d(u_i, T) - d(u_i, W \cup U)$. By $Z(a, b)$ we denote the maximum path length such that we have no reduction rule if $q_i = a$ and $q_n = b$. If $n > Z(q_1, q_n)$ then we have a reduction rule. We have found the following values for $Z(a, b)$: $Z(0, 0) = 1$, $Z(1, 0) = Z(0, 1) = 3$, $Z(2+, 0) = Z(0, 2+) = 4$, $Z(1, 1) = Z(2+, 1) = Z(1, 2+) = 5$, $Z(2+, 2+) = 6$.*

**Theorem 2.** *Consider a planar graph $G = (V, E)$ where none of the reduction rules can be applied. Let $A \subseteq V$ be a minimum FVS of size $k^*$. Then $|V| \leq 112k^*$.*

## 8   Conclusions

In this paper, we gave a polynomial time algorithm for kernelization of planar graphs. Such algorithms are also useful as a preprocessing step, i.e., the algorithm can be applied in a setting where $k$ is not a priory bounded, and then functions as a preprocessing heuristic. Van Dijk [35] has experimentally evaluated the kernelization algorithm for general graphs from [7]. These results are very promising: on several graphs taken from applications, the kernelization gives a significant size reduction, and the kernelization algorithm appears to be very fast.

All rules in this paper are local, and operate on specific subgraphs containing at most 9 nodes if we use the improved Twig-Rules. This implies that *one* automated rule that checks every connected subgraph of size at most 9 can replace them all. The algorithm on general graphs for the cubic kernel [7] uses two rules that are not local. It would be interesting to see if a cubic kernel could be obtained using just local rules. Also, safeness of the rules does not rely on planarity; only the proof of the bound does. It would be interesting to see if the result can be generalized to larger classes of graphs, e.g., graphs embeddable on a fixed surface or graphs avoiding a minor.

Another interesting result would be linear kernel for FEEDBACK VERTEX SET on general graphs. Also a smaller constant for the linear kernel presented in this paper would be of interest.

## References

[1] Alber, J., Fellows, M.R., Niedermeier, R.: Polynomial-time data reduction for dominating sets. J. ACM 51, 363–384 (2004)

[2] Bafna, V., Berman, P., Fujito, T.: A 2-approximation algorithm for the undirected feedback vertex set problem. SIAM J. Disc. Math. 12, 289–297 (1999)

[3] Bar-Yehuda, R., Geiger, D., Naor, J., Roth, R.M.: Approximation algorithms for the vertex feedback set problem with applications to constraint satisfaction and Bayesian inference. In: Proceedings of the 5th Annual ACM-SIAM Symposium on Discrete Algorithms, SODA 1994, pp. 344–354 (1994)

[4] Becker, A., Bar-Yehuda, R., Geiger, D.: Randomized algorithms for the loop cutset problem. J. Artificial Intelligence Research 12, 219–234 (2000)

[5] Becker, A., Geiger, D.: Optimization of Pearl's method of conditioning and greedy-like approximation algorithms for the vertex feedback set problem. Artificial Intelligence 83, 167–188 (1996)

[6] Bodlaender, H.L.: On disjoint cycles. Int. J. Found. Computer Science 5(1), 59–68 (1994)

[7] Bodlaender, H.L.: A cubic kernel for feedback vertex set. In: Thomas, W., Weil, P. (eds.) STACS 2007. LNCS, vol. 4393, pp. 320–331. Springer, Heidelberg (2007)

[8] Burrage, K., Estivill-Castro, V., Fellows, M.R., Langston, M.A., Mac, S., Rosamond, F.A.: The undirected feedback vertex set problem has a poly($k$) kernel. In: Bodlaender, H.L., Langston, M.A. (eds.) IWPEC 2006. LNCS, vol. 4169, pp. 192–202. Springer, Heidelberg (2006)

[9] Chen, J., Fomin, F.V., Liu, Y., Lu, S., Villanger, Y.: Improved algorithms for the feedback vertex set problems. In: Dehne, F., Sack, J.-R., Zeh, N. (eds.) WADS 2007. LNCS, vol. 4619, pp. 422–433. Springer, Heidelberg (2007)

[10] Chen, J., Liu, Y., Lu, S., O'Sullivan, B., Razgon, I.: A fixed-parameter algorithm for the directed feedback vertex set problem. In: Proceedings STOC 2008 (to appear, 2008)

[11] Chudak, F., Goemans, M., Hochbaum, D., Williamson, D.: A primal–dual interpretation of two 2-approximation algorithms for the feedback vertex set problem in undirected graphs. Operations Research Letters 22, 111–118 (1998)

[12] Dehne, F.K.H.A., Fellows, M.R., Langston, M.A., Rosamond, F.A., Stevens, K.: An $O(2^{O(k)}n^3)$ FPT algorithm for the undirected feedback vertex set problem. In: Wang, L. (ed.) COCOON 2005. LNCS, vol. 3595, pp. 859–869. Springer, Heidelberg (2005)

[13] Demaine, E.D., Hajiaghayi, M.: Bidimensionality: New connections between FPT algorithms and PTASs. In: Proceedings of the 16th Annual ACM-SIAM Symposium on Discrete Algorithms, SODA 2005, pp. 590–601 (2005)

[14] Dorn, F.: Dynamic programming and fast matrix multiplication. In: Azar, Y., Erlebach, T. (eds.) ESA 2006. LNCS, vol. 4168, pp. 280–291. Springer, Heidelberg (2006)

[15] Dorn, F., Penninkx, E., Bodlaender, H.L., Fomin, F.V.: Efficient exact algorithms on planar graphs: Exploiting sphere cut branch decompositions. In: Brodal, G.S., Leonardi, S. (eds.) ESA 2005. LNCS, vol. 3669, pp. 95–106. Springer, Heidelberg (2005)

[16] Downey, R.G., Fellows, M.R.: Fixed-parameter tractability and completeness. Congressus Numerantium 87, 161–178 (1992)

[17] Downey, R.G., Fellows, M.R.: Parameterized Complexity. Springer, Heidelberg (1998)

[18] Festa, P., Pardalos, P.M., Resende, M.G.C.: Feedback set problems. In: Handbook of Combinatorial Optimization, Amsterdam, The Netherlands, vol. A, pp. 209–258. Kluwer, Dordrecht (1999)

[19] Flum, J., Grohe, M.: Parameterized Complexity Theory. Springer, Heidelberg (2006)

[20] Fomin, F.V., Gaspers, S., Knauer, C.: Finding a minimum feedback vertex set in time $O(1.7548^n)$. In: Bodlaender, H.L., Langston, M.A. (eds.) IWPEC 2006. LNCS, vol. 4169, pp. 183–191. Springer, Heidelberg (2006)

[21] Fomin, F.V., Thilikos, D.M.: New upper bounds on the decomposability of planar graphs. J. Graph Theory 51, 53–81 (2006)

[22] Garey, M.R., Johnson, D.S.: Computers and Intractability, A Guide to the Theory of NP-Completeness. W.H. Freeman and Company, New York (1979)

[23] Goemans, M.X., Williamson, D.P.: Primal-dual approximation algorithms for feedback problems in planar graphs. Combinatorica 17, 1–23 (1997)

[24] Guo, J., Gramm, J., Hffner, F., Niedermeier, R., Wernicke, S.: Compression-based fixed-parameter algorithms for feedback vertex set and edge bipartization. Journal of Computer and System Sciences 72(8), 1386–1396 (2006)

[25] Guo, J., Niedermeier, R.: Invitation to data reduction and problem kernelization. ACM SIGACT News 38, 31–45 (2007)

[26] Hackbusch, W.: On the feedback vertex set problem for a planar graph. Computing 58, 129–155 (1997)

[27] Kanj, I.A., Pelsmajer, M.J., Schaefer, M.: Parameterized algorithms for feedback vertex set. In: Downey, R.G., Fellows, M.R., Dehne, F. (eds.) IWPEC 2004. LNCS, vol. 3162, pp. 235–248. Springer, Heidelberg (2004)

[28] Kloks, T., Lee, C.M., Liu, J.: New algorithms for k-face cover, k-feedback vertex set, and k-disjoint cycles on plane and planar graphs. In: Kučera, L. (ed.) WG 2002. LNCS, vol. 2573, pp. 282–295. Springer, Heidelberg (2002)

[29] Mehlhorn, K., Näher, S.: LEDA: A Platform for Combinatorial and Geometric Computing. Cambridge University Press, Cambridge (1995)

[30] Niedermeier, R.: Invitation to fixed-parameter algorithms. Universität Tübingen, Habilitation Thesis (2002)

[31] Raman, V., Saurabh, S., Subramanian, C.R.: Faster fixed parameter tractable algorithms for undirected feedback vertex set. In: Bose, P., Morin, P. (eds.) ISAAC 2002. LNCS, vol. 2518, pp. 241–248. Springer, Heidelberg (2002)

[32] Raman, V., Saurabh, S., Subramanian, C.R.: Faster algorithms for feedback vertex set. In: Proceedings 2nd Brazilian Symposium on Graphs, Algorithms, and Combinatorics, GRACO 2005. Electronic Notes in Discrete Mathematics, vol. 19, pp. 273–279 (2005)

[33] Razgon, I.: Exact computation of maximum induced forest. In: Arge, L., Freivalds, R. (eds.) SWAT 2006. LNCS, vol. 4059, pp. 160–171. Springer, Heidelberg (2006)

[34] Stamm, H.: On feedback problems in planar digraphs. In: Möhring, R.H. (ed.) WG 1990. LNCS, vol. 484, pp. 79–89. Springer, Heidelberg (1991)

[35] van Dijk, T.: Fixed parameter complexity of feedback problems. Master's thesis, Utrecht University (2007)

# Parameterized Chess

Allan Scott and Ulrike Stege

Department of Computer Science, University of Victoria

**Abstract.** It has been suggested that the parameterized complexity class AW[*] is the natural home of $k$-move games, but to date the number of problems known to be in this class has remained small. We investigate the complexity of SHORT GENERALIZED CHESS—the problem of deciding whether a chess player can force checkmate in the next $k$ moves. We show that this problem is complete for AW[*].

## 1 Games as Combinatorial Problems

When considering games as cominatorial problems, we ask whether the next player to move has a *winning strategy*—a strategy which is guaranteed to win the game for him regardless of his opponent's moves. Parameterized complexity approaches games by considering *k-move* or *short* games [1], which ask whether the game has a winning strategy that takes at most $k$ moves. To date there are three AW[*]-completeness results for short games: SHORT NODE KAYLES [1], SHORT GENERALIZED GEOGRAPHY [1], and SHORT PURSUIT [6]. All proofs used reductions from QUANTIFIED BOOLEAN $t$-NORMALIZED FORMULA SATISFIABILITY or its unitary variant. One short game, RESTRICTED ALTERNATING HITTING SET, has been shown to be in FPT when using an additional parameter [1]. The games PEBBLE GAME, PEG GAME and CAT AND MOUSE are known to be XP-complete when parameterized by the number of movable pieces available to the players [2,5]. Here, we investigate the complexity of SHORT GENERALIZED CHESS. GENERALIZED CHESS, is known to be EXPTIME-complete [4]. These forms of chess are *generalized* in that they are played on a $n \times n$ board.[1] Formally, a *position* is a tuple containing all information pertinent to a move in the game. A *move* is a transition from one position to another. Each game consists of a discrete sequence of *turns*. On each turn, one player makes a move. Since the concepts of move and turn are closely related in chess, we use the two terms interchangably.

The remainder of this article is organized as follows. In Section 2, we discuss the class AW[*]. In Section 3 we formally define SHORT GENERALIZED CHESS. In sections 4–5 we prove the parameterized membership and hardness of SHORT GENERALIZED CHESS.

---

[1] If we consider how to find a winning strategy for chess exactly as played–namely on an $(8 \times 8)$-chessboard–the number of possible positions is bounded by a constant, i.e. $(6 \cdot 2 + 1)^{64}$. This is since every space can either be occupied by one of 6 kinds of pieces in either black or white, or remain empty. Thus we can evaluate the game in constant time.

M. Grohe and R. Niedermeier (Eds.): IWPEC 2008, LNCS 5018, pp. 172–189, 2008.

# 2    The Class AW[*]

Although many short generalized games can easily be shown to be in PSPACE and XP, in their monograph [3], Downey and Fellows suggest that the "natural home" of $k$-move games is the class AW[*]. In support of this conjecture, we prove that SHORT GENERALIZED CHESS is in AW[*] by reducing it to QUANTIFIED BOOLEAN $t$-NORMALIZED FORMULA SATISFIABILITY (which is complete for AW[*] [3]).

QUANTIFIED BOOLEAN $t$-NORMALIZED FORMULA SATISFIABILITY (QBTNFSAT)

*Instance*: Positive integers $r$, $k_1, \ldots, k_r$; a sequence $s_1, \ldots, s_r$ of pairwise disjoint sets of boolean variables; a boolean formula $F$ over the variables $s_1 \cup \ldots \cup s_r$. $F$ consists of $t$ alternating layers of conjunctions and disjunctions with negations applied only to variables ($t$ is a fixed constant).

*Parameters*: $r$, $k_1, \ldots, k_r$.

*Question*: Does there exists a size-$k_1$ subset $t_1$ of $s_1$ such that for every size-$k_2$ subset $t_2$ of $s_2$, there exists a size-$k_3$ subset $t_3$ of $s_3$ such that ... (quantifiers continue to alternate between universal and existential for all $r$ quantifiers) ... such that, when the variables in $t_1, \ldots, t_r$ are all set to *true* and all other variables are set to *false*, formula $F$ is *true*?

We use a shorthand to describe these subset weights, where the subscript of the quantifier gives the associated Hamming weight. The following example uses the variables given in the definition above.

$$\underset{k_1}{\exists} s_1 \; \underset{k_2}{\forall} s_2 \; \underset{k_3}{\exists} s_3 \ldots \underset{k_r}{\forall} s_r : F$$

QBTNFSAT places requirements on the formatting of the formula. In particular, there are restrictions with an impact on the reductions below. First, only literals may be negated. Further, the formula must be $z$-normalized for some constant $z$ regardless of the input.[2]

# 3    The Game of SHORT GENERALIZED CHESS

Generalized chess is played by two players (players **I** and **II**) who alternate taking turns moving pieces on a *chessboard*, an $(n \times n)$-grid of squares.

SHORT GENERALIZED CHESS

*Input*: An $(n \times n)$-chessboard position[3], a positive integer $k$.

*Parameter*: $k$

---

[2] That is, there are at most $z$ alternating layers of *and* and *or*-operators.

[3] A *chessboard position* includes the position of every piece (including captured pieces), the turn number, and a flag for each king and rook indicating whether that piece has moved yet (essential information to castling, Rule 6).

*Question*: Starting from the given position, does player **I** have a strategy to force a win within the next $k$ moves?[4]

The following provides a basic outline of the rules of chess, though we assume familiarity with the basic movement rules for each piece. For complete rules, see [7].

1. SHORT GENERALIZED CHESS is played with six types of *pieces: pawns, rooks, knights, bishops, queens, and kings.* Each piece belongs to one of the two players. To distinguish which pieces belong to which player, player **I** and **II**'s pieces are coloured white and black respectively. Each player starts the game with $n$ pawns, two rooks, two knights, $n-6$ bishops, one queen, and one king, though pawns may be promoted to other captured pieces later in the game.

2. A chess piece occupies exactly one square at a time, and a square may only be occupied by one piece at a time. The exception is auxiliary square $S_0$ which holds all pieces captured during the game.

3. On her turn, a player must move exactly one of her pieces (except in the case of castling, Rule 6) from the square it currently occupies to another square (which must be empty or occupied by an opposing piece), in a manner legal for the piece being moved . If the square entered is occupied by an opposing piece, then that piece is *captured* and moved to $S_0$.

4. A player is said to be *in check* if her king could be captured by an opposing piece with a single legal[5] move. A player cannot make a move that would leave her in check. If, on her turn, a player is in check and every legal move available leaves her in check, then she is in *checkmate* and loses the game.

5. If a pawn reaches the opponent's end of the board (row 1 or $n$) it may be *promoted*; the pawn may be replaced with any captured piece (other than a pawn). The pawn itself is moved to $S_0$.

6. On her turn, instead of moving a single one of his pieces, a player may choose to *castle* her king with one of her rooks provided that she has not yet moved either piece that game, all the squares between them are empty, the move does not put her king in check, and no square the king would pass through is in check. When castling, the king moves two squares towards the rook and then the rook is advanced to the first square in which it is on the other side of the king.

## 4    Parameterized Membership of SHORT GENERALIZED CHESS

We show that SHORT GENERALIZED CHESS is in AW[*] by reducing it to QBT-NFSAT: we create a QBTNFSAT-formula that captures the rules of chess and applies them to the initial position of the given input instance. To simulate

---

[4] A move in chess usually consists of a player moving one of his pieces, except in the case of a capture, castling (cf. Rule 3 and Rule 6), or promotion, when two pieces are moved.

[5] A *legal* move is a move not breaking any of the rules, while any other move is *illegal*.

the alternating turn-structure of chess, we utilize the alternating quantifiers of the QBTNFSAT-instance. There is a natural correspondence between a QBTNFSAT-formula and a winning strategy for player **I**, since there exists a winning strategy for player **I** iff there exists a move for **I** such that for every move **II** makes there exists a move for **I** such that for every move **II** makes … there exists a move that results in a win for **I**. As such, the existential quantifiers in the QBTNFSAT-formula correspond to player **I**'s moves, and the universal quantifiers correspond to player **II**'s moves. Enforcing the rules of chess, testing the winning condition, and setting the initial position is done by formula $F$ of the QBTNFSAT-instance that we construct. For our purposes, a player can win either by achieving checkmate against the opponent, or because the opponent breaks one of the rules of chess. We remark that, to enforce checkmate rules, the formula used for our reduction actually simulates $k + 2$ moves rather than just $k$. Any test if a rule has been broken or the game has been won has to handle the possibility that the game has already ended.

## 4.1   Encoding Positions

Let $S_{x,y}$ correspond to the square on the chessboard at row $x$ and column $y$, where $x$ and $y$ are positive integers between 1 and $n$. If $a = S_{i,j}$, then we denote $(i, j)$ also with $(a_r, a_c)$. Further, $board \leftarrow \{S_{1,1}, \ldots, S_{n,n}\}$, $white$ and $black$ are the sets of pieces belonging to player **I** and **II** respectively, and $pieces \leftarrow white \cup black$. Note that $S_0 \notin board$. We next consider encoding the positions with the following set of variables:

$$v_{p,a,t} \leftarrow \begin{cases} true & : \quad p \in pieces \text{ is on } a \in board \cup \{S_0\} \text{ on turn } t \\ false & : \quad p \in pieces \text{ is not on } a \in board \cup \{S_0\} \text{ on turn } t \end{cases}$$

Unfortunately, this natural approach to encoding positions is not parameterized. The number of chess pieces in a game is polynomial in $n$ (i.e., $\leq 4n$), and this encoding scheme sets exactly one variable $true$ for each piece on each turn. As each turn corresponds to a single quantifier, this requires that we introduce $n$ into the Hamming weights on those quantifiers. These weights are all parameters; introducing $n$ into any one of them prevents the reduction from preserving the parameter.

To avoid this problem we record only the changes from turn to turn rather than the positions themselves. Given an existing position, a chess move is described using at most four changes to that position. Two changes always occur because a piece leaves one square ($a$) and enters another ($b$). If the move entails a capture then another two changes occur; the captured piece leaves $b$ and enters $S_0$. These two observations are sufficient to describe every chess move but two: castling and pawn promotion. However, both these actions entail exactly two pieces moving. In castling (Rule 6) a rook and a king move simultaneously, while we encode pawn promotion by moving the pawn off the board and another piece from $S_0$ into what would be the pawn's destination square (Rule 5). As such, these moves are described with exactly four changes each. We define the variables for this change-of-position encoding scheme.

$$x_{p,a,t} \leftarrow \begin{cases} true & : \quad v_{p,a,t} \neq v_{p,a,t-1} \\ false & : \quad v_{p,a,t} = v_{p,a,t-1} \end{cases}$$

Using $x_{p,a,t}$, we derive the values of the original $v_{p,a,t}$-variables using the formulas

$$v_{p,a,t} = \bigvee_{S \in Y} \left( \bigwedge_{i \in S} x_{p,a,i} \wedge \bigwedge_{i \in \{1,2,\ldots,t\} - S} \neg x_{p,a,i} \right), \text{ where } Y \text{ is the set of all even-sized}$$

subsets of $\{1,2,\ldots,t\}$ if $v_{p,a,0}$ is *true* and the set of all odd-sized subsets of $\{1, 2, \ldots, t\}$ otherwise. The position on turn 0 corresponds to the initial position given as input to our SHORT GENERALIZED CHESS-instance. Thus, in using these formulas we have implicitly encoded the initial position into the QBTNFSAT-instance.

This formula is a brute-force test of all possible move sequences that result in $v_{p,a,t} = true$. Each *true* variable in $x_{p,a,u}, u \in \{1,\ldots,t\}$, implies that the value of $v_{p,a,u}$ has been inverted w.r.t. the previous turn; an even number of inversions preserves the initial value, while an odd number flips it.[6] For both these formulas, the number of clauses is bounded by $2^k$ (the number of all possible bit sequences) and thus is in FPT.

If at some point we need to negate one of the $v_{p,a,t}$-values, we simply use DeMorgan's rule to receive its negation $\overline{v_{p,a,t}}$. This enables us to encode the board itself. However, just looking at the board does not tell us whether a piece has been moved at some time in the past—information which is necessary to enforce the rules of castling (Rule 6, App. A). We introduce flags to aid the enforcement of these rules. For each $t$, we define a set $CF_t \leftarrow \{c_t^{W\ell}, \overline{c_t^{W\ell}}, c_t^{Wr}, \overline{c_t^{Wr}}, c_t^{B\ell}, \overline{c_t^{B\ell}}, c_t^{Br}, \overline{c_t^{Br}}\}$ where flag $c_x^{ab}$ is *true* iff on turn $x$ castling is allowed on the left ($b = \ell$) or right ($b = r$) side for **I** ($a = W$) or **II** ($a = B$). The overlined flags carry the opposite *true/false* value of the corresponding non-overlined ones, so that exactly four of the variables in this set are always *true*. This means that each quantifier of our QBTNFSAT-instance has a Hamming weight of 8—4 bits for the pieces and another 4 for the castling flags.

We now define $J_t \leftarrow \{x_{p,a,t} : p \in pieces, a \in board \cup \{S_0\}\} \cup \{y_{t,0}, y_{t,1}\}$. $y_{t,0}$ and $y_{t,1}$ are "sink" variables which are set to true if only two changes to the board positions occur on a turn. They have no other purpose. With this we can define the quantifier sets for our QBTNFSAT-instance, which are pair-wise disjoint because each uses a different turn $t$.

$$\exists_8 (J_1 \cup CF_1) \, \forall_8 (J_2 \cup CF_2) \ldots \forall_8 (J_t \cup CF_t)$$

## 4.2   The Winning Condition

Before we define player **I**'s winning condition $W$, we define several sets that we refer to throughout the reduction. We first define the sets *current* and *opponent*.

---

[6] Hence why we look for an even number when the value was initially *true* and an odd number when false.

For odd $t$, let *current* $\leftarrow$ *white* and *opponent* $\leftarrow$ *black*. For even $t$, let *current* $\leftarrow$ *black* and let *opponent* $\leftarrow$ *white*. We also use the set *pawns* of all pawns. Other pieces that we refer to individually are: $WK$, the white king, $BK$, the black king, $WR^\ell$, the left white rook, $WR^r$, the right white rook, $BR^\ell$, the left black rook, and $BR^r$, the right black rook.

$$W \leftarrow \bigvee_{0 \leq t \leq k} (v_{BK,0,t} \wedge \overline{v_{WK,0,t}})$$

Rather than testing whether a position is checkmate for either player, we test whether a king has been captured on the following turn. This simple test has the advantage that it also handles the rules regarding check, as shown below. Player **I**'s winning condition as presented in $W$ is simply to capture $BK$ before $WK$ is captured, as $WK$ being captured previously would imply that black had already won.

**Lemma 1.** *The rules regarding check and checkmate over the next $t$ turns resolve under optimal play to the winning condition of capturing the opponent's king within the next $t + 2$ turns.*

### 4.3    Formula $F$

We use $F$—the formula for the QBTNFSAT-instance produced by the reduction—to enforce the rules of chess, including winning condition $W$. Intuitively, $F$ is *true* in exactly two cases: if no rules were broken and **I** won, or if **II** broke a rule. $F$ is false otherwise. To handle the rules of chess, we define $R_{all}$ and $R_{black}$. $R_{all}$ tests that all the rules were followed, $R_{black}$ tests for each of **II**'s turns that either **II** followed the rules or the game is over.[7] Both formulas use formula $L_t$. Intuitively, $L_t$ is *true* iff the change in position from turn $t - 1$ to $t$ describes a legal move. The formal definition of $L_t$ appears in the next subsection.

$$R_{all} \leftarrow \bigwedge_{1 \leq t \leq k} (L_t \vee v_{BK,0,t-1} \vee v_{WK,0,t-1})$$

$$R_{black} \leftarrow \bigwedge_{1 \leq t \leq \frac{k}{2}} \left( L_{2t} \vee v_{BK,0,2t-1} \vee v_{WK,0,2t-1} \vee \bigvee_{1 \leq s \leq t} \overline{L_{2s-1}} \right)$$

We now combine these formulas resulting in formula $F \leftarrow (R_{all} \wedge W) \vee \overline{R_{black}}$. $F$ is satisfied if both players play a legal game that ends with **I** as the winner, or if **II** moves illegally before the game ends.

### 4.4    Testing for Broken Rules

We define $L_t \leftarrow \overline{M_t} \wedge P_t \wedge C_t \wedge E_t \wedge \overline{D_t} \wedge K_t$, where $M_t$ tests for illegal movements, $P_t$ tests whether the path of moving pieces is clear, $C_t$ tests that the capture

---

[7] Here, the game is over because either king was captured or because **I** moved illegally on a previous turn.

rules have been followed, $E_t$ tests that every piece exists exactly once this turn, $D_t$ tests if two pieces were moved simultaneously, and $K_t$ maintains flags for handling castling moves. We elaborate on each of these sub-formulas.

**Range of Movement ($M_t$).** $M_t$ tests for illegal movements. To handle movement rules we introduce:

$$\Delta(p, a, b) \leftarrow \begin{cases} true : \text{square } b \text{ is within } p\text{'s range of movement from square } a \\ false : \text{square } b \text{ is not within } p\text{'s range of movement from square } a \end{cases}$$

A piece's *range of movement* is considered to be its available moves under ideal conditions (e.g. there are no pieces in the way). A pawn's range of movement includes the diagonal movements that a pawn can make only if the move captures an opposing piece.

$$M_t \leftarrow \bigvee_{\substack{p \in current \\ a,b \in board \\ \Delta(p,a,b)=false}} (v_{p,a,t-1} \wedge v_{p,b,t}) \vee \bigvee_{\substack{p \in opponent \\ a,b \in board}} (v_{p,a,t-1} \wedge v_{p,b,t})$$

$$\vee \bigvee_{\substack{p \in current \cap pawns \\ a,b \in board, a_c \neq b_c \\ \Delta(p,a,b)=true}} \left( v_{p,a,t-1} \wedge v_{p,b,t} \wedge \bigwedge_{q \in opponent} \overline{v_{q,b,t-1}} \right)$$

**Lemma 2.** $M_t$ *is* true *iff on turn $t$ a piece is moved illegally.*

**Path of Movement ($P_t$).** $P_t$ tests whether the path of every moving piece, except the knight's, is clear. We define a path $path(a, b)$ on the chessboard using two squares $a, b \in board$ with $a$ and $b$ belonging to the same column, row, or diagonal of the chessboard. The squares visited by moving from $a$ to $b$ along the associated row, column, or diagonal are on the path between $a$ and $b$.

$$P_t \leftarrow \bigwedge_{\substack{p \in current \\ a,b \in board}} \left( \overline{v_{p,a,t-1}} \vee \overline{v_{p,b,t}} \vee Q_{a,b,t} \wedge \bigwedge_{\substack{q \in pieces \\ q \neq p}} \overline{v_{q,b,t}} \right)$$

$$\wedge \bigwedge_{\substack{p \in current \cap pawns \\ a,b \in board, a_c = b_c}} \left( \overline{v_{p,a,t-1}} \vee \overline{v_{p,b,t}} \vee \bigwedge_{q \in opponent} \overline{v_{q,b,t-1}} \right) \text{ with}$$

$$Q_{a,b,t} \leftarrow \bigwedge_{\substack{p \in pieces \\ d \in path(a,b)}} \overline{v_{p,d,t}} \text{ if } a \text{ and } b \text{ share a row, column, or diagonal.}$$

Otherwise $Q_{a,b,t}$ is *true*.

**Lemma 3.** *If squares $a$ and $b$ are the same row, column, or diagonal, then $Q_{a,b,t}$ is* true *iff all the squares between (but not including) $a$ and $b$ along the associated row, column, or diagonal are unoccupied on turn $t$.*

If $a$ and $b$ do not share a row, column, or diagonal, then either the movement is illegal and rejected by $M_t$, or the piece moved is a knight and can legally jump over other pieces. In either case, $Q_{a,b,t}$ being empty and trivially true is correct. $M_t$ tests for illegal piece movements, and $D_t$ ensures that only one piece is moved on the board at once. Thus, it is sufficient for $P_t$ to ensure that pieces move through empty squares.

**Lemma 4.** $P_t$ is true iff all pieces legally moved on turn $t$ are moved through empty squares and end their move in a square that is not occupied by a same-colored piece on turn $t - 1$.

**Capturing ($C_t$).** $C_t$ tests that if piece $p$ arrived this turn in square $a$, occupied by a opposing piece $q$ last turn, then $q$ is moved to $S_0$. Similarly, if an opposing piece $q$ arrived in $S_0$ this turn, then whichever square $q$ occupied last turn is now occupied by a piece controlled by the current player.

$$C_t \leftarrow \bigwedge_{\substack{p \in current \\ q \in opponent \\ a \in board}} (\overline{v_{p,a,t}} \lor \overline{v_{q,a,t-1}} \lor v_{q,0,t}) \land \bigwedge_{q \in opponent} \left( \overline{x_{q,0,t}} \lor \bigvee_{\substack{p \in current \\ a \in board}} (v_{p,a,t} \land v_{q,a,t-1}) \right)$$

$$\land \bigwedge_{p \in opponent} (\overline{v_{p,0,t-1}} \lor v_{p,0,t}) \land \bigwedge_{p \in current} \left( \overline{v_{p,0,t-1}} \lor v_{p,0,t} \lor \bigvee_{\substack{i \in \{1,...,n\} \\ q \in pawns \cap current}} U_{p,q,t,i} \right) \land$$

$$\bigwedge_{p \in current \cap pawns} \left( v_{p,0,t-1} \lor \overline{v_{p,0,t}} \lor \bigvee_{\substack{i \in \{1,...,n\} \\ q \in current - pawns \cap current}} U_{q,p,t,i} \right) \land$$

$$\bigwedge_{p \in current - pawns} (v_{p,0,t-1} \lor \overline{v_{p,0,t}}) \text{ with}$$

$$U_{p,q,t,i} \quad \leftarrow \quad \left( v_{q,S_{i,n-1},t-1} \land v_{q,0,t} \land v_{p,S_{i,n},t} \land \bigwedge_{\substack{r \in pieces \\ r \neq p}} \overline{v_{r,S_{i,n},t}} \right)$$

$C_t$ does not deal with the move that caused the capture, as that is handled by $M_t$ and $P_t$.

**Lemma 5.** $C_t = true$ iff all pieces that were captured on turn $t$ moved to $S_0$, no opponent's pieces moved to $S_0$ on turn $t$ that were not captured on turn $t$, and no pieces left $S_0$ on turn $t$ except by pawn promotion.

**Every Piece Exists Exactly Once ($E_t$).** $E_t \leftarrow O_t \wedge \overline{T_t}$ tests that every piece is in exactly one position on turn $t$, but not in two positions.

$$O_t \leftarrow \bigwedge_{p \in pieces} \left( \bigvee_{a \in board \cup \{0\}} v_{p,a,t} \right), \quad T_t \leftarrow \bigvee_{\substack{p \in pieces \\ a,b \in board \cup \{0\} \\ a \neq b}} (v_{p,a,t} \wedge v_{p,b,t})$$

**Lemma 6.** $E_t$ *is true iff on turn $t$, for every $p \in pieces$ there is exactly one $a \in board \cup \{0\}$: $v_{p,a,t}$ true.*

**Pieces Moving Simultaneously ($D_t$).** $D_t$ tests if two pieces were moved simultaneously on the board. Each quantifier has Hamming weight 8. Exactly four elements in the quantified sets are used up by the castling flags–four remain. The four bits available allow two pieces to move in the event of capture, castling, or promotion. Otherwise, we restrict the players from moving two board pieces simultaneously.

$$D_t \leftarrow \bigvee_{\substack{p,q \in pieces \\ a,b,c,d \in board \\ a \neq b, c \neq d, p \neq q}} \left( x_{p,a,t} \wedge x_{p,b,t} \wedge x_{q,c,t} \wedge x_{q,d,t} \wedge \neg Y_t^W \wedge \neg Y_t^B \right) \text{ with}$$

$$Y_t^W \leftarrow (c = WK \wedge d = WR^\ell \wedge v_{WK,S_{1,\frac{n}{2}},t-1} \wedge v_{WK,S_{1,\frac{n}{2}-2},t} \wedge v_{WR^\ell,S_{1,1},t} \wedge v_{WR^\ell,S_{\frac{n}{2}-1,1},t} \wedge c_{t-1}^{W\ell} \wedge \overline{c_t^{W\ell}}$$

$$\wedge \overline{c_t^{Wr}} \wedge A_{S_{1,\frac{n}{2}},t} \wedge A_{S_{1,\frac{n-1}{2}},t}) \vee (c = WK \wedge d = WR^r \wedge v_{WK,S_{1,\frac{n}{2}},t-1} \wedge v_{WK,S_{1,\frac{n}{2}+2},t} \wedge v_{WR^r,S_{1,1},t}$$

$$\wedge v_{WR^r,S_{\frac{n}{2}+1,1},t} \wedge c_{t-1}^{Wr} \wedge \overline{c_t^{W\ell}} \wedge \overline{c_t^{Wr}} \wedge A_{S_{1,\frac{n}{2}},t} \wedge A_{S_{1,\frac{n+1}{2}},t}) \text{ and}$$

$$A_{s,t} \leftarrow \bigwedge_{p \in opponent} \left( \bigwedge_{\substack{a \in board:\Delta(p,a,s) \\ a \neq s}} (\overline{v_{p,a,t}} \vee \overline{Q_{a,s,t}}) \right).$$

$Y_t^B$ is identical to $Y_t^W$, except $B$ is substituted for $W$ and the squares are adjusted appropriately. Tests $c = WK \wedge d = WR^\ell$ ensure we consider the correct pieces. The next four tests involving $v$ ensure the pieces are engaged in a castling motion. The next three tests (involving $c$) ensure that the castling flags are maintained. The last two tests (involving $A$) handle the rule that a castling must not pass through a square in check. $A_{s,t}$ is *true* iff square $s$ is not threatened on turn $t$ by a piece in *opponent*. This explains the first clause; the other is identical, except that $\ell$ is replaced with $r$ and squares are on the right instead of left.

**Lemma 7.** $D_t$ *is true iff there exist two pieces $p, q$ which both moved on turn $t$ without starting or ending their moves in $S_0$, and the move is not a legal castling.*

**Castling Maintenance ($K_t$).** $K_t \leftarrow K_t^W \wedge K_t^B$ tracks castling flags. Remember that a chessboard position includes whether any given king or rook had been moved from its starting square. $K_t$ tracks this information.

$$K_t^W \leftarrow (c_t^{W\ell} \vee \overline{c_t^{W\ell}}) \wedge (c_t^{Wr} \vee \overline{c_t^{Wr}}) \wedge (\neg c_t^{W\ell} \vee \neg \overline{c_t^{W\ell}}) \wedge (\neg c_t^{Wr} \vee \neg \overline{c_t^{Wr}}) \wedge (\neg \overline{c_{t-1}^{W\ell}} \vee c_t^{W\ell}) \wedge (\neg \overline{c_{t-1}^{Wr}} \vee c_t^{Wr})$$

$$\wedge \ (x_{WR^\ell, S_{1,1}, t} \vee \overline{c_t^{W\ell}}) \wedge (x_{WR^r, S_{1,n}, t} \vee \overline{c_t^{Wr}}) \wedge (\neg x_{WK, S_{1,\frac{n}{2}}, t} \vee (\overline{c_t^{W\ell}} \wedge \overline{c_t^{Wr}}))$$

$$\wedge \ (\overline{c_{t-1}^{W\ell}} \vee c_t^{W\ell} \vee t \text{ is odd }) \wedge (\overline{c_{t-1}^{Wr}} \vee c_t^{Wr} \vee t \text{ is odd })$$

$K_t^B$ is defined as $K_t^W$, except $B$ is substituted for $W$, the starting positions are updated accordingly, and $t$ must be even instead of odd.

**Lemma 8.** $K_t$ *is true iff the states of the castling flags have been maintained properly.*

## 4.5 Correctness of the Reduction

**Theorem 1.** $L_t$ *is true iff the differences between position* $\{p \in \text{pieces}, a \in \text{board} \cup \{0\} : v_{p,a,t}\}$ *on turn $t - 1$ and position* $\{p \in \text{pieces}, a \in \text{board} \cup \{0\} : v_{p,a,t-1}\}$ *on turn $t$ describe a legal chess move.*

*Proof.* $E_t$ ensures that there is exactly one of each piece, preventing pieces from spontaneously appearing or disappearing. Consequently, pieces can only move from one position to another. $M_t$ ensures that any pieces moved on the board are moved legally, and $P_t$ ensures that the paths of those moves are clear. $C_t$ ensures that any piece which should be captured is indeed moved to $S_0$, that no pieces are moved to $S_0$ without being captured, and that any pieces which are captured do not return to the board. $D_t$ ensures that only one piece moves on the board (excluding captures and castling), and $K_t$ ensures that the castling flags are maintained properly. Finally, the hamming weight of 4 (8 total – 4 for castling flags), combined with the fact that we have only two "sink" variables means that at least one move must take place.

**Lemma 9.** *Formula $F$ is 10-normalized.*

**Theorem 2.** $F$ *is true iff* **I** *has a winning strategy that takes at most $t$ turns to execute.*

*Proof.* If **I** has a winning strategy and plays legally, then either **II** plays legally ($R_{ALL}$ is *true*) and **I** wins the game ($W$ is *true*) or **II** makes an illegal move ($R_{even}$ is false). In either case, $F$ is satisfied. If **I** does not have a winning strategy, then in all instances where **II** plays legally ($R_{even}$ is *true*), either **I** does not win ($W$ is false) or else **I** makes an illegal move ($R_{ALL}$ is false).

# 5   Hardness of Short Generalized Chess

We show that SHORT GENERALIZED CHESS is AW[*]-hard by a reduction from UNITARY QUANTIFIED BOOLEAN $t$-NORMALIZED FORMULA SATISFIABILITY. Using this unitary version of QBTNFSAT allows us to set the state of the variables for an entire quantifier with the position of a single chess piece.

> UNITARY QUANTIFIED BOOLEAN $t$-NORMALIZED FORMULA SATISFIABILITY (U-QBTNFSAT)
> *Instance:* A positive integer $r$; a sequence $s_1, \ldots, s_r$ of pairwise disjoint sets of boolean variables; a boolean formula $X$, involving variables $s_1 \cup \ldots \cup s_r$, that consists of $t$ alternating layers of conjunctions and disjunctions with negations applied only to variables ($t$ is a fixed constant).
> *Parameter:* $r$
> *Question:* Does there exist a variable $t_1$ of $s_1$ such that for every variable $t_2$ of $s_2$, there exists a variable $t_3$ of $s_3$ such that ... (alternating qualifiers) such that, when the variables $t_1, \ldots, t_r$ are made *true* and all other variables are made false, formula $X$ is *true*?

This problem is complete for AW[t] ($=$ AW[*]) [3]. We choose $t = 2$, putting $F$ in conjunctive normal form. In our reduction, we create a chess-instance that uses a large block of pawns arranged in a checkerboard pattern at the center of the board (*walls of pawns*). We carve out paths from the inside of this checkerboard, through which the white queen, several black bishops, and two black rooks are able to move. The white queen then guarantees capture of $BK$ within $k$ turns iff the U-QBTNFSAT-instance is not satisfiable.

We use a set of *variable diagonals* for our chess-instance to represent the variable assignment. Each U-QBTNFSAT-variable has one corresponding diagonal, and the game simulates setting values for these variables by moving black bishops into the diagonals that are "true". Due to the unitary nature of U-QBFTNFSAT, each diagonal represents every other variable in its subset as *false*; if a variable $t \in S$ is *true*, then for $u \in S, u \neq t$, $u$ is *false* because the Hamming weight of $S$ is 1.

The variable diagonals are crossed by a set of vertical paths (*clauseways*). The intersections between variable diagonals and clauseways are constructed s.t. if variable $t = true$ implies that clause $C$ is satisfied, then a black bishop patrolling the corresponding diagonal blocks the white queen's progress through the clauseway corresponding to $C$. If otherwise assigning $t = true$ does not satisfy $C$, then the white queen captures a black bishop by blocking the associated clauseway without any repercussions. Thus, player **I** has only enough turns to capture $BK$ if all $k$ bishops in the chosen clauseway can be captured. Any bishop that cannot be captured forces **I** to spend extra turns moving the queen along an alternate route.[8]

Before we present the detailed gadgets of the reduction, we note important properties of the position that we maintain: The only white piece free to move

---

[8] This makes it impossible to reach the king in the number of turns allotted.

is the queen. Other pieces are able to move only if player **II** does not move as intended and attacks a white pawn. Consequently, player **I** is unable to win the game if her queen is captured. If the white queen moves as the reduction intends, then on every turn player **II** is forced to *move in reply*[9]; failing to react to **I**'s move[10] results in player **I** being able to capture *BK*. Thus, we can guarantee certain constructs of black pieces to be *unassailable*. That is, if player **I** attacks these pieces with her queen, player **II** can capture the queen and prevent **I** from winning the game. For example, consider a black pawn with a bishop in the diagonally adjacent square below and to the right of it. If player **I**'s queen takes the black bishop, then the black pawn can capture **I**'s queen, and vice versa. Because these two pieces protect each other the white queen cannot capture either. Furthermore, this protection can be extended to any other black piece in a square that either of these first two pieces could attack; the white queen cannot capture these other protected pieces without first capturing the piece that protects themselves, but we know that these pieces are unassailable. The protection can be extended recursively; this allows us to eliminate the possibility that the white queen attempts to tunnel through the pawn walls.

**Walls of Pawns.** Pawns have unique properties: they can be blocked from moving forward by an opposing piece, and they cannot move diagonally unless attacking a piece. With these two properties, we can create a situation where no moves are possible simply by placing a single white pawn one square above a single black pawn on an otherwise-empty chessboard (Fig. 1).

We extend this observation by arranging an infinite number of pawns in a checkerboard pattern. No pawn can move to the next square up or down because that square is occupied by a pawn of the opposite color. No pawn can move diagonally because those squares are occupied by pawns of the same color (Fig. 1).

Given such a checkerboard pattern, we are able to remove pawns while still preserving the property that no moves or captures are possible. The most basic way to do so is by removing an adjacent pair of pawns consisting of a white pawn one square above a black pawn. Then the pawns above and below along the same column are still blocked, and the pawns on either side cannot move diagonally because there are no pieces in the newly-emptied squares to capture. We can further carve out paths by using repeated deletions. Fig. 2 demonstrates how we can cut vertically, diagonally, and horizontally so long as the topmost deleted pawn in any column in black and the bottommost is white.

**Assignment (Crooked Path) Gadget.** This gadget enables the two players to alternate setting values for the variables of the U-QBTNFSAT-instance. It does this by having the queen start at the top of a crooked path (as can be seen in Fig. 1, *l*). This path alternates between vertical and diagonal segments (see

---

[9] That is, player **I** will win unless player **II** chooses a move (from a very limited set) that can alleviate the threat.

[10] That is, moving an unrelated piece elsewhere on the board.

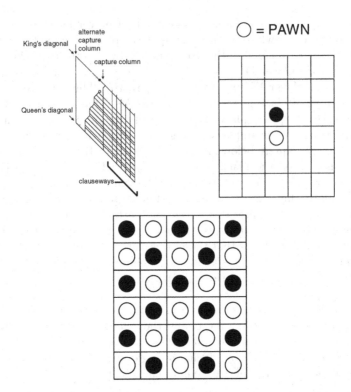

**Fig. 1.** (*l*) An overview of the chessboard as built by the reduction. (*r*) Deadlocked pawns. (*b*) Pawns laid out in a checkerboard arrangement are also deadlocked.

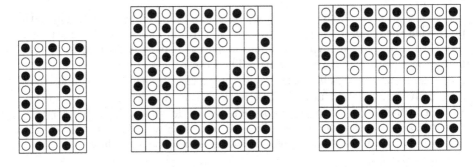

**Fig. 2.** (*l*) Pawns in a checkerboard arrangement with a column segment removed. (*m*) A diagonal removed. The pawns are still deadlocked. (*r*) An example of battlement cuts.

Fig. 3 (*l*)). In each segment, exactly one quantifier is handled. The game simulates setting the values for the variables in the same order as the U-QBTNFSAT instance.

Universal variables are set by player **I**. Each segment of the crooked path intersects with all the assignments diagonals corresponding to variables handled by the universal quantifier corresponding to that segment. When the queen moves to the end of a variable diagonal, the available black bishop must move to block the queen from moving down the diagonal into the capture column. A second black bishop protects the first (Fig. 3 ($r$)).

Existential variables are set by **II**. At the end of each segment is a horizontal path leading to a diagonal which intersects the perpendicular variable diagonals (as above for universal variables). A black bishop is placed in this diagonal such that the queen can attack it using the horizontal path. When the queen reaches the end of a segment, it can attack this bishop and gain entry to the capture column unless **II** moves the bishop. Since the bishop is forced to move, **II** can protect one of the variable diagonals (Fig. 4 ($l$)).

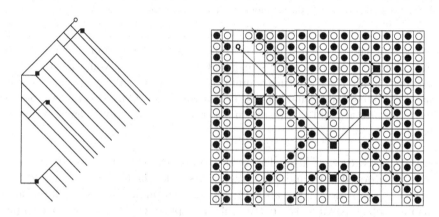

**Fig. 3.** ($l$)The layout of the assignment gadget. This shows only two segments. Additional segments are added to the end as necessary. ($r$) Interaction in the variable assignment gadget. Player **II** has responded to **I**'s attack on the column. Arrows indicate possible moves by **II** to capture the queen if **I** attacks the gadget.

**Lemma 10.** *Player* **II** *must react to player* **I** *in the assignment gadget.*

**Lemma 11.** *Player* **II** *cannot make a move in the assignment gadget to prevent* **I** *from using the capture column.*

**Lemma 12.** *Player* **I** *cannot win by circumventing the assignment gadget.*

**Capture Column.** The capture column is to the right of the assignment gadget, and intersects every variable diagonal. The top of the column connects to the king's diagonal through a short diagonal. The bottom of the capture column intersects with the queen's diagonal in the one-way gadget, explained below.

**Alternate Capture Column.** There exists a second, alternate capture column on the left side of the assignment gadget. This column intersects only the king's

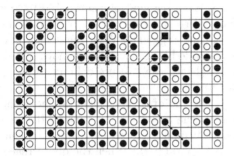 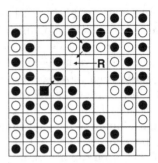

**Fig. 4.** (*l*) Player **II** has moved the bishop out of the queen's path and set the value of an existential variable in the process. (*r*) A gadget to allow **II** to block the alternate capture column. Once the rook is moved next to the two black pawns, the diagonal is blocked and all the black pieces blocking it are protected.

diagonal at the top, and the queen's diagonal at the bottom. A small gadget (Fig. 4) located in the queen's diagonal allows **II** to prevent the queen from entering the alternate capture column. In fact, **II** must do this when the queen enters the queen's diagonal because otherwise **I** wins. Once this route is blocked, the only route available to the queen is through the one-way gadget and towards the clauseways.

**One-Way Gadget.** This gadget is placed between the white queen's exit point from the crooked path and the entry points of the clauseways. It enables **II** to protect the capture column after **I** has passed through. This is accomplished by placing a black pawn at the bottom of the capture column in the queen's path, with a white one beneath it. When the queen captures the black pawn, the black rook takes the white pawn. If the queen now enters the capture column the rook can capture it.

**Lemma 13.** *Player* **II** *must react to player* **I** *in the one-way gadget.*

**Lemma 14.** *Player* **II** *cannot prevent* **I** *from using the capture column while the queen is in the assignment gadget.*

**Lemma 15.** *Player* **I** *cannot circumvent the one-way gadget.*

**Clauseways.** A clauseway is a narrow vertical path that crosses every variable diagonal. For each variable and each clause, if setting the given variable to true would satisfy that clause then a black bishop can safely protect the intersection between the given variable diagonal and clauseway. That is, a queen moving up this clauseway would be forced to move around the bishop rather than capture it.

To create this property, we use a wall construct that is slightly different from the checkered pawn configuration. The intersections between the vertical clauseways and variable diagonals produce parallellogram-shaped walls. We use a

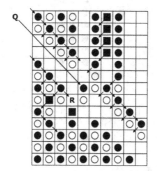
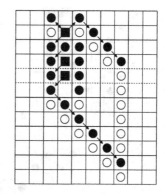

**Fig. 5.** (*l*) Interaction in the one-way gadget. The queen is about to move in and take the black pawn. **II** responds by taking the white pawn below with the rook, guarding the capture column. (*r*) The parallelogram gadget.

different construct for these specific walls, illustrated in Fig. 5. The advantage of these is that their height can be adjusted arbitrarily by extending the segment within the dotted lines. Thus, we can protect or not protect the bishop simply by moving the bottom of the parallelogram sitting above and to the right of the intersection (Fig. 6).

**Lemma 16.** *Player* **II** *must react to* **I** *in the clauseways.*

**Lemma 17.** *Once the queen enters the clauseways, player* **I** *cannot circumvent the clauseways.*

**King's Diagonal.** *BK* is located in a diagonal that runs parallel to the variable diagonals. It is intersected by every clauseway, the capture column, and the alternate capture column. Fig. 6 shows the king in the diagonal. The king itself cannot move because any move would put it in check. If the white queen is placed on the specific diagonal occupied by the king the result is checkmate. The king cannot move out of check, and none of the black pieces can attack the queen in the king's diagonal.[11]

**Correctness of the Reduction**

**Theorem 3.** *Player* **I** *has a winning strategy iff X is not satisfiable.*

*Proof.* If *X* is not satisfiable, then for any variable assignment **II** chooses, there is a variable assignment available for **I** that results in at least one unsatisfied clause. Once this has been achieved, **I** can move her queen to the clauseway representing the unsatisfied clause and have enough turns to capture all the bishops and checkmate the king. **II** cannot prevent this: we know how **II** must

---

[11] Ignoring the possibility of **I** purposely placing the queen so that a pawn could capture it.

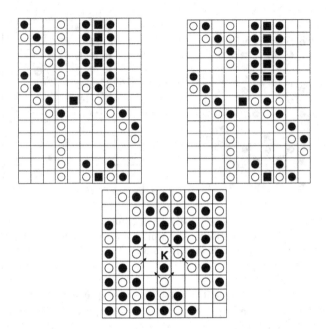

**Fig. 6.** Parallelograms to (*l*) leave the bishop vulnerable to the queen and (*r*) protect the bishop. *(b)* The king's position in the king's diagonal. Note that a white piece can attack any square the king can move to.

respond within each gadget, and that failing to respond appropriately results in a victory for **I**. If $X$ is satisfiable, then **I** loses if it plays the game as expected[12] because **II** can ensure that every clause has at least one satisfied literal and thus every clauseway will have at least one bishop that can block the white queen for more than one turn. However, we have already proved that **I** cannot win by attempting to circumvent any of the gadgets, and **I** cannot attack the pawn walls to escape these gadgets. Thus if **II** plays optimally, **I** cannot win within $k$ turns.

**Board Size.** The size of the board needed by the reduction is tied to the number of pawns used. We can bound the number of pawns needed by estimating the size of the parallelogram produced by the reduction. The height is $n+c+2$ diagonals, where $n$ is the number of variables and $c$ is the number of clauses in the input instance. Each diagonal is 12 squares high. The width of the board is $c$ clauseways (7 squares), 2 additional vertical paths (5 squares), and the assignment gadget, which cannot be wider than parallelogram is high otherwise it would exceed the bounds of the parallelogram. Thus, the parallelogram is $12(n+c+2)+8$ squares high and $19c + 12n + 42$ squares wide. The product gives us a bound on the number of pawns.

---

[12] E.g. setting variables and then looking for a clear clauseway.

# References

1. Abrahamson, K., Downey, R., Fellows, M.: Fixed-parameter tractability and completeness IV: On completeness for W[P] and PSPACE analogues. A. of Pure and Applied Logic 73, 235–276 (1995)
2. Adachi, A., Iwata, S., Kasai, T.: Some combinatorial game problems require $\omega(n^k)$ time. J. ACM 31(2), 361–376 (1984)
3. Downey, R., Fellows, M.: Parameterized Complexity. Springer, Heidelberg (1999)
4. Fraenkel, A., Lichtenstein, D.: Computing a perfect strategy for $n \times n$ chess requires time exponential in $n$. LNCS 115, 278–293 (1981)
5. Kasai, T., Adachi, A., Iwata, S.: Classes of pebble games and complete problems. SIAM J. Comput. 8(4), 574–586 (1979)
6. Scott, A.: Short pursuit-evasion. Texts in Algorithmics 7: Algorithms and Complexity in Durham 2006, 141–152 (2006)
7. FIDE Handbook (Online Version): Chess rules,
   http://www.fide.com/official/handbook.asp?level=EE101

# The Time Complexity of Constraint Satisfaction

Patrick Traxler[*]

Institute of Theoretical Computer Science, ETH Zürich, Switzerland
`patrick.traxler@inf.ethz.ch`

**Abstract.** We study the time complexity of $(d, k)$-CSP, the problem of deciding satisfiability of a constraint system $\mathcal{C}$ with $n$ variables, domain size $d$, and at most $k$ variables per constraint. We are interested in the question how the domain size $d$ influences the complexity of deciding satisfiability. We show, assuming the Exponential Time Hypothesis, that two special cases, namely $(d, 2)$-CSP with bounded variable frequency and $d$-UNIQUE-CSP, already require exponential time $\Omega(d^{c \cdot n})$ for some $c > 0$ independent of $d$. UNIQUE-CSP is the special case for which it is guaranteed that every input constraint system has at most 1 satisfying assignment.

## 1 Introduction

In this work we study the time complexity of the NP-complete Constraint Satisfaction Problem (CSP). We are interested in the following question: *What makes CSP hard to solve?* Besides being NP-hard – already $(3, 2)$-CSP and $(2, 3)$-CSP are NP-hard – many algorithms and heuristics for CSP slow down with increasing domain size $d$. It is however not clear that CSP effectively becomes harder with increasing $d$.

A promising result [10,1] is that we can solve $d$-COL, the $d$-Graph Colorability Problem, in time $2^{\tilde{n}} \cdot \text{poly(input-size)}$, where $\tilde{n}$ is the number of vertices of the input graph. Such a result is however not known for $(d, k)$-CSP or the special cases $(d, 2, 3d^2)$-FREQ-CSP and $d$-UNIQUE-CSP.

- $(d, k, f)$-FREQ-CSP is the $(d, k)$-CSP for which every input constraint system has maximum variable frequency $f$.
- $(d, k)$-UNIQUE-CSP is the $(d, k)$-CSP for which every input constraint system is guaranteed to have at most 1 satisfying assignment. Without any restriction on the constraint size we have $d$-UNIQUE-CSP.

We provide precise definitions in Section 2. We now introduce some definitions to state our results. We call an algorithm a $2^{c \cdot n}$-randomized algorithm iff its running time is bounded by $2^{c \cdot n} \cdot \text{poly(input-size)}$ and its error probability is at most $1/3$. Let

$$c_{d,k} := \inf\{c : \exists 2^{c \cdot n}\text{-randomized algorithm for } (d, k)\text{-CSP}\}.$$

[*] This work was supported by the Swiss National Science Foundation SNF under project 200021-118001/1.

M. Grohe and R. Niedermeier (Eds.): IWPEC 2008, LNCS 5018, pp. 190–201, 2008.

Define $c_{d,k,f}^{FQ}$ and $c_{d,k}^{UQ}$ analogously for $(d, k, f)$-FREQ-CSP and $(d, k)$-UNIQUE-CSP. Let $c_{d,\infty} := \lim_{k\to\infty} c_{d,k}$.

The variant of the *Exponential Time Hypothesis* (ETH) we assume here states that $c_{2,3} > 0$, i.e., 3-SAT is exponentially hard. It is straight forward to apply the results from [7] to show that $c_{3,2} > 0$ iff $c_{2,3} > 0$. In this work we improve on the lower bound $c_{d,2} > 0$, assuming ETH.

**Theorem 1.** *If ETH holds, there exists $c > 0$ such that for all $d \geq 3$*

$$c \cdot \log(d) \leq c_{d,2,3d^2}^{FQ}$$

*(where $c$ depends on $c_{3,2}$.)*

Theorem 1 strongly contrasts the time complexity of $d$-COL for which we know a $2^{\tilde{n}}$-algorithm [10,1]. Such an algorithm is however unlikely to exist for $(d, 2, 3d^2)$-FREQ-CSP because its existence implies that ETH fails.

The second special case of $(d, k)$-CSP we study is $d$-UNIQUE-CSP.

**Theorem 2.** *For all $d \geq 2$, it holds that*

$$c_{2,\infty} \cdot \lfloor \log(d) \rfloor \leq c_{d,\infty}^{UQ}.$$

Theorem 2 roughly says that the unique case is already the hardest one. Note that the currently best upper bound for $c_{2,\infty}$ is 1.

*Motivation.* The motivation for our results comes from the design and analysis of exponential time algorithms. We usually fix some natural parameter like the number of variables $n$ and try to find some small $c$ such that we can solve CSP in time $O(c^n)$. The best known upper bound $(d(1-1/k)+\varepsilon)^n \cdot \text{poly(input-size)}$, $\varepsilon > 0$, for $(d, k)$-CSP [12] is achieved by Schöning's algorithm and it was improved to, omitting the polynomial factor, $d!^{n/d}$ for $(d, 2)$-CSP [4], to $1.8072^n$ for $(4, 2)$-CSP [3], and to $1.3645^n$ for $(3, 2)$-CSP [3]. The problem of maximizing the number of satisfied constraints of a $(d, 2)$-constraint system is considered in [15]. Our results say that the dependency on $d$ of these algorithms comes close to the best possible.

Studying the special case $(d, 2, 3d^2)$-FREQ-CSP is motivated by the observation that algorithms for CSP are also analyzed w.r.t. to the number of constraints $m$ instead of $n$. This is in particular the case if optimization variants of CSP are considered. See [13] for such an algorithm and also for further references. A $(d, 2)$-constraint system in which every variable has maximum frequency $3d^2$ has at most $3d^2n$ constraints. Our results imply therefore limitations of algorithms which are analyzed w.r.t. $m$ (Corollary 2).

The second special case we study, $d$-UNIQUE-CSP, is motivated by the use of randomness. The expected running for finding a satisfying assignment of a constraint system with $s > 0$ satisfying assignments is roughly $d^n/s$. A considerable improvement, namely $(2^n/s)^{1-1/k}$, exists for $k$-SAT [2]. It also seems likely that the algorithms in [12,4] become faster if many satisfying assignments are present. Our results say that $d$-UNIQUE-CSP still has increasing complexity w.r.t. to $d$ and that CSP can only become easier if $s$ is large enough. The observed dependency of randomized algorithms on $d$ and $s$ seem therefore to be unavoidable.

*Related Work.* This work builds upon a series of papers [6,7,2] which mainly deal with SAT and special cases of SAT like $k$-SAT. A central question is: What makes SAT hard to solve? This question is motivated by the observation that many algorithms and heuristics for SAT work better on instances with special properties. For example, there exists a $1.324^n$-randomized algorithm for 3-SAT [8], whereas the best algorithms for SAT still take $2^n$ steps in the worst case. In [6] it was shown, assuming ETH, that for every $k$ there exists $k' > k$ such that $c_{2,k} < c_{2,k'}$. In other words, $k$-SAT becomes harder with increasing $k$. Our results, Theorem 1 and 2, are of the same kind. It is however not clear how to adapt the techniques in [6] to our problem. In particular, Impagliazzo & Paturi [6] ask if a similar result as theirs holds for $d$-COL. Our approach is indeed different from theirs. They use the concept of a forced variable whereas we work with a different technique of partitioning variables (see Lemma 1).

In [9] the exponentially hard instances of the Maximum Independent Set (MIS) problem with respect to the maximum degree were identified. It was shown that if there exists a subexponential time algorithm for MIS with maximum degree 3, then there exists one for MIS (which would contradict ETH). One part of the proof of this theorem is a sparsification lemma for MIS. In the proof of our Lemma 2, a sparsification lemma for $(d, 2)$-CSP, we apply the same technique as there. We prove a new sparsification lemma for $(d, 2)$-constraint systems because we want good bounds. The sparsification lemma in [7] could also be used. But it gives much worse bounds.

Calabro et al. [2] proved that $c_{2,k} \leq c_{2,k}^{UQ} + O(\log^2(k)/k)$ (Lemma 5 of [2]). We generalize and improve this to $c_{d,k} \leq c_{d,k}^{UQ} + O(\log(dk)/k)$ (see Section 4). Calabro et al. [2] concluded $c_{2,\infty}^{UQ} = c_{2,\infty}$ from their result. This theorem generalizes to $c_{d,\infty}^{UQ} = c_{d,\infty}$ by the previous relation. We remark that our proof is different from theirs although the dependency on $k$ is similar. Calabro et al. adapt the isolation lemma from [14] whereas we build upon the isolation lemma from [11]. In particular, we need a new idea to apply a generalization of the isolation lemma from [11] in our situation (see Lemma 4).

*Overview of Work.* In Section 2 we introduce the constraint satisfaction problem we study in this work. In Section 3 we prove Theorem 1 and in Section 4 we prove Theorem 2.

## 2   Preliminaries

A $(d, k)$-*constraint system* $\mathcal{C}$ consists of a set of values $\Sigma$ with $|\Sigma| = d$, called the *domain* of $\mathcal{C}$, and a set of constraints of the form

$$C := \{x_1 \neq s_1, x_2 \neq s_2, ...\}$$

with $|C| \leq k$, $x_i$ being some variable and $s_i \in \Sigma$. We often identify $C$ with the set of constraints and denote by $\mathrm{Dom}(C)$ the associated domain $\Sigma$. Let $\mathrm{Var}(C)$ denote the set of variables occurring in $C$. Unless stated otherwise $n := |\mathrm{Var}(C)|$. We call a mapping $a : \mathrm{Var}(C) \rightarrow \mathrm{Dom}(C)$ an *assignment*. A constraint $C \in \mathcal{C}$ is *satisfied* by an assignment $a$ iff there exists some $(x \neq s) \in C$ such that $a(x) \neq s$. A constraint system $\mathcal{C}$ is *satisfied* iff every $C \in \mathcal{C}$ is satisfied. We denote by $\mathrm{Sat}(\mathcal{C})$ the set of all satisfying assignments of $\mathcal{C}$. The *Constraint Satisfaction Problem* $(d, k)$-CSP is the problem of deciding if a satisfying assignment for a given $(d, k)$-constraint system exists. The $(d, k, f)$-FREQ-CSP is the special case of $(d, k)$-CSP with maximum variable frequency $f$, that is, we require that every variable occurs at most $f$ times in an input $(d, k)$-constraint system, and $(d, k)$-UNIQUE-CSP is the special case for which an input $(d, k)$-constraint system is guaranteed to have at most 1 satisfying assignment.

The $(2, k)$-CSP is the $k$-Boolean Satisfiability Problem ($k$-SAT). The $(d, 2)$-CSP is a generalization of the $d$-Graph Colorability Problem ($d$-COL). For seeing this, consider the following example.

*Example 1.* Let $G = (U, E)$ be a graph. For $\{u, v\} \in E$ the constraints

$$\{u \neq 1, v \neq 1\}, \{u \neq 2, v \neq 2\}, ..., \{u \neq d, v \neq d\}$$

are in $\mathcal{C}$. Set $\mathrm{Dom}(\mathcal{C}) := \{1, ..., d\}$. Then $\mathcal{C}$ is satisfiable iff $G$ is $d$-colorable. $\square$

## 3    Binary Sparse CSP (Proof of Theorem 1)

The proof of Theorem 1 consists of three steps. We show first how to reduce the number of variables by increasing the domain size (Lemma 1). We need this lemma to provide a relation between $(d, k)$-CSP and $(d', k)$-CSP for $d'$ larger than $d$. This is the core of our result that CSP has increasing complexity w.r.t. $d$. Then, we show in the second step how to transform a $(d, 2)$-constraint system into a sparse $(d, 2)$-constraint system, i.e., we prove a sparsification lemma for $(d, 2)$-constraint systems (Lemma 2). Our transformation can be carried out in subexponential time. Combining both lemmas we are finally able to prove Theorem 1.

At the end of this section we point out how our result relates to algorithms which are analyzed w.r.t. the number of constraints.

**Lemma 1.** *Let $r \in \mathbb{N}$, $r > 0$. For every $(d, k)$-constraint system $\mathcal{C}$ over $n$ variables there exists a satisfiability equivalent $(d^r, k)$-constraint system $\mathcal{C}'$ over $n' := \lceil \frac{n}{r} \rceil$ variables which is computable in time $d^{rk} \cdot \mathrm{poly}(|\mathcal{C}|)$.*

*Proof.* The idea of our algorithm is to group the variables in groups of size $r$ and replace every group of variables by a new variable. We need the following definition. Let $U \subseteq \mathrm{Var}(\mathcal{C})$ and $D$ be a constraint. Define $\mathrm{Nonsat}_U(D)$ to be the set of all assignments $a : U \rightarrow \mathrm{Dom}(\mathcal{C})$ which do not satisfy $D$. Our algorithm gets as input a $(d, k)$-constraint system $\mathcal{C}$ and outputs a $(d^r, k)$-constraint system $\mathcal{C}'$. It works as follows.

Compute a partition of pairwise disjoint subsets $P_1, ..., P_t$ of $\mathrm{Var}(\mathcal{C})$ such that $|P_i| = r$ for $1 \leq i < t$ and $1 < |P_t| \leq r$. Extend $P_t$ with new variables such that $|P_t| = r$. Find new variables $y_1, ..., y_t$, i.e., variables which are not in $\mathrm{Var}(\mathcal{C})$. Set $\mathcal{C}' \leftarrow \mathcal{C}$ and $\mathrm{Dom}(\mathcal{C}') \leftarrow \mathrm{Dom}(\mathcal{C})^r$, i.e., $\mathrm{Dom}(\mathcal{C}')$ is the set of all strings of length $r$ with symbols from $\mathrm{Dom}(\mathcal{C})$. For every $C \in \mathcal{C}$ and every $1 \leq i \leq t$: if $\mathrm{Var}(C) \cap P_i \neq \{\}$ then replace all the variables of $\mathrm{Var}(C) \cap P_i$ in $C$ in the following way. Let $D \subseteq C$ be the set of all inequalities with variables from $\mathrm{Var}(C) \cap P_i$. Add the constraint $C' \leftarrow (C \setminus D) \cup \{y_i \neq b\}$ to $\mathcal{C}'$ for every $b \in \mathrm{Nonsat}_{P_i}(D)$. Remove $C$ from $\mathcal{C}'$.

We claim that $\mathcal{C}$ is satisfiable iff $\mathcal{C}'$ is satisfiable. Let $a \in \mathrm{Sat}(\mathcal{C})$. For $y_i$ we define $a'(y_i) := a(x_1') \cdot ... \cdot a(x_r')$ with $\{x_1', ..., x_r'\} = P_i$, i.e., $a'(y_i)$ is the concatenation of the values of variables in $P_i$. Let $C' \in \mathcal{C}'$ and $C \in \mathcal{C}$ be the corresponding constraint $C'$ emerged from. Since $a$ satisfies $C$ there exists some inequality $x \neq s \in C$ satisfied by $a$, i.e., $a(x) \neq s$. Assume $x \in P_i$. Then $x \neq s \in D$, $D$ as in the algorithm. This implies that $y_i \neq b \in C'$ for some $b \in \mathrm{Nonsat}_{P_i}(D)$. Since $a'(y_i) \neq b$ for all $b \in \mathrm{Nonsat}_{P_i}(D)$ (because of $a(x) \neq s$) it follows that $C'$ is satisfied by $a'$ and therefore $a' \in \mathrm{Sat}(\mathcal{C}')$.

For the other direction, assume that $a \notin \mathrm{Sat}(\mathcal{C})$ for all assignments $a$ of $\mathcal{C}$. We have to show that $a' \notin \mathrm{Sat}(\mathcal{C}')$ for all assignments $a'$ of $\mathcal{C}'$. For $x \in \mathrm{Var}(\mathcal{C})$ and $x \in P_i$, $1 \leq i \leq t$, we define $a(x) := (a'(y_i))(x)$. The assignment $a$ is well defined because $P_1, ..., P_t$ is a partition of $\mathrm{Var}(\mathcal{C})$. Also note that we consider $a'(y_i)$ here as an assignment of the form $P_i \to \mathrm{Dom}(\mathcal{C})$. We know that there exists some $C \in \mathcal{C}$ which is not satisfied by $a$. This implies that there exists some $C' \in \mathcal{C}'$ which is not satisfied by $a'$ and which emerged from $C$. For seeing this, choose in the construction of $C'$ the partial assignment $b \in \mathrm{Nonsat}_{P_i}(D)$ according to $a$, i.e., choose $b$ such that $b(x) = a(x)$ for all $x \in P_i$.

It holds that $n' = t = \lceil \frac{n}{r} \rceil$. Note that the length of some assignment in $\mathrm{Nonsat}_{P_i}(D)$ is $r$ and we introduce therefore $d^r$ new values. The old values are not used any longer. The running time is polynomial in the input size with the exception of enumerating $\mathrm{Nonsat}_{P_i}(D)$ which takes time $O(d^r \cdot |\mathcal{C}|)$ and we may have to do this for every variable in a constraint of size at most $k$. This yields $O(d^{rk} \cdot |\mathcal{C}|)$.                                                                    □

The following result is a direct implication of this lemma.

**Corollary 1.** *For constant $d$, $d'$ and $k$ with $d' \geq d$. It holds that $c_{d',k} \geq \lfloor \log_d(d') \rfloor \cdot c_{d,k}$.*

To prove Lemma 2 we will use algorithm $\mathrm{SPARSIFY}_\varepsilon$ defined in Figure 1. The idea of our algorithm is similar to one of the many backtracking algorithms for the Maximum Independent Set Problem (MIS), namely, branching on vertices with large degree first. Johnson & Szegedy [9] applied the same technique to prove a sparsification lemma for MIS. Let $\varepsilon > 0$ and $K_{\varepsilon,d} := \frac{d}{\varepsilon \cdot \log(d)} \cdot \log(d^{(\varepsilon/d)}/(d^{(\varepsilon/d)} - 1))$. $\mathrm{SPARSIFY}_\varepsilon$ uses the procedure $\mathrm{SUBS}(\mathcal{C})$ which searches in $\mathcal{C}$ for constraints of the form $\{x \neq s\}$ and removes all $C \in \mathcal{C}$ with $|C| \geq 2$ and $(x \neq s) \in C$. It also uses the operation $\mathcal{C}^{[x \mapsto f]}$ by which all constraints $C \in \mathcal{C}$ with $(x \neq f') \in C$, $f' \neq f$, and all inequalities $x \neq f$ get removed from $\mathcal{C}$. Let $\mathrm{freq}(\mathcal{C}, x, s)$ be the number of times $x \neq s$ occurs in $\mathcal{C}$.

---

**Input:** a $(d, 2)$-constraint system $\mathcal{C}$.
**Output:** a list $\mathcal{L}$ of $(d, 2)$-constraint systems.
1. if there exists $x \in \mathrm{Var}(\mathcal{C})$ and $s \in \mathrm{Dom}(\mathcal{C})$ s.t. $\mathrm{freq}(\mathcal{C}, x, s) > \lceil K_{\varepsilon,d} \rceil$, then
2.     call $\mathrm{SPARSIFY}_\varepsilon(\mathrm{SUBS}(\mathcal{C}^{[x \mapsto s]}))$;
3.     call $\mathrm{SPARSIFY}_\varepsilon(\mathrm{SUBS}(\mathcal{C} \cup \{\{x \neq s\}\}))$;
4. else output $\mathcal{C}$;

---

**Fig. 1.** Algorithm $\mathrm{SPARSIFY}_\varepsilon$

**Lemma 2.** *Let $\mathcal{C}$ be a $(d, 2)$-constraint system and $\varepsilon > 0$. $\mathrm{SPARSIFY}_\varepsilon$ enumerates with polynomial delay a list $\mathcal{L}$ of $(d, 2)$-constraint systems which has the following properties:*

1. *(Correctness) it holds that $\mathcal{C}$ is satisfiable iff there exists some satisfiable $\mathcal{C}' \in \mathcal{L}$,*
2. *(Bounded frequency) for all $\mathcal{D} \in \mathcal{L}$, $x \in \mathrm{Var}(\mathcal{D})$, and $s \in \mathrm{Dom}(\mathcal{D})$:*

$$\mathrm{freq}(\mathcal{D}, x, s) \leq \lceil K_{\varepsilon,d} \rceil,$$

3. *(Size) $|\mathcal{L}| \leq d^{\varepsilon \cdot n}$.*

*Proof.* To see the correctness of algorithm $\mathrm{SPARSIFY}_\varepsilon$ note that $\mathcal{C}$ is satisfiable iff $\mathrm{SUBS}(\mathcal{C}^{[x \mapsto s]})$ or $\mathrm{SUBS}(\mathcal{C} \cup \{\{x \neq s\}\})$ is satisfiable. The bounded frequency property holds because of the branching rule. It remains to prove the last property. $\mathrm{SPARSIFY}_\varepsilon$ branches on pairs $(x, s)$ with $\mathrm{freq}(\mathcal{C}, x, s) > \lceil K_{\varepsilon,d} \rceil$. There are at most $d \cdot n$ such pairs. Let $n'$ be the number of these pairs in $\mathcal{C}$ and $t(n')$ be the size of the search tree induced by $\mathrm{SPARSIFY}_\varepsilon$. If we can show that $t(n') \leq d^{(\varepsilon/d) \cdot n'}$, then $|\mathcal{L}| \leq d^{\varepsilon \cdot n}$. For $n' \leq \lceil K_{\varepsilon,d} \rceil$ we can assume that $t(n') \leq d^{(\varepsilon/d) \cdot n'}$ holds. Now assume that the induction hypothesis $t(i) \leq d^{(\varepsilon/d) \cdot i}$ holds for $i \leq n' - 1$. $\mathrm{SPARSIFY}_\varepsilon$ removes either at least 1 or at least $\lceil K_{\varepsilon,d} \rceil$ pairs according to the two cases of the branching rule. In the first case, $\mathrm{SUBS}(\mathcal{C} \cup \{\{x \neq s\}\})$ yields a constraint system in which $\{x \neq s\}$ occurs once. No superset of $\{x \neq s\}$ occurs in $\mathrm{SUBS}(\mathcal{C} \cup \{\{x \neq s\}\})$. In the second case, the constraint system $\mathcal{C}^{[x \mapsto s]}$ contains $\mathrm{freq}(\mathcal{C}, x, s)$ new constraints of size 1 with no superset in $\mathrm{SUBS}(\mathcal{C}^{[x \mapsto s]})$. Hence $t(n') \leq t(n' - 1) + t(n' - \lceil K_{\varepsilon,d} \rceil)$ which is by the induction hypothesis at most $d^{(\varepsilon/d) \cdot n' - (\varepsilon/d)} + d^{(\varepsilon/d) \cdot n' - (\varepsilon/d) \cdot \lceil K_{\varepsilon,d} \rceil} \leq d^{(\varepsilon/d) \cdot n'} \cdot (d^{-(\varepsilon/d)} + d^{-(\varepsilon/d) \cdot K_{\varepsilon,d}})$. By the definition of $K_{\varepsilon,d}$: $d^{-(\varepsilon/d)} + d^{-(\varepsilon/d) \cdot K_{\varepsilon,d}} = 1$. □

*Proof (of Theorem 1).* We apply Lemma 2 to a $(d, 2)$-constraint system $\mathcal{C}$ with fixed $\varepsilon = \gamma := c_{3,2}/(4 \log(3))$ and get a list $\mathcal{L}$ of constraint systems. Every $\mathcal{C}' \in \mathcal{L}$ has maximum frequency $K_{\gamma,d} \cdot d$. Let $K_\gamma := \lceil \gamma^{-2} \rceil$. Then, $K_{\gamma,d} \leq K_\gamma \cdot d$. $K_{\gamma,d} \leq \gamma^{-2} d$ simplifies to $y - \frac{\ln(d)^2}{d} \frac{1}{y} \leq \ln(e^y - 1)$ with $y := \gamma \frac{\ln(d)}{d}$. Note that

$0 < \gamma \leq 1/4$ by the definition of $\gamma$ and therefore we can assume $0 < y \leq \frac{\ln(d)}{4d}$. The function $f(y) := \ln(e^y - 1) - y + \frac{\ln(d)^2}{d} \frac{1}{y}$ takes the minimum in $y = \frac{\ln(d)}{4d}$. The claim follows from $f(\frac{\ln(d)}{4d}) \geq 0$ for all $d \geq 3$. To reduce the maximum frequency to $3d^2$, we introduce for every variable $x$ new variables $x^{(1)}, ..., x^{(K_\gamma)}$. We can express that $x^{(i)}$ has exactly the same value as $x^{(i+1)}$ with at most $d^2$ constraints, namely, with all constraints $\{x^{(i)} \neq s_1, x^{(i+1)} \neq s_2\}$, $s_1 \neq s_2$. We add all constraints for $1 \leq i \leq K_\gamma - 1$ to $\mathcal{C}'$ and replace every occurrence of $x$ in such a way that for all $x \in \text{Var}(\mathcal{C}')$, $s \in \text{Dom}(\mathcal{C}')$: $\text{freq}(\mathcal{C}', x, s) \leq 3d$. The number of variables is at most $K_\gamma \cdot n$. Using Corollary 1 we get the relation $c_{d,2} \geq \lfloor \log_3(d) \rfloor \cdot c_{3,2}$. Thus

$$c_{d,2,3d^2}^{FQ} \cdot K_\gamma + \gamma \cdot \log(d) \geq c_{d,2} \geq \lfloor \log_3(d) \rfloor \cdot c_{3,2},$$

and $c_{d,2,3d^2}^{FQ} \geq \lfloor \log_3(d) \rfloor \cdot c_{3,2}/(2K_\gamma)$. This completes the proof of Theorem 1. $\square$

As a direct consequence we get a lower bound for

$$e_{d,k} := \inf\{c : \exists 2^{c \cdot m}\text{-randomized algorithm for } (d,k)\text{-CSP}\}$$

where $m$ is the number of constraints.

**Corollary 2.** *If ETH holds, there exists $c > 0$ such that for all $d \geq 3$: $e_{d,2} \geq c \cdot \log(d)/d^2$.*

Note that $e_{d,2} \leq 2 \cdot \log(d)/d$ since we can remove every variable $x$ which occurs less than $d$ times (because then there is a remaining value we can assign to $x$ to satisfy every constraint $x$ occurs in). Hence, we may assume $|\mathcal{C}| \geq d/2 \cdot n$. Enumerating all possible assignments of the $n$ variables yields the claimed upper bound.

Lemma 2 and the transformation afterwards give an upper bound of $3d^2$ on the variable frequency and actually the bound $\text{freq}(\mathcal{C}, x, s) \leq 3d$. Let $p_x$ be the number of possible values of $x$, that is, $d$ minus the number of constraints of size 1 in which $x$ occurs. Since we expect in the worst case that $p_x = \Omega(d)$ for $x \in V$ the following result suggests that this upper bound comes close to the best possible. For example, an improvement of $\text{freq}(\mathcal{C}, x, s) \leq \sqrt{d}$ seems to be questionable.

**Proposition 1 ([5]).** *Let $\mathcal{C}$ be a $(d, 2)$-constraint system and define $p_{\min} := \min_{x \in \text{Var}(\mathcal{C})} p_x$. Then $\mathcal{C}$ is satisfiable, if for all $x \in \text{Var}(\mathcal{C})$ and $s \in \text{Dom}(\mathcal{C})$: $\text{freq}(\mathcal{C}, x, s) \leq \frac{p_{\min}}{2}$.*

## 4   Unique CSP (Theorem 2)

The proof of Theorem 2 consists of four steps. Our goal is to prove the relation

$$c_{2,k} \cdot \lfloor \log(d) \rfloor \leq c_{d,k} \leq c_{d,k}^{UQ} + O\left(\frac{\log(dk)}{k}\right) \quad \text{(Corollary 3).}$$

Taking the limit $k \to \infty$ proves Theorem 2. In this section we prove the upper bound on $c_{d,k}$. The lower bound follows from Corollary 1, Section 3. The first step in our proof of the upper bound is a generalization of the isolation lemma from [11]. We generalize this lemma from the boolean to the non-boolean case (Lemma 3). We can however not apply this lemma directly to prove our upper bound. In a second step, we therefore show how to use it to get an isolation lemma (Lemma 4) which fits our needs. The crucial difference between the isolation lemma from [11] and Lemma 4 is that we can encode the random linear equations from Lemma 4 by a $(d, k)$-constraint system. This is done in the third step (Lemma 5). Finally, we can put it all together (Corollary 3) and prove Theorem 2.

We conclude this section with a remark on our main technical contribution, Lemma 4.

We start with a generalization of Lemma 1 from [11]. The lemma there states that if $S \subseteq \{0,1\}^n$ is non-empty, then the probability that $S$ has a unique minimum w.r.t. a random weight function is at least $1/2$. Our result works for non-empty $S \subseteq \{0, ..., d-1\}^n$.

**Lemma 3.** *Let $n, c \in \mathbb{N}$ and $S \subseteq \{0, ..., d-1\}^n$, $S \neq \{\}$. Choose $w_i$, $1 \le i \le n$, independently and uniformly from $\{1, ..., c\}$. Define a random weight function $w : \{0, ..., d-1\}^n \to \mathbb{N}$ as $w : a \mapsto \sum_{i=1}^n w_i \cdot a_i$. It holds that*

$$\Pr_w(S \text{ has a unique minimum w.r.t. } w) \ge 1 - n \cdot \frac{\binom{d}{2}}{c}.$$

*Proof.* Let $1 \le i \le n$ and $0 \le l \le d-1$. Define $S_{i,l} := \{a \in S : a_i = l\}$ and

$$M_{i,l} := \begin{cases} \min_{a \in S_{i,l}} w(a) - l \cdot w_i & \text{if } S_{i,j} \neq \{\} \\ 0 \end{cases}.$$

Denote by $E_i$ the event that $\exists 0 \le j < k \le d-1 : M_{i,j} + j \cdot w_i = M_{i,k} + k \cdot w_i$. For any $i$ it holds that

$$\Pr_w(E_i) = \Pr_w(\exists 0 \le j < k \le d-1 : (M_{i,j} - M_{i,k})/(k-j) = w_i) \le$$

$$\binom{d}{2} \Pr_w((M_{i,j} - M_{i,k})/(k-j) = w_i) \le \binom{d}{2}/c.$$

Here, we used the union bound and the fact that

$$\Pr_w((M_{i,j} - M_{i,k})/(k-j) = w_i) = \begin{cases} \frac{1}{c} & \text{if } (M_{i,j} - M_{i,k})/(k-j) \in \{1, ..., c\} \\ 0 \end{cases}$$

($w_i$ is chosen independently of $w_1, ..., w_{i-1}, w_{i+1}, ..., w_n$). Applying the union bound we get (*)

$$\Pr_w(\exists 0 \le i \le n : E_i) \le n \cdot \frac{\binom{d}{2}}{c}.$$

Finally, assume that there exist $a \neq b \in S$ which take the minimum value w.r.t. $w$. Since $a \neq b$ there exists $1 \leq i \leq n$ such that $a_i \neq b_i$ and $M_{i,a_i} + a_i \cdot w_i = M_{i,b_i} + b_i \cdot w_i$. This can happen with probability at most $n \cdot \binom{d}{2}/c$ because of (*). Hence, the probability that $S$ has a unique minimum w.r.t. $w$ is at least $1 - n \cdot \binom{d}{2}/c$. □

The random weight function $w$ depends on $n$ variables. This makes it at the first sight useless for our needs since we can not encode it as a constraint system in subexponential time. We can however apply it iteratively as we will see in the proof of the following lemma.

**Lemma 4.** *Let $d \geq 2$, $k \geq 1$, and $S \subseteq \{0, ..., d-1\}^n$ be non-empty. There exists a polynomial time computable set $\mathcal{L}$ of $\lceil \frac{n}{k} \rceil$ random linear equations, each depending on at most $k$ variables, such that*

$$\Pr_{\mathcal{L}}(|S \cap \mathrm{Sol}_d(\mathcal{L})| = 1) \geq 2^{-O(n\frac{\log(dk)}{k})},$$

*where $\mathrm{Sol}_d(\mathcal{L})$ is the set of solutions of $\mathcal{L}$ in $\{0, ..., d-1\}^n$.*

*Proof.* We employ Lemma 3. Let $c := 2 \cdot \binom{d}{2} \cdot k$. Independently and uniformly choose $w_i$ from $\{1, ..., c\}$ for all $i$. We define $\mathcal{L}$ to be the set of linear equations

$$\sum_{i=1+j \cdot k}^{(1+j) \cdot k} w_i \cdot x_i = r_j$$

for $0 \leq j \leq t-1$ and $\sum_{i=1+t \cdot k}^{n} w_i \cdot x_i = r_t$, where $r_0, ..., r_t$ are chosen uniformly at random from $\{0, ..., c \cdot (d-1) \cdot k\}$.

For simplicity we assume that $n$ is a multiple of $k$, i.e., there exists $i$ such that $n = ik$. We prove by induction over $i$ that the $i$ equations in $\mathcal{L}$ have a unique solution in $S$ with probability at least $(2c(d-1)k+2)^{-i}$. If $i = 1$, then the probability that $S$ has a unique minimum w.r.t. $w_1, ..., w_k$ is at least $1/2$ by Lemma 3 and the probability of guessing the right value $r_1$ is at least $1/(c(d-1)k+1)$; together at least $1/(2c(d-1)k+2)$. Now, assume the induction hypothesis holds for $i-1$. Let $S' := \{a_{n-k+1}...a_n : a \in S\}$. The probability that the corresponding equation in the variables $x_{n-k+1}, ..., x_n$ has a unique solution in $S'$ is at least $1/(2c(d-1)k+2)$. Let $a_{n-k+1}...a_n$ be this solution and $S'' := \{b \in S : b_{n-k+1} = a_{n-k+1}, ..., b_n = a_n\}$. By the induction hypothesis the probability that the first $i-1$ equations have a unique solution in $S''$ is at least $(2c(d-1)k+2)^{-i-1}$. Since $w_1, ..., w_n$ and $r_0, ..., r_{i-1}$ are chosen uniformly and independently the overall success probability is at least $(2c(d-1)k+2)^{-i}$. Hence, $\Pr_{\mathcal{L}}(|S \cap \mathrm{Sol}_d(\mathcal{L})| = 1)$ is greater or equal than

$$(2c(d-1)k+2)^{-n/k-1} \geq (4d^3k^2)^{-n/k-1} \geq 2^{-O(n\frac{\log(dk)}{k})}. \qquad \square$$

The next lemma states the simple but important fact that we can encode the random linear equations from the previous lemma as a $(d, k)$-constraint system.

**Lemma 5.** *Let $C$ be a constraint system with domain size $d$ over $n$ variables. There exists a $(d, k)$-constraint system $C'$ over $n$ variables computable in time $d^k \cdot \text{poly}(|C|)$ such that if $C$ is satisfiable, then $C \cup C'$ has exactly one satisfying assignment with probability at least $2^{-O(n \cdot \frac{\log(dk)}{k})}$. Moreover, $|C'| \leq d^k \cdot (n/k+1)$.*

*Proof.* Let $\mathcal{L}$ be as in Lemma 4. For every equation in $\mathcal{L}$ we can enumerate all assignments of the $k$ variables in $d^k$ steps. Thus, we can encode a single equation as a $(d, k)$-constraint system in polynomial time. We define $C'$ to be the set of these at most $d^k \cdot (n/k+1)$ constraints. Now, let $S := \text{Sat}(C)$. If $C$ is unsatisfiable, then $C \cup C'$ is unsatisfiable. Otherwise, $S \neq \{\}$. The probability that $C \cup C'$ has exactly one satisfying assignment is as in Lemma 4.    □

We are now at a point where we can prove Theorem 2. It follows from the following corollary by taking the the limit $k \to \infty$.

**Corollary 3.** *For all $d \geq 2$ and $k \geq 2$, it holds that $c_{2,k} \cdot \lfloor \log(d) \rfloor \leq c_{d,k} \leq c_{d,k}^{UQ} + O(\frac{\log(dk)}{k})$.*

*Proof.* The relation $c_{2,k} \cdot \lfloor \log(d) \rfloor \leq c_{d,k}$ follows from Corollary 1. To prove $c_{d,k} \leq c_{d,k}^{UQ} + O(\log(dk)/k)$ we apply Lemma 5 $2^{O(n \cdot (\log(dk)/k))}$ times and test every time if $C \cup C'$ is satisfiable using an algorithm $A$ of time complexity $2^{(c_{d,k}+\delta) \cdot n}$, $\delta \geq 0$, for $(d, k)$-UNIQUE-CSP. If $A$ once accepts, $C$ gets accepted, otherwise rejected.

□

In the proof of Lemma 5 we used the fact that we can encode the solutions of a random linear equation as a constraint system without changing the number of variables. The opposite is however not true. Therefore we may say that Lemma 4 is stronger than Lemma 5. In particular, if we want to obtain similar relations as in Corollary 3 for other problems, Lemma 4 is appropriate if the problem at hand allows a compact encoding of the solutions of a random linear equation. This is for example the case for Binary Integer Programming. We also remark here that for the proof of Theorem 1 it is not necessary that $C'$ has constant constraint size $k$. For example, $k = \sqrt{n}$ suffices. To give an example of a situation where it is necessary that $C'$ has constant constraint size $k$ we prove Corollary 4.

**Corollary 4.** *ETH holds iff $c_{3,2}^{UQ} > 0$.*

*Proof.* Let $C$ be a $(3, 2)$-constraint system over $n$ variables and $\varepsilon > 0$. Make $k = k(\varepsilon)$ large enough such that $O(\log(dk)/k) < \varepsilon$. Applying Lemma 5 we get a constraint system $C \cup C'$. The constraints in $C'$ have size at most $k$ and $|C'| \leq n \cdot (3^k+k)/k$. By introducing $K \leq n \cdot (3^k+k)$ new variables we can transform $C'$ into a $(3, 2)$-constraint system $C''$ with the same number of satisfying assignments. Set $C''$ to $C'$. Replace every $\{x_1 \neq s_1, ..., x_l \neq s_l\} \in C''$ with $l > 2$ by $\{x_1 \neq s_1, y \neq 1\}$, $\{x_2 \neq s_2, y \neq 2\}$, $\{x_3 \neq s_3, ..., x_l \neq s_l, y \neq 3\}$. Here, $y$ is a new variable not used before. We add constraints to $C''$ which say that $x_1 \neq s_1$ implies $y \neq 2$ and that

$x_1 \neq s_1$ implies $y \neq 3$. Hence, if the inequality $x_1 \neq s_1$ is satisfied $y$ is forced to be 1. The constraints are $\{x_1 \neq s_1^1, y \neq 2\}$, $\{x_1 \neq s_1^2, y \neq 2\}$, $\{x_1 \neq s_1^1, y \neq 3\}$ and $\{x_1 \neq s_1^2, y \neq 3\}$ where $\{s_1, s_1^1, s_1^2\} = \mathrm{Dom}(\mathcal{C})$. Next we add constraints to $\mathcal{C}''$ which say that $x_2 \neq s_2$ implies $y \neq 3$. In the case that the inequality $x_1 \neq s_1$ is not satisfied but $x_2 \neq s_2$ is $y$ is forced to be 2. The constraints are $\{x_2 \neq s_2^1, y \neq 3\}$, and $\{x_2 \neq s_2^2, y \neq 3\}$ where $\{s_2, s_2^1, s_2^2\} = \mathrm{Dom}(\mathcal{C})$. In the last case that $x_1 \neq s_1$ and $x_2 \neq s_2$ are not satisfied $y$ is forced to be 3. We repeat this step until every constraint has size at most 2. In every step the size of one constraint is reduced by one and exactly one variable is used. Hence, we need $K \leq n \cdot (3^k + k)$ new variables. We conclude that $c_{3,2} \leq (3^{k(\varepsilon)} + k(\varepsilon)) \cdot c_{3,2}^{\mathrm{UQ}} + \varepsilon$ for every $\varepsilon > 0$. If $c_{3,2}^{\mathrm{UQ}} = 0$, then $c_{3,2} \leq \varepsilon$ for every $\varepsilon > 0$. A contradiction to ETH. $\qquad\square$

## Acknowledgments

Thanks to Robert Berke for pointing out [5].

## References

1. Björklund, A., Husfeldt, T.: Inclusion–exclusion algorithms for counting set partitions. In: Proc. of the 47th Annual IEEE Symposium on Foundations of Computer Science, pp. 575–582 (2006)
2. Calabro, C., Impagliazzo, R., Kabanets, V., Paturi, R.: The complexity of unique $k$-SAT: An isolation lemma for $k$-CNFs. In: Proc. of the 18th Annual IEEE Conference on Computational Complexity, pp. 135–141 (2003)
3. Eppstein, D.: Improved algorithms for 3-coloring, 3-edge-coloring, and constraint satisfaction. In: Proc. of the 12th Annual ACM-SIAM Symposium on Discrete Algorithms, pp. 329–337 (2001)
4. Feder, T., Motwani, R.: Worst-case time bounds for coloring and satisfiability problems. J. Algorithms 45(2), 192–201 (2002)
5. Haxell, P.E.: A condition for matchability in hypergraphs. Graphs and Combinatorics 11, 245–248 (1995)
6. Impagliazzo, R., Paturi, R.: On the complexity of $k$-SAT. J. Computer and System Sciences 62(2), 367–375 (2001)
7. Impagliazzo, R., Paturi, R., Zane, F.: Which problems have strongly exponential complexity? J. Computer and System Sciences 63(4), 512–530 (2001)
8. Iwama, K., Tamaki, S.: Improved upper bounds for 3-SAT. In: Proc. of the 15th Annual ACM-SIAM Symposium on Discrete Algorithms, pp. 328–329 (2004)
9. Johnson, D.S., Szegedy, M.: What are the least tractable instances of max independent set? In: Proc. of the 10th Annual ACM-SIAM Symposium on Discrete Algorithms, pp. 927–928 (1999)
10. Koivisto, M.: An $O(2^n)$ algorithm for graph coloring and other partitioning problems via inclusion–exclusion. In: Proc. of the 47th Annual IEEE Symposium on Foundations of Computer Science, pp. 583–590 (2006)
11. Mulmuley, K., Vazirani, U.V., Vazirani, V.V.: Matching is as easy as matrix inversion. Combinatorica 7(1), 105–113 (1987)

12. Schöning, U.: A probabilistic algorithm for $k$-SAT and constraint satisfaction problems. In: Proc. of the 40th Annual Symposium on Foundations of Computer Science, pp. 410–414 (1999)
13. Scott, A.D., Sorkin, G.B.: An LP-Designed Algorithm for Constraint Satisfaction. In: Azar, Y., Erlebach, T. (eds.) ESA 2006. LNCS, vol. 4168, pp. 588–599. Springer, Heidelberg (2006)
14. Valiant, L.G., Vazirani, V.V.: NP is as easy as detecting unique solutions. Theoretical Computer Science 47(1), 85–93 (1986)
15. Williams, R.: A new algorithm for optimal 2-constraint satisfaction and its implications. Theoretical Computer Science 348(2–3), 357–365 (2005)

# A Tighter Bound for Counting Max-Weight Solutions to 2SAT Instances

Magnus Wahlström

Max-Planck-Institut für Informatik, Saarbrücken, Germany
wahl@mpi-inf.mpg.de

**Abstract.** We give an algorithm for counting the number of max-weight solutions to a 2SAT formula, and improve the bound on its running time to $\mathcal{O}(1.2377^n)$. The main source of the improvement is a refinement of the method of analysis, where we extend the concept of compound (piecewise linear) measures to multivariate measures, also allowing the optimal parameters for the measure to be found automatically. This method extension should be of independent interest.

## 1 Introduction

From a computational complexity point of view, the problem class #P of problems where you want to know the number of solutions to some problem in NP is a very difficult one. The class was proposed by Valiant in the 1970's [14], and it was later proved that the so-called polynomial hierarchy is contained in $P^{\#P}$ [12] (i.e. that a polynomial-time algorithm for any #P-complete problem would allow us to solve any problem in the polynomial hierarchy in polynomial time; in fact, a single query to the algorithm would suffice). #P-complete problems include the counting counterparts of both NP-complete problems such as 3SAT (counting counterpart #3SAT) and problems that are in P. The problem considered in this paper, #2SAT, is an example of the latter: the "decision variant" 2SAT is a well-known polynomial problem, while the counting version #2SAT is #P-complete [9, 13]. However, despite the apparent difficulty of the class, individual #P-complete problems can be solved in reasonable exponential time; for instance, the bound $\mathcal{O}^*(1.2377^n)$[1] for #2SAT is significantly faster than any bound for solving 3SAT (for which the best bounds are a probabilistic algorithm with a bound of $\mathcal{O}^*(1.3238^n)$ by Iwama and Tamaki [8], and a deterministic algorithm with a bound of $\mathcal{O}^*(1.473^n)$ by Brueggemann and Kern [1]).

One of the first algorithms for a counting problem came in the early 1960's with Ryser's [11] $O(n^2 2^n)$ time algorithm for counting the number of perfect matchings in a bipartite graph (also known as computing the *permanent* of a 0/1 matrix). Previous work on the #2SAT problem with better bounds than $\mathcal{O}^*(2^n)$ includes results by Dubois [4], Zhang [17], Littman *et al.* [10], Dahllöf, Jonsson and Wahlström [2, 3], and Fürer and Kasiviswanathan [7]. In terms

---

[1] $\mathcal{O}^*(\cdot)$, $\Theta^*(\cdot)$, etc, signify that polynomial factors are ignored.

M. Grohe and R. Niedermeier (Eds.): IWPEC 2008, LNCS 5018, pp. 202–213, 2008.
© Springer-Verlag Berlin Heidelberg 2008

of the actual bounds, the bound $\mathcal{O}^*(1.3247^n)$ appeared in [2]; later, a complete rewrite of the algorithm for the journal version produced the bound $\mathcal{O}^*(1.2561^n)$ [3], where the method of compound measures was introduced (under the name of piecewise linear measures); and the previously best bound $\mathcal{O}^*(1.2461^n)$ [7] was produced by a more detailed version of the analysis in [3].

The bound $\mathcal{O}^*(1.2377^n)$ produced in this paper is the result of a further improvement of the analysis through compound measures, this time introducing multi-variate compound measures, which are a combination of compound measures with the multi-variate recurrences of Eppstein [5]. Analysis through compound measures allows us to model that the behaviour of the algorithm varies depending on certain parameters on the instance, in this case the average degree, i.e. that the behaviour of the algorithm is non-uniform. Apart from earlier #2SAT$_w$ publications, this type of analysis has been applied to the problem of finding a solution to SAT instances $F$ with a bounded $\ell(F)/n(F)$ value [15]. Combining the method with Eppstein's method for solving multi-variate recurrence improves the quality of the bound, and allows us to automate the bound calculations.

The paper is structured into Sect. 2 on preliminaries, Sect. 3 describing the improved compound measures, Sect. 4 defining the problem precisely and giving the algorithm, Sect. 5 providing the analysis of upper bounds for maximum degree four, and finally Sect. 6 containing the general upper bound.

Major portions of this paper appeared as Chapt. 7 in the author's thesis [16], but the material has not been published in any refereed publication. Some proofs have been omitted due to lack of space; these can usually be found in [16].

## 2    Preliminaries

A variable, in this paper, can take the values true or false, referred to as 1 resp. 0; a *literal* of a variable is either the unnegated literal $v$, having the same truth value as the variable, or the negated literal $\bar{v}$, with the opposite truth value; a *clause* is a disjunction of $(a \vee b)$ of literals, referred to as a *k-clause* if it is a disjunction on $k$ literals; and a 2SAT formula $F = (a \vee b) \wedge (\bar{a} \vee c) \wedge \ldots$ is a conjunction of clauses where every clause contains at most two literals. A *model* for a 2SAT formula $F$ is an assignment to all its variables that satisfies the formula (i.e. at least one literal in every clause is true). When we write $\tilde{v}$, this refers to either literal $v$ or $\bar{v}$.

We talk of the *graph* of a 2SAT formula $F$. This is mainly an analogous term: we consider a graph where we have one vertex for every variable in $F$, and one edge $(a, b)$ for every 2-clause $(\tilde{a}, \tilde{b})$ in $F$, where $\tilde{v}$ is $v$ or $\bar{v}$, and let terms such as *connected component* and *subgraph* hold the meaning they would have in this graph. Therefore, the *degree* $d(x)$ of a variable $x$ is the number of clauses where it occurs, and $d(F)$ is the maximum degree of any variable in $F$. $Var(F)$ is the set of variables occurring in $F$, $n(F) = |Var(F)|$ is the number of variables, and $n_i(F)$ is the number of variables of degree $i$. A variable of degree $i$ is called an

$i$-variable; a 1-variable is rather called a *singleton*. A *heavy* variable has degree at least three. We also use $\ell(F)$ for the total length of $F$, i.e. $\ell(F) = \sum_i i n_i(F)$.

Other graph terms used in the paper include the concept of a *neighbourhood*: for a variable $v$, the *(open) neighbourhood* $N(v)$ of $v$ is the set of all variables $w$ such that there exists a 2-clause $(\tilde{v} \vee \tilde{w})$ in $E$ (i.e. the definition is identical to the neighbourhood of the vertex $v$ in the graph of the formula). The *closed neighbourhood* $N[v]$ is defined as $N(v) \cup \{v\}$.

# 3   On Analysis by Compound Measure

A *complexity measure* $\mu(F)$ is a function assigning a non-negative complexity value to any possible instance $F$ of some problem. They are used for estimating upper bounds on the running times of branching algorithms; see e.g. the survey by Fomin et al. [6].

For a branching algorithm and a complexity measure $\mu(F)$, the *branching number* of a particular branching from an instance $F$ to subinstances $F_1, \ldots, F_k$, with $\mu(F) - \mu(F_i) = \delta_i$, is $\tau(\delta_1, \ldots, \delta_k)$, defined as the unique positive root of $\sum_i x^{-\delta_i} = 1$.

We say that $\mu(F)$ is a *well-behaved* measure for the algorithm if the following hold:

1. $\mu(F) \geq 0$ for all possible $F$;
2. $\mu(F) = 0$ only for cases the algorithm solves in a given polynomial time;
3. $\mu(F') \leq \mu(F)$ if the algorithm, when applied to $F$, may apply a reduction replacing $F$ by $F'$; and
4. $\mu(F') < \mu(F)$ if the algorithm, when applied to $F$, may perform a branching where $F'$ is one of the branches.

If this is the case, then the running time of the algorithm will be in $\mathcal{O}^* \left( c^{\mu(F)} \right)$, where $c$ is the largest possible branching number that can occur in the algorithm. A *hard case* is a case with a branching number identical to the maximum.

Under a model of an algorithm as a set of possible branchings, of which any branching can be applied at any time, and of an instance as a point in $\mathbb{Z}^d$ (i.e. a set of integer attributes, e.g. $n_i(F)$ for $i \leq d$), Eppstein [5] gives a method for finding a *weight-based measure* $\mu(I) = \mathbf{w} \cdot I$ (e.g. $\mu(F) = \sum_i w_i n_i$) such that the resulting bound for the worst-case behaviour is tight to within a polynomial factor (under a further restriction that the instance agrees with a certain *target vector* $t$; see the paper for details).

However, for the algorithm in this paper (and for other algorithms as well), the basic assumption of the model is false: not all branchings can be used in all cases. In this paper, we use a model where along with every branching $B$ that could possibly be a hard case in the above sense, we associate a *highest average degree* $p_B$, such that if $\ell(F)/n(F) \geq p_B$, then the branching $B$ will not be used (since better cases are found). We analyse this using *compound measures*. Compound measures were used in previous publications for #2SAT$_w$ [3, 7], but here, we introduce multi-variate compound measures, and show how to use Eppstein's method to find the optimal weights for them.

Our compound measures in this paper follow the pattern

$$\mu(F) = \begin{cases} \mu_0(F) & \text{If } \ell(F)/n(F) \leq p_0 \\ \mu_1(F) & \text{If } p_0 < \ell(F)/n(F) \leq p_1 \\ \cdots & \\ \mu_t(F) & \text{If } \ell(F)/n(F) > p_{t-1} \end{cases}$$

where $p_i$ is an increasing sequence selected from the $p_B$, referred to as the *pivot points*, and each $\mu_i(F)$ is a weight-based measure, specifically $\mu_i(F) = \sum_j w_{i,j} n_j(F)$; our measure $\mu(F)$ is *piecewise linear*. The region from $p_i$ to $p_{i+1}$ is referred to as *section $p_i$ to $p_{i+1}$*. Two additional things must hold for $\mu$:

$$\mu_i(F) \geq \mu(F) \text{ for every possible } F, i \tag{1}$$
$$\mu_i(F) = \mu_{i+1}(F) \text{ when } \ell(F)/n(F) = p_i \tag{2}$$

(i.e. the measure is concave and continuous). Note that by (1), $\mu(F) - \mu(F') \geq \mu(F) - \mu_i(F')$ for any $i$, so we do not need to know the value of $\ell(F')/n(F')$ to find $\Delta\mu(F)$. By this, we can derive

$$\mu_{i+1}(F) = \mu_i(F) - \alpha_i(\ell(F) - p_i n(F))$$

for $\alpha_i \geq 0$; $\alpha_i$ is the *pivot amount* at pivot point $i$.

Note that for the worst-case behaviour, $\mu_i(F)$ for $i < t$ is of only secondary interest, since $\ell(F)/n(F) \leq p_i(F)$ must hold, limiting the possible number of highest-degree variables. What we are interested in is $\mu_t(F)$, and the values of $\alpha_i$ that show that our bound is correct. The values of $w_{i,j}$ can be derived from this:

$$w_{t-k,j} = w_{t,j} + \sum_{i=t-k}^{t-1} (j - p_i)\alpha_i.$$

Now, for every branching $B$, we have one case for every measure $\mu_i$ as long as $p_B \geq p_i$, and this case can be expressed in terms of $w_{t,j}$ and $\alpha_i$ using the previous formula. Finally, select the target vector to be the situation $n_d(F) = n(F)$ for the maximum considered degree $d$.

**Lemma 1.** *The adaption to Eppstein's method that is described above will produce the best possible compound measure for a given set of degree-bounded branchings $(B, p_B)$ and preselected pivot points $p_i$.*

*Proof.* There is a one-to-one correspondence between values of $w_{t,j}$ and $\alpha_i$ on the one hand, and all values $w_{i,j}$ on the other, so the optimality guarantee of Eppstein is enough. $\qquad\square$

Finally, we point at two things that are missing. First, we have not described how to select pivot points. In general, we have no answer, but for the branchings in this paper, where $(\Delta\ell(F))/(\Delta n(F)) \geq p_B$ for every branch of every branching $B$, it can be shown that only the values $p_B$ need to be used. Second, there is no lower bound guaranteeing optimality of the resulting bound relative to the model we have described.

## 4    Problem Definition and Algorithm

In its basic form, the problem #2SAT consists of a 2SAT instance $F$, and the question is how many solutions $F$ has. Here, we extend the problem with weights, and ask for the number of solutions with maximum solution weight (i.e. the number of solutions with weights identical to the maximum possible weight of any solution to $F$)[2]: each instance consists of a 2SAT formula $F$ along with a *weight vector* $\mathbf{w}$ and a *cardinality vector* $\mathbf{c}$, assigning a weight $w(\tilde{v})$ resp. a cardinality $c(\tilde{v})$ to each literal $\tilde{v}$ of each variable $v$ occurring in $F$. The weight of a model $M$ of $F$ is

$$\mathcal{W}(M) = \sum_{l \text{ is true in } M} w(l);$$

that is, the sum of $w(v)$ for every variable that is true in $M$ and $w(\bar{v})$ for every variable that is false in $M$. In the same manner, let the cardinality of a model $M$ be

$$\mathcal{C}(M) = \prod_{l \text{ is true in } M} c(l).$$

Again, $l$ ranges over literals, not over variables. The solution to the instance is the maximum weight of any model $M$ for $F$, and the sum of the cardinalities of all true models $M$ of $F$.

We use cardinality vectors in order to enable the use of multiplier reduction (described below); the most natural instances to the problem may be where the cardinality of every literal is 1, so that we ask simply for the number of max-weight models.

Because of the weight and cardinality vectors, there is some bookkeeping involved in performing an assignment to a formula and propagating its direct implications (e.g. from a 2-clause $(a \vee b)$ and an assignment $a = 0$, a further assignment $b = 1$ is derived). We will understand the term *recursively branch on* $x$ to include handling all details of weight, cardinality, and propagation; thus, we will also always assume that the formula we are dealing with contains only 2-clauses. A precise description, if required, can be found in earlier publications [3, 16].

Our main algorithm $\mathrm{C}(F, \mathbf{c}, \mathbf{w})$, taking a formula $F$, a cardinality vector $\mathbf{c}$, and a weight vector $\mathbf{w}$, is defined as Algorithm 1 later in this section. All references to $\mathrm{C}(\cdots)$ in the following definitions are references to this algorithm. First, we give a definition that is used when selecting a variable to branch on.

---

[2] The most immediate uses of the algorithm are probably to either find the number of solutions in total, or the maximum possible weight of a solution (both of which being problems for which this algorithm is the fastest known), but note that the max-weight requirement can be used to implement non-2SAT constraints, e.g. the gadget $(\bar{x} \vee \bar{u}), (y \vee \bar{u}), (\bar{x} \vee \bar{v}), (\bar{y} \vee \bar{v}), (z \vee \bar{v})$ with $w(l) = 1$ iff $l \in \{x, u, v\}$, where $u$ and $v$ only occur in these clauses, implements $(x \vee y \vee z)$.

**Definition 1.** *In a formula $F$ with average degree $\ell(F)/n(F) = k$, the associated average degree of a variable $x$ in $F$ is $\alpha(x)/\beta(x)$, where:*

$$\alpha(x) = d(x) + |\{y \in N(x) \mid d(y) < k\}| \tag{3}$$

$$\beta(x) = 1 + \sum_{\{y \in N(x) \mid d(y) < k\}} 1/d(y) \tag{4}$$

The following lemma shows the use of this definition in the algorithm.

**Lemma 2 (Lemma 6 of [3]).** *Let $F$ be a non-empty formula such that $\ell(F)/n(F) = k$. There exists some variable $x \in Var(F)$ with $d(x) \geq k$ with associated average degree at least $k$.*

We also use a reduction called *multiplier reduction*. When a formula $F$ consists of two parts $F_1$ and $F_2$, with $|Var(F_1) \cap Var(F_2)| = 1$ (say, the variable $v$), then multiplier reduction applies. We then need only two pieces of information from $F_2$: the number and weight of max-weight solutions when $v = 1$ resp. $v = 0$. We can find these numbers through calls to $C(F_2[v = 1])$ and $C(F_2[v = 0])$ and incorporate them into $w(\tilde{v}), c(\tilde{v})$, to then proceed to calculate $C(F_1)$ with the modified values. This will be referred to as *removing $F_2$ by multiplier reduction*.

Finally, we provide the algorithm. Note that though the analysis is split into several parts, using different measures, these parts are only different ways of analysing this same algorithm.

## Algorithm 1 $C(F, \mathbf{c}, \mathbf{w})$

1. If $F = \emptyset$, then return $(1, 0)$. If $\emptyset \in F$, then return $(0, 0)$.
2. If $F$ is not connected, then return $(c, w)$ where $c = \prod_{i=0}^{j} c_i$, $w = \sum_{i=0}^{j} w_i$, and $(c_i, w_i) = C(F_i, \mathbf{c}, \mathbf{w})$ for the connected components $F_0, \ldots, F_j$.
3. If multiplier reduction applies, then apply it, removing the lightest part (as measured by the complexity measure).
4. If $d(F) \in \{3, 4\}$, then let $x$ be a variable of maximum degree, secondarily maximising the associated average degree $\alpha(x)/\beta(x)$.
   (a) If there exists a set of two heavy variables $\{y, z\}$, $y, z \notin N[x]$, whose removal leaves $F$ disconnected and leaves $N(x)$ in a non-heaviest component, then recursively branch on $y$.
   (b) Otherwise, recursively branch on $x$.
5. Let $x$ be a variable of maximum degree, which if possible does not have only neighbours of degree $d(x)$, and recursively branch on $x$.

## 5    Analysis: Maximum Degree 4

In this section, we will give upper bounds for the running time of the algorithm in cases where $d(F) \leq 4$. The bounds of this section are given using the method described in Sect. 3. We begin with an observation for the case $d(F) = 2$.

**Table 1.** Possible cases when branching on a 3-variable

| Degrees of neighbours | Highest average degree | Branching (case 4b) |
|---|---|---|
| $(2,2,2)$ | $6/2.5 = 2.4$ | $\tau(12w_l + 4w_n, 12w_l + 4w_n)$ |
| $(2,2,3)$ | $5/2 = 2.5$ | $\tau(10w_l + 3_w n, 18w_l + 6w_n)$ |
| $(2,3,3)$ | $4/1.5 \approx 2.67$ | $\tau(8w_l + 2w_n, 16w_l + 5w_n)$ |
| $(3,3,3)$ | $3/1 = 3$ | $\tau(6w_l + w_n, 16w_l + 4w_n)$ |

**Lemma 3.** *The algorithm $C$ applied to a formula $F$ with $d(F) \leq 2$ runs in polynomial time.*

We will now give the bound for the case $d(F) = 3$. For reference, and to illustrate the process, the possible neighbourhoods of a heavy variable, with their respective average degree guarantees, are given in Table 1. The measure is based on the attributes $\ell(F)$ and $n(F)$, rather than $n_2(F)$ and $n_3(F)$, since they are equivalent when there are only two attributes, and the former is somewhat easier to work with.

**Table 2.** Component measures $w_l \ell(F) + w_n n(F)$ for maximum degree 3

| Section | $w_l$ | $w_n$ | Time |
|---|---|---|---|
| 2 to 2.4 | 0.25 | $-0.5$ | $\mathcal{O}^*\left(2^{0.1n}\right) \subset \mathcal{O}^*\left(1.0718^n\right)$ |
| 2.4 to $2 + 2/3$ | 0.185373 | $-0.344895$ | $\mathcal{O}^*\left(2^{0.1495n}\right) \subset \mathcal{O}^*\left(1.1092^n\right)$ |
| $2 + 2/3$ to 3 | 0.155985 | $-0.266527$ | $\mathcal{O}^*\left(2^{0.2015n}\right) \subset \mathcal{O}^*\left(1.1499^n\right)$ |

**Lemma 4.** *For a formula $F$ with $d(F) \leq 3$, algorithm $C$ runs in $\mathcal{O}^*\left(1.1499^n\right)$ time.*

*Proof.* The components of the compound measure for this case are on the form $\mu_a(l,n) = w_l l + w_n n$, with the parameters of the measures given in Table 2. It may seem strange that $w_n < 0$ for these components, but this can be translated into the form $\sum_i w_i n_i$ with $w_i = iw_l + w_n$, in which case $w_i \geq 0$ for every $i \geq 2$, which also shows that $\mu_a(l,n)$ is non-increasing over every reduction (and since $\mu_a(F) \geq \mu(F)$, so is $\mu(F)$). For cases 2 and 3 of the algorithm, note that since $\mu_a$ is linear, $\mu(F) = \mu_a(F) = \sum_i \mu_a(F_i) \geq \sum_i \mu(F_i)$; the time used is dominated by the time spent on the heaviest component.

Also, when estimating $\Delta f$, this means that our underestimations are safe unless the formula we compare against contains a singleton (since $w_1 < 0$ for some sections of $f$). Specifically, $\Delta f$ can be described as $w_2$ for every removed 2-variable, $w_3$ for every removed 3-variable, and $w_l$ for every variable that has had its degree reduced from 3 to 2.

Next, consider case 4a. We can see that in both branches we will have removed all of $N[x]$ plus the variable $y$, and at least two heavy variables will have been reduced to light variables (since multiplier reduction does not apply).

Both branches get a reduction of at least $(S(x) + 5)w_l + 5w_n$, where $S(x) = \sum_{y \in N[x]} d(y)$, which will compare favourably to the results of using case 4b, and will never result in a worse branching.

**Table 3.** Component measures $\sum_i w_i n_i(F)$ for maximum degree 4

| Section | $w_2$ | $w_3$ | $w_4$ | Time |
|---------|-------|-------|-------|------|
| 2–3 | 0.045443 | 0.201428 | 0.324788 | $\mathcal{O}^*\left(1.1499^n\right)$ |
| 3–3.2 | 0.084777 | 0.201428 | 0.285454 | $\mathcal{O}^*\left(1.1634^n\right)$ |
| 3.2–3.5 | 0.092882 | 0.202779 | 0.280051 | $\mathcal{O}^*\left(1.1822^n\right)$ |
| 3.5–3.75 | 0.097593 | 0.204349 | 0.278481 | $\mathcal{O}^*\left(1.1975^n\right)$ |
| 3.75–4 | 0.107950 | 0.208788 | 0.277001 | $\mathcal{O}^*\left(1.2117^n\right)$ |

The worst cases of the algorithm remain. These branchings are given in Table 1, in terms of generic weights $w_l$ and $w_n$, since they do not depend upon the particular measure associated with the current section, and the branchings can be verified without great difficulty. Using the measures from Table 2, it can be verified that every branching has a branching number of at most 2 in a section where it is applicable (e.g. the first measure gives branching numbers of at most 2 for every case, while in the final section up to 3, only the 3-regular case is applicable).

We see that the time is indeed in $\mathcal{O}^*\left(2^{\mu_3(F)}\right)$ for the $\mu_3(F)$ given in Table 2, and the total worst time is $\mathcal{O}^*\left(1.1499^n\right)$, as given in the table.   □

Now, we present the analysis of the case when $d(F) = 4$. For this case, the multiple attributes-version of the analysis is used, with component measures $\sum_i w_i n_i(F)$, as explained in Sect. 3. We use $\Delta w_i = w_i - w_{i-1}$ to simplify expressions of branchings. The weights of the measure are given in Table 3. These weights were calculated automatically according to the approach described in Sect. 3, with resulting non-zero pivot points at average degrees 3, 3.2, 3.5, and 3.75 (the amount of pivot at the other potential pivot points was found to be zero in an optimal solution). The component measure for section 2–3 coincides with the top-most component measure for $d(F) = 3$: $0.155985\ell(F) - 0.266527n(F)$ results in $w_2 = 2w_l - w_n = 0.045443$ and $w_3 = 3w_l - w_n = 0.201428$. The automatic weight calculation also guarantees that the choice of weights and pivoting strategy is optimal. The bound achieved for $d(F) = 4$ is $\mathcal{O}^*\left(1.2117^n\right)$.

**Lemma 5.** *The weights of Table 3 form a correct compound measure.*

**Lemma 6.** *For a formula $F$ with $d(F) = 4$, $C(F)$ runs in time $\mathcal{O}^*\left(1.2117^n\right)$.*

*Proof.* We refer again to Table 3 for a definition of the weights in the compound measure. The measure is clearly well-behaved. The branching depends on the neighbourhood of the variable that is chosen; the explicit table has been cut, but the cases are similar to those for $d(F) = 3$ (and easier to find). We will prove that these are the worst-case branchings shortly, but first we consider case 4a:

if case 4a is used, then $N[x]$ and $y$ are removed in both branches, and at least two variables decrease their degree, which can be adjusted for the parity of $\ell(F)$. In the heavy branch of the maximally unbalanced branching, $N[x]$ is removed and at least three variables get their degrees decreased, likewise adjusted for the parity of $\ell(F)$. We see that in case 4a, both branches will be at least as heavy as the heaviest possible branch of case 4b. Thus, we only consider case 4b in the following.

When removing only the variable $x$ and repeatedly applying cases 1 and 2 of the algorithm, if any variable gets its degree reduced to $0^3$ or ends up in a non-biggest connected component (even if this happens after subsequent applications of case 2), then $x$ is a cut-vertex and multiplier reduction applies to $F$. Also, obviously, if any variable gets its degree reduced to 1, then multiplier reduction applies and one more reduction of the degree of some variable occurs. Thus, when $x$ has $k$ light neighbours and only $x$ is removed in one branch, the total reduction is at least $w_4 + kw_2 + k\Delta w_4$ plus the reductions in degree of the other neighbours of $x$, for the light branch of the maximally unbalanced branching. When some other variable is assigned, this does not apply, though we do know that in total, there are at least three variables outside of $N[x]$ that have links to variables in $N(x)$.

The case when the neighbours of $x$ are not all removed in the same branch does not provide a hard case; this is in agreement with the general principle that $\tau(a-d, b+d) > \tau(a,b)$ when $a \le b$ and $d \ge 0$, and can be proven by going through the cases in a simple though lengthy manner.

All that remains are the cases of a neighbourhood of degrees $(d_1, d_2, d_3, d_4)$ where in one branch only the branching variable disappears (producing $\Delta\mu = w_4 + \sum_i \Delta w_{d_i}$) and in the other, $N[x]$ disappears (producing $\Delta\mu = \sum_{y \in N[x]} w_{d(y)} + k\Delta w_4$, for $k = 3$ or $k = 4$ according to the parity of the sum of degrees of $N[x]$). Again, it can be verified that each such case has a branching number of at most 2 in every section where it is applicable. The total worst-case time for the $d(F) = 4$ case, as stated, is $\mathcal{O}^*\left(2^{w_4 n}\right)$ for the final value of $w_4 = 0.277001$, or $\mathcal{O}^*\left(1.2117^n\right)$.    □

**Table 4.** Weights for $d(F) > 4$ analysis

| $w_2$ | $w_3$ | $w_4$ | $w_5$ | $w_6$ |
|---|---|---|---|---|
| 0.115507 | 0.208788 | 0.277001 | 0.301245 | 0.307612 |

## 6    Analysis: The General Case

With $d(F) > 4$, the effects of a changing average degree seem to be less important than the number of variables removed. The analysis is performed in terms of a standard weight-based measure $\mu(F) = \sum_i w_i n_i(F)$, whose weights are given in

---

[3] By "reduced to 0" we mean that all neighbours of the variable are removed, and we do not include when a variable is removed by multiplier reduction.

Table 4. Note that while the values of $w_3$ and $w_4$ are the same as in the topmost measure for the $d(F) = 4$ analysis, the value of $w_2$ is increased to get a better worst-case branching number. This inequality is no problem, since the degree of a variable never increases by the application of a reduction: once $d(F) < 5$, the case $d(F) > 4$ does not appear in any subinstance.

The hard cases will be one case with a smallest-possible neighbourhood ($d(x) = 5$ and $N(x)$ is 2-regular), and the two cases with biggest-possible neighbourhoods (for $d(x) = 5$ and $d(x) = 6$). Since the latter two cases have average degree limits of 5 resp. 6, a compound measure would not be the right tool for this analysis.

**Lemma 7.** *Using $\mu(F) = \sum_i w_i n_i(F)$ with the weights given in Table 4, the running time of $C$ for a formula $F$ with $d(F) \leq 6$ is in $\mathcal{O}^* \left(2^{\mu(F)}\right)$.*

*Proof.* If $d(F) < 5$, then see Sect. 5 (note that the weights $w_2, \ldots, w_4$ give a bound that is consistent with that for section 3.75–4, which is in turn a valid bound for all cases with $d(F) \leq 4$). As before, the application of case 4a guarantees a reduction in both branches that is at least as high as the reduction in the heavy branch of the maximally unbalanced branching, and all cases with a branching number of 2 appear in case 4b with the maximally unbalanced branching. Providing a list of all branching numbers would require several pages; such lists are omitted for space, but the claim can be verified by a simple computer program. The cases of $k$-regular neighbourhoods with $d(x) = k$ are avoided as far as possible in case 5 of the algorithm, and as a result these cases happen at most once each in every path through the branching tree: they only apply if the $k$-variables form a regular connected component, and since no reduction creates a new occurrence of any variable in the formula, any $k$-regular connected component that appears in some subsequent subcase of some $k$-regular formula $F$ must occur as a subformula in $F$, which is impossible. Since these cases occur at most once in every path of the tree, they contribute only to the polynomial part of the running time.

No case with a more balanced branching has a higher branching number than 2; the proof for this is also omitted. We see that all cases have a branching number of at most 2. $\qquad\square$

**Theorem 1.** *The algorithm $C$ counts the number of max-weight models for a formula $F$ in time $\mathcal{O}^* (1.2377^n)$.*

*Proof.* If $d(F) \leq 6$, then this follows from Lemma 7. Otherwise, we can perform a quick analysis in terms of $n(F)$: the measure $n(F)$ is a well-behaved measure for the algorithm and since $d(F) \geq 7$, the branching number for case 5 is at worst $\tau(1, 8) < 1.2321$. $\qquad\square$

Finally, we make a brief note on lower bounds for the algorithm. While it would be good as a reference point to have strong lower bounds to match the upper bounds on the algorithm's running time, e.g. through presenting a class of instances for which we can prove that the algorithm, through making poor choices compatible with the algorithm description, can require $\Omega^* (c^n)$ time for some $c$, actually producing instances for which we can confidently predict the execution process proves very difficult. The best we are able to present in this paper is a type of instance of maximum degree 3 taking $\Omega^*(1.1048^n)$ time, to contrast with our upper bound $\mathcal{O}^*(1.1499^n)$ for the same situation.

For this purpose, build an instance $I_k$ as a form of warped $k$-rung "ladder": Use variables $x_i$ and $y_i$ for $1 \leq i \leq k$, and clauses $(y_i \vee y_{i+1})$ for all $1 \leq i < k$, $(x_i \vee x_{i+2})$ for all $1 \leq i \leq k - 2$, and $(x_i \vee y_i)$ for all $1 \leq i \leq k$. We show that $\Omega^*(1.1048^n)$ is consistent with the algorithm description for this type of instance.

**Lemma 8.** *The algorithm C, applied to an instance $I_k$ with $n = 2k$ variables, can take $\Omega^*(\tau(6,8)) > \Omega^*(1.1048^n)$ time.*

*Proof.* It is consistent with the algorithm description that it branches on the leftmost $y_i$ whose every neighbour is of degree 3 (in the original instance $I_k$, this would be $y_3$). When $y_i = 0$, all variables $x_j$ and $y_j$ for $j \geq i + 2$ remain, making the next branching candidate $y_{i+4}$; when $y_i = 1$, at least all variables $x_j$ and $y_j$ with $j > i$ remain, making the next branching candidate $y_{i+3}$. Following this process until $i \geq k - 3$ forms a recursion tree with, asymptotically, $\tau(3,4)^k$ lowest-level instances. $\qquad\square$

# 7   Conclusions

We have shown how to integrate analysis by non-uniform, piecewise linear measures, as used in previous $\#2\mathrm{SAT}_w$ bounds [3, 7], with the multi-variate recurrence approach by Eppstein [5], thereby combining the ability of the former to model algorithms whose behaviour varies depending on parameters of the input, with the good bounds and the automatability of the bound calculation of the latter. On the one hand, we have used this to give a tighter upper bound of $\mathcal{O}^*(1.2377^n)$ on the time required for solving $\#2\mathrm{SAT}_w$. On the other, we would like to point out that the question of the tightness of the resulting bound is unresolved. In other words, under what conditions will the bound $\mathcal{O}^*(c^\mu)$ produced by the method correctly describe the worst-case behaviour of a model of degree-bounded branchings as described in Sect. 3, or of other related variants?

# References

[1] Brueggemann, T., Kern, W.: An improved deterministic local search algorithm for 3-SAT. Theoretical Computer Science 329(1–3), 303–313 (2004)

[2] Dahllöf, V., Jonsson, P., Wahlström, M.: Counting satisfying assignments in 2-SAT and 3-SAT. In: H. Ibarra, O., Zhang, L. (eds.) COCOON 2002. LNCS, vol. 2387, pp. 535–543. Springer, Heidelberg (2002)

[3] Dahllöf, V., Jonsson, P., Wahlström, M.: Counting models for 2-SAT and 3-SAT formulae. Theoretical Computer Science 332(1–3), 265–291 (2005)

[4] Dubois, O.: Counting the number of solutions for instances of satisfiability. Theoretical Computer Science 81, 49–64 (1991)

 [5] Eppstein, D.: Quasiconvex analysis of backtracking algorithms. In: Proceedings of the 15th annual ACM-SIAM symposium on Discrete algorithms (SODA 2004), pp. 788–797 (2004)
 [6] Fomin, F.V., Grandoni, F., Kratsch, D.: Some new techniques in design and analysis of exact (exponential) algorithms. Bulletin of the EATCS 87, 47–77 (2005)
 [7] Fürer, M., Kasiviswanathan, S.P.: Algorithms for counting 2-SAT solutions and colorings with applications. Electronic Colloquium on Computational Complexity (ECCC) 5(033) (2005)
 [8] Iwama, K., Tamaki, S.: Improved upper bounds for 3-SAT. In: Proceedings of the 15th Annual ACM-SIAM Symposium on Discrete Algorithms (SODA 2004), p. 328 (2004)
 [9] Kozen, D.: The design and analysis of algorithms. Springer, New York (1992)
[10] Littman, M., Pitassi, T., Impagliazzo, R.: On the complexity of counting satisfying assignments. In: The working notes of the LICS 2001 workshop on Satisfiability (2001)
[11] Ryser, H.J.: Combinatorial Mathematics. The Mathematical Association of America, Washington (1963)
[12] Toda, S.: PP is as hard as the polynomial-time hierarchy. SIAM Journal on Computing 20(5), 865–877 (1991)
[13] Valiant, L.: The complexity of enumeration and reliability problems. SIAM Journal of Computing 8, 410–421 (1979)
[14] Valiant, L.G.: The complexity of computing the permanent. Theoretical Computer Science 8, 189–201 (1979)
[15] Wahlström, M.: An algorithm for the SAT problem for formulae of linear length. In: Brodal, G.S., Leonardi, S. (eds.) ESA 2005. LNCS, vol. 3669, pp. 107–118. Springer, Heidelberg (2005)
[16] Wahlström, M.: Algorithms, measures, and upper bounds for satisfiability and related problems. Linköping Studies in Science and Technology, PhD Dissertation no. 1079 (2007), http://urn.kb.se/resolve?urn=urn:nbn:se:liu:diva-8714
[17] Zhang, W.: Number of models and satisfiability of sets of clauses. Theoretical Computer Science 155, 277–288 (1996)

# Exact Algorithms for Edge Domination*

Johan M.M. van Rooij and Hans L. Bodlaender

Institute of Information and Computing Sciences, Utrecht University,
P.O. Box 80.089, 3508 TB Utrecht, The Netherlands
{jmmrooij, hansb}@cs.uu.nl

**Abstract.** In this paper we present a faster exact exponential time algorithm for the edge dominating set problem. Our algorithm uses $O(1.3226^n)$ time and polynomial space. The algorithm combines an enumeration approach based on enumerating minimal vertex covers with the branch and reduce paradigm. Its time bound is obtained using the measure and conquer technique. The algorithm is obtained by starting with a slower algorithm which is refined stepwise. In this way a series of algorithms appears, each one slightly faster than the previous, resulting in the $O(1.3226^n)$ time algorithm.

The techniques also gives faster exact algorithms for: minimum weight edge dominating set, minimum (weight) maximal matching, matrix domination and the parametrised version of minimum weight maximal matching.

**Keywords:** edge dominating set, minimum maximal matching, exact algorithms, exponential time algorithms, measure and conquer.

## 1   Introduction

Research on exponential time algorithms for finding exact solutions to NP-hard problems dates back to the sixties and seventies. Some natural problems such as independent set [20,22], colouring [13] and Hamiltonian circuit [10] have been studied for a long time, while for other problems such as dominating set [7,18,24], treewidth [25] and feedback vertex set [19] exact exponential algorithms with non-trivial running times date from only recently. There is a renewed interest in these algorithms, also visible in a recent series of surveys [8,11,21,26,27].

In this paper, we consider the minimum edge dominating set problem. This problem is identical to the problem of finding a minimum dominating set in a line graph. While both the minimum edge dominating set problem and the minimum dominating set problem are NP-hard [28], in some ways the problem restricted to line graphs is easier. For instance, minimum dominating set is hard to approximate [4], while minimum edge dominating set is constant-factor approximable [1]. Also from the parametrised point of view, minimum dominating set most likely is not fixed parameter tractable (it is W[2]-complete [2]), while

---

* This research was partially supported by project BRICKS (Basic Research for Creating the Knowledge Society).

minimum edge dominating set is fixed parameter tractable [5]. In the setting of exact exponential time algorithms, it also seems that the edge dominating set problem is somewhat easier; the currently best known time bound for an exact algorithm for minimum dominating set is $O(1.5063^n)$ [24], while in this paper we present an $O(1.3226^n)$ time algorithm for minimum edge dominating set.

The first exact algorithm for minimum edge dominating set is from 2005 due to Randerath and Schiermeyer [18] and has a running time of $O(1.4423^m)$. Raman et al. [17] improved this to $O(1.4423^n)$ and recently Fomin et al. [6] obtained an algorithm using $O(1.4082^n)$ time.

In this paper we combine the idea of enumerating minimal vertex covers in order to compute (variants of) minimum edge dominating set with reduction rules and matching techniques. In this way we improve upon the currently fastest algorithms for these problems. The time bound for our algorithm is tightened considerably by analysing it with measure and conquer. Furthermore this approach allows us to apply *design by measure and conquer*. This enables us to create a series of improved algorithms with even smaller upper bounds on their running times, similar to [24].

## 2    Preliminaries

Let $G = (V, E)$ be an $n$-node undirected simple graph. Let $G[V']$ be the subgraph of $G$ induced by a subset $V' \subseteq V$. Let $N(v)$ be the open neighbourhood of a vertex $v \in V$, $N[v]$ be the closed neighbourhood of $v \in V$ $(N[v] = N(v) \cup \{v\})$, and $N_{V'}(v), N_{V'}[v]$ be the open, respectively closed, neighbourhoods of $v$ in $G[V']$. $N_{V'}(V'')$ is an extension of this notation to neighbourhoods of $V'' \subseteq V$: $N_{V'}(V'') = (\bigcup_{v \in V''} N_{V'}(v)) \setminus V'', N_{V'}[V''] = \bigcup_{v \in V''} N_{V'}[v]$. For a subset of the vertices $V' \subseteq V$, we define the $V'$-degree of $v \in V$ to be the degree of $v$ in $G[V']$.

An *edge dominating set* is a subset $D \subseteq E$ such that every edge $e \in E$ is dominated by an edge $f \in D$, where $f$ dominates $e$ if $e$ and $f$ have a common end point. We consider the *minimum edge dominating set problem*: given a graph $G$, find an edge dominating set of minimum cardinality. For an edge weight function $\omega : E \to \mathbb{R}_{\geq 0}$ the *minimum weight edge dominating set problem* is: given a graph $G$, find an edge dominating set $D$ of minimum total weight. Also we consider the *minimum (weight) maximum matching problem*; this problem is equivalent to *minimum (weight) independent edge dominating set*, where an edge dominating set $D$ is independent if no two edges in $D$ are incident to the same vertex.

## 3    Using Minimal Vertex Covers for Edge Dominating Set

We start by showing how vertex covers and matchings can be used to compute minimum edge dominating sets. This leads to a reduction rule, a special treatment of graphs of low degree, and an algorithm that improves upon the currently fastest algorithm for this problem. We use:

*Property 1.* If $D \subseteq E$ is an edge dominating set in $G = (V, E)$, then $C = \{v \in V | \exists_{e \in D} \ v \in e\}$ is a vertex cover in $G$.

*Proof.* For each $e \in E$, there is an edge $f \in D$ that dominates $e$, i.e. $e$ and $f$ have an end point $v$ in common. Because $v \in C$, $C$ is a vertex cover.     □

For a vertex cover $C$ define $D_C$ to be the minimum edge dominating set containing $C$ in its set of endpoints. For given $C$, $D_C$ can be computed in the following way: compute a maximum matching $C'$ in $G[C]$ and add an edge for every unmatched vertex in $C$. $|D_C|$ is minimum because $|C'|$ is maximum.

Since every vertex cover contains a minimal vertex cover, we can use this in a simple algorithm for minimum edge dominating set: compute $D_C$ for all minimal vertex covers $C$ in $G$, and return one of minimum cardinality. This can be done in $O(1.4423^n)$ time, since minimal vertex covers can be enumerated with polynomial delay [12,14] and their number is bounded by $3^{n/3}$ [15].

This number is tight: consider the family of graphs consisting of $l$ triangles. Finding a minimum edge dominating set in a triangle, however, is trivial. This observation can be generalised: we obtain Rule 1 used in Algorithm 1.

Algorithm 1 continuously keeps track of a partitioning of the vertices of $G$ in three sets: a set $C$ of vertices that must become part of the vertex cover, a set $I$ of vertices that may not become part of the vertex cover (they are in the complementing independent set), and a set $U$ of vertices, which we call the set of *undecided* vertices. We denote such a state by the four-tuple $(G, C, I, U)$.

**Rule 1**

> **if** $G[U]$ contains a connected component $H$ that is a clique **then**
> > let $\tilde{G} = (V \cup \{v\}, E \cup \{\{u, v\} | u \in H\})$; $\tilde{C} = C \cup H \cup \{v\}$; $\tilde{U} = U \backslash H$
> > recursively solve $(\tilde{G}, \tilde{C}, I, \tilde{U})$ and obtain the edge dominating set $D$
> > **if** $D$ contains two distinct edges $\{u, v\}, \{v, w\}$ incident to $v$ **then**
> > > **return** $(D \backslash \{\{u, v\}, \{v, w\}\}) \cup \{\{u, w\}\}$
> > **return** $D \backslash \{\{u, v\}\}$, where $\{u, v\}$ is the unique edge in $D$ incident to $v$

*Proof of Correctness.* After the recursive call the extra vertex $v$ is incident to one or two edges in $D$, since $v \in \tilde{C}$ and if there are more edges incident to $v$ then two such edges can be replaced by the edge joining the other endpoints: this gives a smaller edge dominating set with $\tilde{C}$ as a subset of the set of endpoints.

All clique edges in the original graph are dominated if at most one clique vertex is not incident to a dominating edge. Therefore both return statements return an edge dominating set in $G$ that contains $C$ within its set of endpoints.

Because $D$ is of minimum cardinality (in $\tilde{G}$) and in both cases the returned set is of cardinality one smaller, it must also be of minimum cardinality (in $G$): if it is not then in both cases we can obtain a smaller edge dominating set in $\tilde{G}$ by adding one edge to the returned set.     □

Algorithm 1 enumerates minimal vertex covers by branching on vertices of maximum degree. It either puts such a vertex in the vertex cover, or in the independent set and hence all its neighbours in the vertex cover. If $G[U]$ is maximum degree two, then the algorithm branches on vertices in cycles until $G[U]$ is a collection of paths. Then it branches on the third vertex of such a path: Rule 1 guarantees that in both branches at least the first three vertices are removed from $U$. We estimate the number of subproblems generated:

**Algorithm 1.** Algorithm for Minimum Edge Dominating Set

---

**Input:** a graph $G = (V, E)$ and sets $C, I, U$ partitioning $V$ (initially $C = I = \emptyset, U = V$)

**Output:** a minimum edge dominating set in $G$

1: exhaustively apply Rule 1
2: **if** a vertex $v$ of maximum degree in $G[U]$ has $U$-degree at least three
    **or** $G[U]$ contains a connected component $H$ which is a cycle (pick $v \in H$) **then**
3:    create two subproblems and solve each one recursively:
4:        1: $(G, C \cup N_U(v), I \cup \{v\}, U \backslash N_U[v])$    2: $(G, C \cup \{v\}, I, U \backslash \{v\})$
5: **else if** $G[U]$ contains a connected component which is a path of length $\geq 4$ **then**
6:    Let $v_1, v_2, v_3, v_4$ be the vertices at an end of the path. Recursively solve:
7:        1: $(G, C \cup \{v_2, v_4\}, I \cup \{v_1, v_3\}, U \backslash \{v_1, v_2, v_3, v_4\})$    2: $(G, C \cup \{v_3\}, I, U \backslash \{v_3\})$
8: **else if** $G[U]$ contains a connected component which is a path of length three **then**
9:    Let $v$ be the middle vertex and recursively solve the subproblems:
10:      1: $(G, C \cup N_U(v), I \cup \{v\}, U \backslash N_U[v])$    2: $(G, C \cup \{v\}, I \cup N_U(v), U \backslash N_U[v])$
11: **else**
12:    compute the candidate edge dominating set $D_C$    {here: $U = \emptyset, C \cup I = V$}
13: **return** the smallest edge dominating set encountered

---

**Lemma 1.** *For Algorithm 1 and $l \geq 4$:*

1. *A cycle component $C_l$ in $G[U]$ generates a maximum of $4^{l/6}$ subproblems.*
2. *A path component $P_l$ in $G[U]$ generates a maximum of $4^{(l-1)/6}$ subproblems.*

*Proof.* (1) Let $P(l), C(l)$ be the number of subproblems generated by Algorithm 1 when dealing with a path or cycle of length $l$ respectively. Derive the values of $P(l)$ and $C(l)$ for $l \leq 4$ directly and consider the recurrence relation:

$$P(1) = P(2) = C(1) = C(2) = C(3) = 1 \quad P(3) = P(4) = C(4) = 2$$

$$\forall_{l \geq 5} : P(l) = P(l-3) + P(l-4) \qquad C(l) = C(l-1) + P(l-3)$$

Let $x$ be the solution to $1 = x^{-3} + x^{-4}$. For $l \geq 4$, $P(l) < x^l$ follows by induction after noting that it holds for $l \in \{4, 5, 6, 7\}$. For $l \geq 10$ we have:

$$C(l) < x^{l-1} + x^{l-3} = x^l(x^{-1} + x^{-3}) = \left( x \sqrt[l]{x^{-1} + x^{-3}} \right)^l < (4^{1/6})^l$$

using the fact that $\sqrt[l]{x^{-1} + x^{-3}}$ is decreasing and smaller than $4^{1/6}$ if $l \geq 10$. Direct computation shows that for $l < 10$: $C(l) \leq 4^{l/6}$.

(2) For $l \geq 8$, $x^{l/(l-1)}$ is decreasing and smaller than $4^{1/6}$, therefore:

$$P(l) < x^l = \left( x^{l/(l-1)} \right)^{l-1} < (4^{1/6})^{l-1}$$

For $4 \leq l \leq 7$: $P(l) \leq 4^{(l-1)/6}$, by direct computation.     □

A classical analysis of Algorithm 1 already improves upon previous results on this problem, using much simpler techniques as for example in [6].

**Theorem 1.** *Algorithm 1 solves the edge dominating set problem in $O(1.3803^n)$ time and polynomial space.*

*Proof.* Correctness follows from the discussion about enumerating vertex covers and the correctness of Rule 1.

Let $u$ be the number of undecided vertices in our problem instance (initially $u = n$), and $S(u)$ be the number of subproblems generated to solve an instance with $|U| = u$. The algorithm follows the following recurrence relation:

$$S(u) \leq \begin{cases} S(u-1) + S(u-4) & \text{branch on a vertex of } U\text{-degree} \geq 3 \\ P(l)S(u-l) & \text{total effect of removing a path of length } l \\ C(l)S(u-l) & \text{total effect of removing a cycle of length } l \end{cases}$$

Here we group all branchings corresponding to the removal of an entire path or cycle from $G[U]$. Let $\alpha$ be the solution to $1 = \alpha^{-1} + \alpha^{-4}$. Now $S(u) \leq \alpha^u$ because of the branching on a vertex of degree at least three. Also when a path or cycle is removed from $G[U]$ by a series of branchings, for each path or cycle of length $l$: $S(u) \leq 4^{l/6}S(u-l) \leq (4^{1/6})^l \alpha^{u-l} < \alpha^l$, by Lemma 1. This gives $S(u) \leq \alpha^u$, which results in the running time of $O(poly(n)\alpha^n)$ or $O(1.3803^n)$. $\qquad\square$

## 4    Design by Measure and Conquer

We can prove a much smaller upper bound on the running time of Algorithm 1 by analysing it with *measure and conquer* [7]. This approach also allows us to create a series of improved algorithms, each with a faster running time than the previous. This is what we call *design by measure and conquer* [24].

For the measure and conquer analysis we need a weight function $w : \mathbb{N} \to [0,1]$ assigning weights $w(d)$ to vertices of degree $d$ in $G[U]$. Instead of counting the number of undecided vertices to measure the progress of our algorithm, we will now use their total weight $k = \sum_{v \in U} w(\deg_{G[U]}(v))$ as a measure of complexity. If we can show that our algorithm runs in $O(\alpha^k)$ time using weight function $w$, it will also run in $O(\alpha^n)$ time, since for any problem instance $k \leq n$.

**Theorem 2.** *Algorithm 1 solves the minimum edge dominating set problem in $O(1.3323^n)$ time and polynomial space.*

*Proof.* Let $w : \mathbb{N} \to [0,1]$ be the weight function assigning weight $w(\deg_{G[U]}(v))$ to vertices $v \in G[U]$. The algorithm removes all vertices of $U$-degree zero, therefore $w(0) = 0$. Let $\Delta w(i) = w(i) - w(i-1)$. Vertices with a larger $U$-degree should be given a larger weight, hence we demand: $\forall_{n \geq 1} \Delta w(n) \geq 0$. Furthermore we impose the non-restricting steepness inequalities, $\forall_{n \geq 1} \Delta w(n) \geq \Delta w(n+1)$.

Consider an instance where the algorithm branches on a vertex $v$ of maximum $U$-degree $d \geq 3$ with $r_i$ neighbours of degree $i$ in $G[U]$ ($d = \sum_{i=1}^d r_i$). Let $d_2$ be a lower bound on the number of vertices at distance two from $v$ in $G[U]$:

$$d_2 = \left( \sum_{i=1}^d (i-1)r_i \pmod 2 \right) \quad \text{except when } d = r_3 = 3 \text{ then:} \quad d_2 = 2$$

This follows from a parity argument: there must be an edge in $G[U]$ with only one endpoint in $N_U[v]$ if $1 \equiv \sum_{i=1}^{d}(i-1)r_i \pmod 2$. Also $N_U[v]$ cannot be a clique by Rule 1, hence if $d = r_d$ we can use $d_2 = 2$.

If $v$ is put in the vertex cover, it is removed from $U$ and the $U$-degrees of all its neighbours in $G[U]$ are decreased by one. If $v$ is placed in the independent set then $N_U[v]$ is removed from $U$, and at least $d_2$ additional vertices have their $U$-degree reduced by at least one. Hence the algorithm recurses on two instances which are reduced $\Delta_{indep}$ and $\Delta_{vc}$ in measured complexity:

$$\Delta_{indep} = w(d) + \sum_{i=1}^{d} r_i w(i) + d_2 \Delta w(d) \qquad \Delta_{vc} = w(d) + \sum_{i=1}^{d} r_i \Delta w(i)$$

Let $S(k)$ be the number of subproblems generated to solve a problem of measured complexity $k$. For all $d \geq 3$ and $(d = \sum_{i=1}^{d} r_i)$ we have a recurrence relation of the form: $S(k) \leq S(k - \Delta_{indep}) + S(k - \Delta_{vc})$. Let $q(w)$ be the functional mapping a weight function to the solution of this entire set of recurrence relations.

By Lemma 1, an $l$-cycle or $l$-path generates a maximum of $4^{l/6}$, respectively $4^{(l-1)/6}$, subproblems. An $l$-cycle has a measured complexity of $l \cdot w(2)$ and a path of length $l$ has measured complexity at least $(l-1) \cdot w(2)$, since $\Delta w(1) \geq \Delta w(2)$ and hence $2w(1) \geq w(2)$. Therefore, in an instance where the vertices in cycle components and path components of length at least four in $G[U]$ have measured complexity $k'$, the removal of these vertices from $U$ by Algorithm 1 results in a maximum of $4^{k'/6w(2)}$ subproblems.

We now look for the optimal weight function $w : \mathbb{N} \to [0,1]$, satisfying the restrictions, such that the following maximum over the worst case behaviours of the different branch cases is minimum. We distinguish between three such cases: the maximum $U$-degree is three or more, cycles and paths of length at least four are removed from $G[U]$, and a path of length three is removed from $G[U]$.

$$S(k) \leq \left( \min_{w:\mathbb{N}\to[0,1]} \max \left\{ q(w), \ 4^{(1/6)w(2)}, 2^{1/(w(2)+2w(1))} \right\} \right)^k \tag{1}$$

By setting $w(i) = 1$ for all $i \geq 4$ we obtain a finite dimensional quasiconvex program [3]. Its solution shows a running time of $O(1.3323^k)$ using weights:

$$w(1) = 0.750724 \qquad w(2) = 0.914953 \qquad \forall_{i \geq 3} \ w(i) = 1$$

Since no subproblems are stored, Algorithm 1 uses polynomial space.     □

We will now stepwise improve upon this result, designing even faster algorithms using *design by measure and conquer* [24]. The idea is the following: the quasiconvex program proving the time bound of Theorem 2 has only a few cases that are tight to the maximum of Equation 1. These cases follow by substituting the optimal weights in the recurrence and can be considered for improvement.

The quasiconvex program of Theorem 2 has the following tight worst cases:

1. $d = 3, r_2 = 2, r_3 = 1$, i.e. in $G[U]$ we have a vertex of maximum $U$-degree 3, with two neighbours of $U$-degree 2 and one neighbour of $U$-degree 3.

**Table 1.** Bounds on the running times of the algorithms in the improvement series

| Strategies: | none | 1 | 1-2 | 1-3 | 1-4 | 1-5 | 1-7 | 1-8 | 1-9 |
|---|---|---|---|---|---|---|---|---|---|
| $O(x^n)$: | 1.3323 | 1.3315 | 1.3296 | 1.3280 | 1.3265 | 1.3248 | 1.3240 | 1.3228 | 1.3226 |
| $\Omega(x^n)$ [23]: | 1.3160 | 1.2968 | 1.2968 | 1.2968 | 1.2968 | 1.2753 | 1.2753 | 1.2753 | 1.2753 |

2. $d = 3, r_3 = 3$: we have a vertex of maximum $U$-degree 3, with three neighbours in $G[U]$ of $U$-degree 3.
3. a connected component in $G[U]$ is a path of length three.

Consider the first case. Let $v$ be the vertex of maximum $U$-degree three, with two neighbours $u_1, u_2 \in U$ of $U$-degree two and one neighbour $u_3 \in U$ of $U$-degree three. In our analysis of Theorem 2, we had a lower bound $d_2$ on the number vertices with distance two from $v$ in $G[U]$; for this case we had $d_2 = 0$. We now consider two subcases.

In the first subcase $v$, $u_1$, $u_2$ and $u_3$ form a connected component in $G[U]$ isomorphic to Subgraph 1 in Figure 1. Algorithm 1 branches on $v$. We modify this now, by instead branching on one of the $U$-degree two vertices, e.g. $u_1$. In both subproblems that are obtained after branching on $u_1$, the vertices of the subgraph that remains in $U$ form a clique in $G[U]$, and so are dealt with by Rule 1. Therefore the entire subgraph disappears from $G[U]$ after one branching step and application of Rule 1, while previously, we had a path of length three in $G[U]$ remaining in one subproblem that required another branching step.

In the second subcase vertices in $U \backslash \{v, u_1, u_2, u_3\}$ are adjacent to $u_1$, $u_2$ and/or $u_3$. If we branch on $v$, then these vertices will have their $U$-degrees reduced by one in one branch, implying a larger progress than estimated in Theorem 2: by a parity argument we can use $d_2 = 2$ as a new lower bound on the number vertices with distance two from $v$ in $G[U]$.

Thus we modify the algorithm and split this case in two subcases in the measure and conquer analysis. The optimum of the new quasiconvex program proves an upper bound on the running time of $O(1.3315^n)$ for this new algorithm.

Arguments, similar to the argument given above can be given in a large number of other case as well. This leads to a series of improvement steps, and a series of algorithms: each algorithm slightly improves upon the previous by introducing a new more efficient branching strategy on a tight case. In each case a number of local configurations around a vertex the previous algorithm would branch on are considered for more efficient branching. This results in an increase of the lower bound $d_2$ for all remaining configurations.

We do not have the space here to explain all these alternative branching strategies in detail: we give these in a schematic manner in Figure 1; see [23] for a more elaborate treatment. The graphs in the figure are ordered on their appearance as a tight case in the algorithms in the series. The $i$-th algorithm uses alternative branching strategies 1 up to $i$. Table 1 contains upper and lower [23] bounds on the running times of the individual algorithms.

**Algorithm 2.** Let Algorithm 2 be the modification of Algorithm 1 using all the alternative branching strategies illustrated in Figure 1.

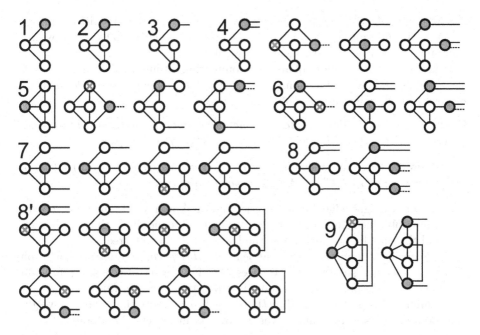

The leftmost vertex in every subgraph corresponds to a vertex we could branch on in Algorithm 1 and grey vertices represent more efficient alternatives. If multiple vertices are grey, simultaneously branch on these vertices generating four or eight subproblems. Crossed vertices represent vertices branched on directly hereafter, but only in the subproblems where this induces extra 1, 2 or 3-cliques. Sometimes small path components remain in a subproblem; these are immediately branched upon also.

Unfinished edges always connect to vertices outside the drawn subgraph, and there are no other edges in $G[U]$ between vertices with at least one drawn endpoint. Dashed edges are optional.

**Fig. 1.** More efficient branching strategies on possible subgraphs of $G[U]$

**Theorem 3.** *Algorithm 2 solves the minimum edge dominating set problem in $O(1.3226^n)$ time and polynomial space.*

*Proof.* Reconsider the quasiconvex program used to prove the running time of Theorem 2 and include new recurrence relations corresponding to the subcases in Figure 1 and modify the values of $d_2$ for previously tight cases.

The solution to this modified quasiconvex program gives a running time of $O(1.3226^n)$ for Algorithm 2 using weights:

$$w(1) = 0.779416 \quad w(2) = 0.920821 \quad w(3) = 0.997106 \quad \forall_{i \geq 4}\, w(i) = 1 \qquad \square$$

## 5   Related Problems

The results of the previous sections also imply faster exact algorithms for a variety of other problems.

The *matrix domination* problem is: given a $m \times n$ matrix $M$ with entries in $\{0, 1\}$, find a minimum subset $S$ of the 1-entries in $M$ such that every row and column of $M$ contains at least one 1-entry in $S$.

Using the transformations of an edge dominating set into a minimum maximal matching [9] and of a matrix domination instance to a bipartite edge dominating set instance [28], one directly obtains:

**Corollary 1.** *The minimum maximal matching problem can be solved by a modification of Algorithm 2 in $O(1.3226^n)$ time and polynomial space.*

**Corollary 2.** *The matrix domination problem can be solved by modification of Algorithm 2 in $O(1.3226^{n+m})$ time and polynomial space.*

For minimum weight edge dominating set and minimum weight maximal matching we have obtained similar results. Hereto we need to modify both the reduction rule and the matching algorithm at the leaves of the search tree.

First consider the *minimum weight generalised edge cover* problem: in a graph $G$ cover a specified subset of the vertices $C \subseteq V$ by a set of edges of minimum total weight. This problem is solvable in cubic time too [16] (also see [5]). An algorithm for this problem can directly be used to compute the minimum weight edge dominating set containing a vertex cover $C$ in its sets of endpoints.

**Theorem 4.** *The minimum weight edge dominating set problem can be solved by a modification of Algorithm 2 in $O(1.3226^n)$ time and polynomial space.*

*Proof.* Consider Algorithm 2 using a polynomial time algorithm for minimum weight generalised edge cover to compute the minimum weight edge dominating set containing a vertex cover, and with Rule 1 replaced by the rules:

**Rule 2.** Put isolated vertices in $G[U]$ in the independent set $I$.

**Rule 3**
    **if** $G[U]$ contains a connected component $H$ that is a clique of size 2 or 3 **then**
        let $H = (V', E')$, $e = \text{argmin}_{e' \in E'} \omega(e')$ and $\tilde{C} = C \cup H \cup \{v\}$; $\tilde{U} = U \backslash H$
        let $\tilde{G} = (V \cup \{v\}, E \cup \{\{u, v\} | u \in H\})$ with $\forall_{u \in H} \omega(\{u, v\}) = \omega(e)$
        recursively solve $(\tilde{G}, \tilde{C}, I, \tilde{U})$ and obtain the edge dominating set $D$
        **if** $D$ contains two distinct edges $f, g$ incident to $v$ **then**
            **return** $(D \backslash \{f, g\}) \cup \{e\}$
        **return** $D \backslash \{f\}$, where $f$ is the unique the edge in $D$ incident to $v$

Correctness is identical to the previous algorithms, only for the correctness of Rule 3 we also need the observation that in a clique of size two or three at least one vertex in any pair of vertices is incident to an edge of minimum weight.

The running time follows from Theorem 3.                     □

For minimum weight maximal matching we need to consider *the minimum weight generalised independent edge cover problem*: given a graph $G$, cover a specified subset of the vertices $C \subseteq V$ by a set of edges of minimum total weight such that no two edges are incident to the same vertex.

---

**Algorithm 3.** Minimum Weight Generalised Independent Edge Cover Algorithm

---

**Input:** a graph $G = (V, E)$ and a subset of its vertices $C \subseteq V$
**Output:** a minimum weight generalised independent edge cover of $C$ in $G$ if one exists

1: **if** $G$ has an odd number of vertices **then**
2:     add a new vertex $v$ to $G$ $(v \notin C)$
3: **for all** $v, w \in V \backslash C, v \neq w$ **do**
4:     add a new edge between $v$ and $w$ to $G$ with zero weight
5: **if** there exists a minimum weight perfect matching $P$ in $G$ **then**
6:     **return** $P$ with all edges between vertices not in $C$ removed
7: **return false**

---

**Proposition 1.** *Algorithm 3 solves the minimum weight generalised independent edge cover problem in polynomial time.*

*Proof.* The returned edge set is a generalised independent edge cover of $C$ in $G$ since it is a matching and it contains all vertices in $C$ in its set of endpoints. The edges in $P$ between vertices in $V \backslash C$ have zero weight and thus the returned set and $P$ have equal total weight. Because all generalised independent edge covers of $C$ in $G$ can be obtained in this way from perfect matchings, the returned set is of minimal weight and false is only returned if no generalised independent edge cover of $C$ exists in $G$.      $\square$

**Theorem 5.** *The minimum weight maximal matching problem can be solved by a modification of Algorithm 2 in $O(1.3226^n)$ time and polynomial space.*

*Proof.* Consider Algorithm 2 using Algorithm 3 to compute the minimum weight maximal matching containing a vertex cover, and with Rule 1 replaced by:

**Rule 4**
    **if** $G[U]$ contains a connected component $H$ which is a clique **then**
       let $\tilde{G} = (V \cup \{v\}, E \cup \{\{u, v\} | u \in H\})$ with $\forall_{u \in H} \; \omega(\{u, v\}) = 0$
       $\tilde{C} := C \cup H; \; \tilde{I} := I \cup \{v\}; \; \tilde{U} := U \backslash H$
       recursively solve $(\tilde{G}, \tilde{C}, \tilde{I}, \tilde{U})$ and obtain the maximal matching $D$
       **if** $D$ contains an edge $e$ incident to $v$ **then**
          **return** $D \backslash \{e\}$
       **return** $D$

Correctness is identical to the previous algorithms, using that Rule 4 is correct because the extra vertex $v$ can only be incident to one edge in $D$, and this edge is of zero weight. The running time follows from Theorem 3.      $\square$

Using different techniques, Proposition 1 also implies the following:

**Proposition 2 ([23]).** *The parametrised minimum weight maximal matching problem, with parameter $k$ can be solved in $O^*(2.4178^k)$[1].*

---

[1] Here we use the $O^*$ notation which suppresses not only constant but all polynomial parts of the running time.

*Proof.* Identical to the treatment of similar problems in [6], using Algorithm 3 if no path decomposition is computed.                                                  □

## 6   Conclusion and Further Research

We have presented faster exact exponential time (and polynomial space) algorithms for minimum edge dominating set and related problems. These algorithms are obtained by using a vertex cover structure on the input graph, special branching strategies and reduction rules applied to simple instances and the iterative improvement of a measure and conquer analysis.

It would be interesting to see if there are more related problems, such as minimum (weight) total edge domination to which our methods can be applied.

We note that our algorithms have their running times expressed in the number of vertices $n$ in the input graph $G$, instead of the number of edges $m$ in $G$. As an interesting research topic, we mention the analysis and design of exact algorithms for edge domination (and other problems), where we focus on the running time as function of the number of edges $m$. See the discussion about complexity parameters in [26].

## References

1. Carr, R., Fujito, T., Konjevod, G., Parekh, O.: A $2\frac{1}{10}$ approximation algorithm for a generalization of the weighted edge-dominating set problem. Journal of Combinatorial Optimization 5, 317–326 (2001)
2. Downey, R.G., Fellows, M.R.: Fixed-parameter tractability and completeness. Congressus Numerantium 87, 161–178 (1992)
3. Eppstein, D.: Quasiconvex analysis of backtracking algorithms. In: Proceedings of the 15th Annual ACM-SIAM Symposium on Discrete Algorithms, SODA 2004, pp. 781–790 (2004)
4. Feige, U.: A threshold of ln $n$ for approximating set cover. J. ACM 45, 634–652 (1998)
5. Fernau, H.: Edge dominating set: Efficient enumeration-based exact algorithms. In: Bodlaender, H.L., Langston, M.A. (eds.) IWPEC 2006. LNCS, vol. 4169, pp. 140–151. Springer, Heidelberg (2006)
6. Fomin, F.V., Gaspers, S., Saurabh, S.: Branching and treewidth based exact algorithms. In: Asano, T. (ed.) ISAAC 2006. LNCS, vol. 4288, pp. 16–25. Springer, Heidelberg (2006)
7. Fomin, F.V., Grandoni, F., Kratsch, D.: Measure and conquer: Domination — a case study. In: Caires, L., Italiano, G.F., Monteiro, L., Palamidessi, C., Yung, M. (eds.) ICALP 2005. LNCS, vol. 3580, pp. 191–203. Springer, Heidelberg (2005)
8. Fomin, F.V., Grandoni, F., Kratsch, D.: Some new techniques in design and analysis of exact (exponential) algorithms. Bulletin of the EATCS 87, 47–77 (2005)
9. Harary, F.: Graph Theory. Addison-Wesley, Reading, MA (1969)
10. Held, M., Karp, R.: A dynamic programming approach to sequencing problems. J. SIAM 10, 196–210 (1962)
11. Iwama, K.: Worst-case upper bounds for kSAT. Bulletin of the EATCS 82, 61–71 (2004)

12. Johnson, D.S., Yannakakis, M., Papadimitriou, C.H.: On generating all maximal independent sets. Information Processing Letters 27, 119–123 (1988)
13. Lawler, E.L.: A note on the complexity of the chromatic number problem. Information Processing Letters 5, 66–67 (1976)
14. Lawler, E.L., Lenstra, J.K., Rinnooy Kan, A.H.G.: Generating all maximal independent sets: NP-hardness and polynomial-time algorithms. SIAM J. Comput. 9, 558–565 (1980)
15. Moon, J.W., Moser, L.: On cliques in graphs. Israel J. Math. 3, 23–28 (1965)
16. Plesník, J.: Constrained weighted matchings and edge coverings in graphs. Disc. Appl. Math. 92, 229–241 (1999)
17. Raman, V., Saurabh, S., Sikdar, S.: Efficient exact algorithms through enumerating maximal independent sets and other techniques. Theory of Computing Systems 42, 563–587 (2007)
18. Randerath, B., Schiermeyer, I.: Exact algorithms for minimum dominating set. Technical Report zaik2005-501, Universität zu Köln, Cologne, Germany (2005)
19. Razgon, I.: Exact computation of maximum induced forest. In: Arge, L., Freivalds, R. (eds.) SWAT 2006. LNCS, vol. 4059, pp. 160–171. Springer, Heidelberg (2006)
20. Robson, J.M.: Algorithms for maximum independent sets. J. Algorithms 7, 425–440 (1986)
21. Schöning, U.: Algorithmics in exponential time. In: Diekert, V., Durand, B. (eds.) STACS 2005. LNCS, vol. 3404, pp. 36–43. Springer, Heidelberg (2005)
22. Tarjan, R.E., Trojanowski, A.: Finding a maximum independent set. SIAM J. Comput. 6, 537–546 (1977)
23. van Rooij, J.M.M., Bodlaender, H.L.: Exact algorithms for edge domination. Technical Report UU-CS-2007-051, Department of Information and Computing Sciences, Utrecht University, Utrecht, The Netherlands (2007)
24. van Rooij, J.M.M., Bodlaender, H.L.: Design by measure and conquer: A faster exact algorithm for dominating set. In: Proc. 24th Symp. Theoretical Aspects of Computer Science, STACS 2008 (2008)
25. Villanger, Y.: Improved exponential-time algorithms for treewidth and minimum fill-in. In: Correa, J.R., Hevia, A., Kiwi, M. (eds.) LATIN 2006. LNCS, vol. 3887, pp. 800–811. Springer, Heidelberg (2006)
26. Woeginger, G.J.: Exact algorithms for NP-hard problems: A survey. In: Jünger, M., Reinelt, G., Rinaldi, G. (eds.) Combinatorial Optimization - Eureka, You Shrink! LNCS, vol. 2570, pp. 185–207. Springer, Heidelberg (2003)
27. Woeginger, G.J.: Space and time complexity of exact algorithms: Some open problems (invited talk). In: Downey, R.G., Fellows, M.R., Dehne, F. (eds.) IWPEC 2004. LNCS, vol. 3162, pp. 281–290. Springer, Heidelberg (2004)
28. Yannakakis, M., Gavril, F.: Edge dominating sets in graphs. SIAM J. Appl. Math. 38, 364–372 (1980)

# Author Index

# Lecture Notes in Computer Science

Sublibrary 1: Theoretical Computer Science and General Issues

For information about Vols. 1– 4681
please contact your bookseller or Springer

Vol. 4863: A. Bonato, F.R.K. Chung (Eds.), Algorithms and Models for the Web-Graph. X, 217 pages. 2007.

Vol. 4860: G. Eleftherakis, P. Kefalas, G. Păun, G. Rozenberg, A. Salomaa (Eds.), Membrane Computing. IX, 453 pages. 2007.

Vol. 4855: V. Arvind, S. Prasad (Eds.), FSTTCS 2007: Foundations of Software Technology and Theoretical Computer Science. XIV, 558 pages. 2007.

Vol. 4854: L. Bougé, M. Forsell, J.L. Träff, A. Streit, W. Ziegler, M. Alexander, S. Childs (Eds.), Euro-Par 2007 Workshops: Parallel Processing. XVII, 236 pages. 2008.

Vol. 4851: S. Boztaş, H.-F.(F.) Lu (Eds.), Applied Algebra, Algebraic Algorithms and Error-Correcting Codes. XII, 368 pages. 2007.

Vol. 4848: M.H. Garzon, H. Yan (Eds.), DNA Computing. XI, 292 pages. 2008.

Vol. 4847: M. Xu, Y. Zhan, J. Cao, Y. Liu (Eds.), Advanced Parallel Processing Technologies. XIX, 767 pages. 2007.

Vol. 4846: I. Cervesato (Ed.), Advances in Computer Science – ASIAN 2007. XI, 313 pages. 2007.

Vol. 4838: T. Masuzawa, S. Tixeuil (Eds.), Stabilization, Safety, and Security of Distributed Systems. XIII, 409 pages. 2007.

Vol. 4835: T. Tokuyama (Ed.), Algorithms and Computation. XVII, 929 pages. 2007.

Vol. 4818: I. Lirkov, S. Margenov, J. Waśniewski (Eds.), Large-Scale Scientific Computing. XIV, 755 pages. 2008.

Vol. 4800: A. Avron, N. Dershowitz, A. Rabinovich (Eds.), Pillars of Computer Science. XXI, 683 pages. 2008.

Vol. 4783: J. Holub, J. Žďárek (Eds.), Implementation and Application of Automata. XIII, 324 pages. 2007.

Vol. 4782: R. Perrott, B.M. Chapman, J. Subhlok, R.F. de Mello, L.T. Yang (Eds.), High Performance Computing and Communications. XIX, 823 pages. 2007.

Vol. 4771: T. Bartz-Beielstein, M.J. Blesa Aguilera, C. Blum, B. Naujoks, A. Roli, G. Rudolph, M. Sampels (Eds.), Hybrid Metaheuristics. X, 202 pages. 2007.

Vol. 4770: V.G. Ganzha, E.W. Mayr, E.V. Vorozhtsov (Eds.), Computer Algebra in Scientific Computing. XIII, 460 pages. 2007.

Vol. 4769: A. Brandstädt, D. Kratsch, H. Müller (Eds.), Graph-Theoretic Concepts in Computer Science. XIII, 341 pages. 2007.

Vol. 4763: J.-F. Raskin, P.S. Thiagarajan (Eds.), Formal Modeling and Analysis of Timed Systems. X, 369 pages. 2007.

Vol. 4759: J. Labarta, K. Joe, T. Sato (Eds.), High-Performance Computing. XV, 524 pages. 2008.

Vol. 4746: A. Bondavalli, F. Brasileiro, S. Rajsbaum (Eds.), Dependable Computing. XV, 239 pages. 2007.

Vol. 4743: P. Thulasiraman, X. He, T.L. Xu, M.K. Denko, R.K. Thulasiram, L.T. Yang (Eds.), Frontiers of High Performance Computing and Networking ISPA 2007 Workshops. XXIX, 536 pages. 2007.

Vol. 4742: I. Stojmenovic, R.K. Thulasiram, L.T. Yang, W. Jia, M. Guo, R.F. de Mello (Eds.), Parallel and Distributed Processing and Applications. XX, 995 pages. 2007.

Vol. 4739: R. Moreno Díaz, F. Pichler, A. Quesada Arencibia (Eds.), Computer Aided Systems Theory – EUROCAST 2007. XIX, 1233 pages. 2007.

Vol. 4736: S. Winter, M. Duckham, L. Kulik, B. Kuipers (Eds.), Spatial Information Theory. XV, 455 pages. 2007.

Vol. 4732: K. Schneider, J. Brandt (Eds.), Theorem Proving in Higher Order Logics. IX, 401 pages. 2007.

Vol. 4731: A. Pelc (Ed.), Distributed Computing. XVI, 510 pages. 2007.

Vol. 4728: S. Bozapalidis, G. Rahonis (Eds.), Algebraic Informatics. VIII, 291 pages. 2007.

Vol. 4726: N. Ziviani, R. Baeza-Yates (Eds.), String Processing and Information Retrieval. XII, 311 pages. 2007.

Vol. 4719: R. Backhouse, J. Gibbons, R. Hinze, J. Jeuring (Eds.), Datatype-Generic Programming. XI, 369 pages. 2007.

Vol. 4711: C.B. Jones, Z. Liu, J. Woodcock (Eds.), Theoretical Aspects of Computing – ICTAC 2007. XI, 483 pages. 2007.

Vol. 4710: C.W. George, Z. Liu, J. Woodcock (Eds.), Domain Modeling and the Duration Calculus. XI, 237 pages. 2007.

Vol. 4708: L. Kučera, A. Kučera (Eds.), Mathematical Foundations of Computer Science 2007. XVIII, 764 pages. 2007.

Vol. 4707: O. Gervasi, M.L. Gavrilova (Eds.), Computational Science and Its Applications – ICCSA 2007, Part III. XXIV, 1205 pages. 2007.

Vol. 4706: O. Gervasi, M.L. Gavrilova (Eds.), Computational Science and Its Applications – ICCSA 2007, Part II. XXIII, 1129 pages. 2007.

Vol. 4705: O. Gervasi, M.L. Gavrilova (Eds.), Computational Science and Its Applications – ICCSA 2007, Part I. XLIV, 1169 pages. 2007.

Vol. 4703: L. Caires, V.T. Vasconcelos (Eds.), CONCUR 2007 – Concurrency Theory. XIII, 507 pages. 2007.

Vol. 4700: C.B. Jones, Z. Liu, J. Woodcock (Eds.), Formal Methods and Hybrid Real-Time Systems. XVI, 539 pages. 2007.

Vol. 4699: B. Kågström, E. Elmroth, J. Dongarra, J. Waśniewski (Eds.), Applied Parallel Computing. XXIX, 1192 pages. 2007.

Vol. 4698: L. Arge, M. Hoffmann, E. Welzl (Eds.), Algorithms – ESA 2007. XV, 769 pages. 2007.

Vol. 4697: L. Choi, Y. Paek, S. Cho (Eds.), Advances in Computer Systems Architecture. XIII, 400 pages. 2007.

Vol. 4688: K. Li, M. Fei, G.W. Irwin, S. Ma (Eds.), Bio-Inspired Computational Intelligence and Applications. XIX, 805 pages. 2007.

Vol. 4684: L. Kang, Y. Liu, S. Zeng (Eds.), Evolvable Systems: From Biology to Hardware. XIV, 446 pages. 2007.

Vol. 4683: L. Kang, Y. Liu, S. Zeng (Eds.), Advances in Computation and Intelligence. XVII, 663 pages. 2007.